RIDING ON A RAY OF LIGHT

NEW CONCEPTS IN THE STUDY OF LIGHT, MATTER AND GRAVITY

KRISHNAGOPAL DHARANI

PARTRIDGE

ISBN:	Hardcover	978-1-5437-0650-5
	Softcover	978-1-5437-0649-9
	eBook	978-1-5437-0648-2

Print information available on the last page.

To order additional copies of this book, contact
Partridge India
000 800 10062 62
orders.india@partridgepublishing.com

www.partridgepublishing.com/india

CONTENTS

PART II: THE NEGENTROPIC MODEL

PREFACE

For over a decade I have been working, rather relentlessly though amateurishly, on the diverse principles that govern the universe both at the cosmic and atomic scales, and I have even been brooding ever since over the knotty problem of understanding nature on a single principle from a general perspective. At various stages in my journey into the cosmos, it appeared strange to me that the workings of nature could be so uncomprehendingly complex, and I have thought, with some naive optimism, that all the vagaries of nature may somehow be explained on one single principle. With this apparent hope at the back of my mind, I have continued to study the current concepts in theoretical science in as simple terms as possible and tried to understand the subtle implications of the diverse principles of nature which govern the universe, and the result of these efforts constitutes the first six big chapters of *Riding on a Ray of Light* which are meant for the benefit of a general science reader.

As I waded through the intricate theories, I found that some of the concepts in theoretical science, even those which are critically acclaimed by the scientific elite of today, are inconsistent and ambiguous. In fact, some of the most cherished concepts of today have a significant number of equally important contestants in the field, and I felt that these conflicts in theoretical science simply reflect upon the inadequacy of current theories in explaining the various phenomena of nature. These theoretical shortcomings have prompted me to search for new avenues in a different direction, which may explain the natural principles in a more coherent manner, and the result of these efforts constitutes the next six chapters of this book.

Writing science, though meritoriously rewarding, is a demanding enterprise – I have spent many laborious years writing this book, which demanded a great deal of musing and mulling over the complex issues accompanied by corrections and recorrections in order to make all the supplied information more pithy and intelligible to the reader, and this whole exercise has left me with the conviction that the ease with which a book of science can be understood is inversely proportional to the ease with which it is written! Moreover, proposing and presenting a new and innovative concept to the scientific community are still harder an exercise. Consequently, though I have written this book amidst moments of academic enthusiasm and exhilaration, I should admit that on several occasions, I was subjected to many shades of pessimism, including periods of despair, dejection, and self-doubt—especially on occasions when my new model conflicted with the established concepts in theoretical science.

But then I assure the reader, with a reasonable confidence, that this book will give him/her a comprehensive understanding of the workings of nature with a brand-new model of the universe which would clear most, if not all, of the major issues in the field of cosmology and fundamental science! And I earnestly hope that if ever this book falls in the hands of a specialist, he/she would make a fresh reading of this work with an unreserved judgement and then consider the new concepts for further analysis and experimentation only if he/she thinks they deserve it!

Finally, I would like to state that I have written *Riding on a Ray of Light* with the chief purpose of upholding the cause of science, and in the same vein, I wish that this work would motivate younger students of science to shun academic dogma whenever possible and to do some critical thinking wherever necessary and to bring about a change in the climate of scientific thought, which is really the backbone of scientific progress.

I wish the reader all the best and a rewarding academic journey through the book!

Krishnagopal Dharani

INTRODUCTION

From a Ray of Light to the Beginnings of the Universe

R *iding on a Ray of Light* is a scientific treatise on the fundamental nature of matter and energy. The book aims to propose a new perspective in the conceptual understanding of the fundamental nature of matter and energy, matter and energy being the two most basic building blocks of the universe. It describes a new model by which a ray of light is generated in nature and a mechanism by which this ray of light gets transformed into matter in the universe. By the way, this new hypothesis examines many intriguing issues related to mass, motion, gravity, and such others at the fundamental level and defines these phenomena in very precise terms.

It is known that many of the fundamental questions related to matter and energy have persisted in the milieu of science as enigmas for ages despite mighty advances in our understanding of nature. Many theoretical scientists, experimental researchers, and even philosopher–scientists, down the centuries, have attempted to resolve the fundamental structure and meaning of matter and energy but with no real success. Not only these, but there are several fundamental scientific issues that are left unanswered despite giant leaps in our understanding of nature both at the atomic and at the cosmic scales – we neither have a complete understanding of the basic principles that design the structure of an atom on a micro scale nor have we got a consummate grasp of the precise workings of the laws that govern the universe on a macro scale. But then here in this book, we have dealt with almost all the fundamental issues of physics in a new theoretical perspective in a systematic manner and arrived at some profound conclusions related to the microcosm as well as the macrocosm.

The reader may see that to understand the new theoretical perspective, we must first have a brief understanding of the established principles in the broad field of theoretical physics, and this understanding would, in fact, let the reader realize that there are several unresolved voids in our current understanding. Equipped with this scientific knowledge and the lacunae, the reader may then proceed to explore new avenues in the workings of nature. The reader may, however, realize that it would be a Herculean task not only to deal with such wide range of topics related to the theoretical principles of the universe under one heading, but it is also a mind-bending exercise to comprehend the new ideas all at one go, and hence the book is conveniently divided into several chapters arranged in two parts—Part I and Part II.

Part I consists of six chapters which outline the established principles of modern theoretical science starting from the general principles of the universe that govern the movement of all celestial bodies to the rules of physics that operate the microscopic world of the atom. *Part II* has another six chapters which systematically analyse the basic tenets of physics and study them in a new perspective, and this analysis would resolve many of the hitherto perplexing secrets of modern physics ranging from the nature of light to the creation of matter, to the origins of the universe, and to many other such diverse physical phenomena of nature. It may interest the reader, at this juncture, to note that the 'Overviews' written at the beginning of each chapter in Part I and Part II are all connected to one another in a chain so that, when read continuously, the reader may get a grasp of the aims and objectives of the book!

From here, a hurried reader may jump over to the chapters, plough through the topics, and start downloading all the scientific cargo and whatnot. But a leisurely reader, on the other hand, if at all we find one such nowadays, may take a break and peek at the following few passages, which would introduce the chief purpose of *Riding on a Ray of Light* and its importance in the arena of modern science.

Riding on a Ray of Light
A New Conceptual Understanding of the Universe

Over the past hundred years, our understanding of the fundamental governing rules of nature has changed—our scientific knowledge has progressed from simple working theories of nature, such as the Newton's laws of motion and gravitation, laws of heat transfer, laws of electricity, Maxwell's theory of electromagnetism and so on; to some complex, and sometimes ambiguous, theories, such as the Einstein's theory of relativity, quantum theory, and the uncertainty principle. Simply put, our science has graduated itself from a simple field of direct observational and experimental discipline to a complex field of weird concepts, probabilities, and speculation.

Part I of *Riding on a Ray of Light* highlights the recent developments in the field of theoretical physics explaining the essence of many modern theories such as the theory of relativity, the quantum theory, the big bang theory, the expanding universe, the arrow of time, and many such others. A general science reader might have come across many of these intriguing theories in some context or the other in his/her rigorous academic studies or during his/her amateur scientific pursuits, but the real meaning of these theories may generally be lost upon the reader mostly because of their complex and cryptic nature. A science reader may even be dissuaded to seriously pursue these theories because of the intimidating maths they involve or perhaps because these theories themselves are based on some other unintelligible hypotheses which muddle up their understanding. Whatever the reason, despite our deep interest in these matters, much of the scientific knowledge the modern physicists boast of remains by and large obscure and esoteric to a common reader.

But there is a joy in understanding the secrets of nature! Many times, as science enthusiasts, we may come across scientific titbits in the form of some scintillating statements in our media such as, 'Einstein proved right once again!' or 'God particle discovered!' or 'Faster-than-light particles found?' and so on. We might have known a great deal about Einstein and of course of his genius, but do we really have a clear theoretical understanding of what he had discovered which made him so sensationally famous? Or we might have studied atoms and electrons in our college curricula and have known even about some wobbly subatomic particles, but do we correctly understand the current model of the atom called the Standard Model? Or from time to time, we might have palpated a lot of excitement in the scientific circles about certain major breakthroughs in the much celebrated theories such as the big bang explosion or the gravitational waves, but do any of these discoveries mean anything to us? Or we might have heard of the mighty supernovas exploding in the far-off skies or the eerie black holes lurking deep in the cosmos, but do we have any theoretical background of their formation? The list of such perplexing phenomena of nature is endless. Though fascinating these theories are, they remain unintelligible to an average science reader as ever! Part I of *Riding on a Ray of Light* gives us a succinct account of all the main theories of theoretical physics in an easily understandable manner.

Apart from such exotic theories, we may sometimes become curious of many commonplace scientific problems that we may encounter in our daily lives. For example, we may ask why heat always flows from hotter to colder bodies, how our sun shines so resplendently and relentlessly for billions of years, why the evening sky looks crimson red, what makes uranium radioactive but not a bar of iron—zillions of such questions! We may have some makeshift answers to some of these questions, but a correct scientific understanding of the various phenomena of nature not only gives us an immense scholastic pleasure, but it would also tickle our scientific zeal to explore further into the unknown. Part I answers many such questions wherever they become relevant in the stride of discussion of many diverse scientific topics covered in the book.

There is yet another set of intriguing questions that have fascinated men for ages such as these: How big is our universe? Is the universe finite and limited, or is it infinite and boundless? How has the universe originated, and what is its ultimate fate? Is there life beyond the earth? Many such seemingly unanswerable questions have troubled many charlatans of the yore as they have plagued the scientific elite of today! Many of such unresolved questions are discussed at the end of each chapter in Part I wherever they become relevant.

However, it must be acknowledged that Part I covers all the established theories from a bird's-eye view with some sweeping generalizations, and if an interested reader finds the information inadequate, he/she may dig deeper into the topics from more learned and masterly sources, which are aplenty in the market. It may be reiterated at this point that the chief purpose of Part I is only to serve as a general guide to the reader to understand the new concepts presented in Part II.

Now we will look into Part II of *Riding on a Ray of Light*. Part II proposes a unique model which effectively answers several fundamental questions in modern theoretical physics which were left unanswered for a long time in the history of science. Hereunder, we will have a brief rundown of the fundamental questions tackled in the book, but before going into that, we will look into the general theoretical approach that is followed here in this book to unravel the secrets of nature.

There are two chief ways by which a scientific problem may be studied—one is by the way of experimentation, and the other is by the way of conceptualization. Scientific experiment is a methodology by which we may observe, manipulate, and interpret the events of nature to answer a question. Experiments form the most important aspect of any research in science. Conceptualization, on the other hand, is the formation of an abstract principle developed in the mind of the researcher in order to answer the question under observation. A concept is a dainty idea which usually collates the results of several ongoing experiments along with the already acquired knowledge in that field to answer the question under observation in the simplest possible terms. Conceptualization is also an important aspect of scientific research. The importance of conceptualization in the field of science is briefly discussed below.

Experiments, in general, provide us with bits of facts about nature, but a concept is born when we arrange these facts in a meaningful way in a larger perspective. Or we may say that conceptualization is the assemblage of several jigsaw pieces of hard factual knowledge into a theoretical big picture. We may even say that without a proper conceptual reconstruction of the available facts, the subtle meaning of a particular observation or experiment remains buried in the avalanche of crude facts of nature. We have a vast supply of examples in the history of science which endorse the importance of unifying concepts arising out of diverse experimental data. We will have a few historical examples here.

Tycho Brahe, an astronomer of the fourteenth century, was a meticulous observer of the night sky, and he had an impeccable collection of vast astronomical data of his time (perhaps owing to the fact that he could remove his false metal nose, which enabled him to align his eyes perfectly to the sextant for astronomical navigation!). But then even with all this enormous data right under his nose, he could not envision the correct model of the universe, and his planetary model was no better than ancient Ptolemy's earth-centred model. Only when this data had fallen into the hands of his astute assistant, Johannes Kepler (after the death of Tycho, who had guarded off his treasure from others until then), did the deeper meaning of these meticulous astronomical data became apparent. Kepler, with a great conceptual insight, analysed Tycho's records and developed his immaculate sun-centred model with three indomitable laws of planetary motion (which are applicable even today), which correctly predicted the celestial orbits of almost all the known planets of that time!

We have several other examples all along the history of science which underline the importance of conceptualization: Rutherford's planetary model of the atom (as we will see in Ch 3) was a great conceptual rearrangement of available facts which has really initiated the golden era of nuclear physics, Einstein's theory of relativity (Ch 1) was another crucial conceptual revamping that has set off an array of big changes in theoretical science, the quantum theory (Ch 2), the Standard Model (Ch 3), and the big bang theory (Ch 5) – these were all important concept-based successes. The reader may find a countless number of such examples etched up in the annals of science in all of its various disciplines!

In Search of a Unified Theoretical Model: Now we will get back to the context of our book. Part II of *Riding on a Ray of Light* is a conceptual work based exclusively on the established scientific facts in the field of modern theoretical physics. It ushers in a working model in theoretical physics called the *negentropic model*, which answers many of the hitherto unanswered fundamental problems in this field. We will have a brief overview of the fundamental principles that are taken up for our study in the book.

For over a hundred years or more, scientists have been looking vainly for a unified theoretical model which could explain all the physical phenomena occurring both at the macrocosm and the microcosm on a single platform, and such a unifying idea has already been propitiously named the *theory of everything*. But then there has been no actual success with any of the models presented so far! However, all the experimental and conceptual data accumulated so far has given us to understand that the first step in this direction would be to unify the four fundamental forces of nature (electromagnetic radiation, gravitational force, strong force, and weak force) into a single theoretical framework (i.e. into a 'unified force'), and this unified force has been the holy grail of physics all along the past century! But then again, despite the hardest efforts of our scientists, none of the theoretical models of today have succeeded in unifying all these four forces. Though three of the four forces could be unified by the quantum theory, the gravitational force has remained stubbornly aloof, refusing to be merged with the rest (we will study all the details in Part I). But then why is this grand merger proved to be so nearly impossible? Do we have to erect some new kind of theoretical edifice in place of the theory of relativity or quantum theory, as is believed by many scientists? Or is it because we need some sort of new physics or some sort of surreal subatomic particle to accomplish this grand merger? Or is the entire academic exercise merely a theoretical mirage which our human mind is simply not designed to decipher?

On the other hand, consider these questions: Could this conceptual hurdle be due to some sort of theoretical misinterpretation of the already existing experimental data we have acquired down the centuries? Or putting it the other way, is there a scientific procedural oversight on our part? In other words, is it still possible for us to rearrange the vast amount of available knowledge in a new way to build up a simple theoretical model? With these questions in the background, we will undertake a studious reorganization of the existing data, starting afresh with the fundamentals and climbing our way up to many complex theories to finally unravel a working model which would stitch up all the pieces in this gigantic jigsaw puzzle into one simple grand design.

However, it may be realized at the outset, in this new approach, that the first and foremost block in theoretical science is our lack of complete understanding of the phenomenon of light. In other words, the first step in our expedition for unification must be to conceptually understand the very nature of photon. And it is shown in Part II that this understanding would allow us to interpret matter, gravity, and our universe in a proper perspective! Thus it may be stated that:

A ray of light is at the centre stage of our theoretical understanding.

In accordance with the above reasoning, we will first study the general features of motion in Ch 7, which actually paves way for a correct conceptual understanding of light. In Ch 8, a mechanism of generation of light is presented in a versatile model called the *photonic negentropic model*. In Ch 9, we will see an extension of this model, aptly called the *fermionic negentropic model*, which unravels the mystery of how photons transform into matter particles. We know that matter, once it takes its birth, exhibits an unimpeachable property of gravity. Ch 10 gives us an account, once again based strictly on the negentropic model, of the origin of gravity at the most fundamental level. In Ch 11, we will present a comprehensive view of the negentropic model, which gives us some tolerably simple explanation of a few of the weird quantum phenomena we have encountered in Part I. And, finally, in Ch 12, we will apply the negentropic model at the cosmic level and see how it could answer some of the most perplexing questions of the cosmos related to black holes, dark matter, dark energy, the beginning and ending of the universe, and several such issues.

Author's Note: Having outlined the purpose of this work, I may conclude this introduction with a brief critique. Firstly, the reader may excuse me for my lack of an academic qualification in laying out this exposition, but I would, rather humbly, submit that I am a professional in another scientific field. And when readers peep into the chapters, they would find the job well done to the general satisfaction of a specialist in this field! However, it is my belief that sometimes, a specialist in one scientific field may be able to look more critically (and perhaps with an advantage of independent perspective) into the unresolved problems of some other specialty, which may really help science to step into new avenues! Nevertheless, I must appeal to the learned reader that throughout the book, I have worked only to elevate the cause of science, and I did not digress, at any stage of the book, from the scientific method.

In the end, I must admit that the negentropic model is only an approximate theory in the sense that it simply indicates a general approach to the problems in theoretical physics. Neither is it a complete model because, though based firmly on empirical grounds, it is not treated with intricate mathematical formulations as it is often required in the modern scientific procedure! And I should also admit that a few inconsistencies might have crept into the hypothesis, but then it may be conceded that such incongruities are inevitable considering the vast amount of academic exercise that is undertaken in this work!

Now off we go into the *Riding on a Ray of Light*!

PART I

THE KNOWN UNIVERSE

CHAPTER 1

The Macrocosm
And the Majestic Theory of Relativity

Overview

The universe, which is exclusively made up of matter and energy, may be studied in two ways – on a large scale, it may be examined as planets, stars, and galaxies; while on the smaller scale, it may be studied as molecules, atoms, and subatomic particles. In other words, we may discuss the universe either as a *macrocosm* or in a *microcosm*.

By the early twentieth century, two prominent theories emerged in the field of modern physics to study the universe—one is the majestic *theory of relativity*, which enshrined the rules that govern the events of macrocosm of which we will learn in this chapter; and the other is the intriguing *quantum theory*, which encased the rules that govern the world of subatomic particles of which we will learn in the next chapter. There is a compelling reason for us to study these two theories at the outset of our journey into the cosmos. The concepts that are presented in these two chapters keep recurring, in one form or the other, in our later discussions throughout the book – hence, the general reader is advised to go slow in these two chapters and assimilate the ideas presented therein. However, the reader may find some of the relativity and quantum concepts a little offbeat because these ideas may run counter to his/her intuition, sometimes driving the reader to a point of proclaiming them as nonsensical or even insane. Nevertheless, the reader must realize that these theories have gracefully stood the test of time amidst criticism and disapprobation, and they have ultimately become the two flagship theories of the twentieth century which formed the basis of further developments in the field of theoretical physics, as we will see throughout the book.

The story of relativity began quite insidiously in the medieval period, chronicled in the notebooks of several great scientists such as Galileo and Newton and finally culminated in a consummate theory in the mind of Einstein in the twentieth century. The intricate theory of relativity is introduced to the reader in this chapter in a stepwise manner, first dealing with the evolution of the 'special theory of relativity' followed by the development of the 'general theory of relativity'. We will then learn about several implications of the theory and finally see how it has acquired the status of a valid theory in the field of theoretical science.

However, in this chapter, we will be dealing with only the essentials of relativity theory, but then by the end of this chapter, hopefully, the reader would be able to comprehend the significance of various relativity jargon such as *space-time, time dilation, twin paradox, gravitational waves*, etc. And of course, the reader may also get a toehold of the meaning of the famous relativity equation, $E = mc^2$.

Now starts our journey into the uncanny world of relativity!

Section A
Introducing Relativity

The theory of relativity is all about moving objects in the universe. The motion of celestial objects in space has always intrigued the scientists, and the early scientists have come up with some interesting questions regarding the nature of motion of objects in space, and the answers to these questions have really become the basis of our modern understanding of the universe, as we will be seeing in the following sections. We will first study motion in a systematic way, and then we will see how generations of scientists have strived to answer and solve the problems related to motion, which has ultimately culminated in the theory of relativity.

We will start our discussion with a simple thought experiment put forth by the French mathematician Henri Poincaré in 1898: you went to bed at one night as usual, and it so happened that suddenly during the night, everything in the universe has become a thousand times larger. This includes *absolutely everything*—the bed you have slept on, the clock by your side, your house, the mountains outside, you yourself, and each of the atoms and molecules in existence, the earth, the sun, the stars, your measuring scale, your telescope—absolutely everything. What of it then? Upon waking up, is there any way of telling that *anything* has changed at all? No way! You could never tell, in this setting, that anything has changed at all. You would still feel much the same as it was before. This means to say that we rely upon some 'fixed standard' to make all our measurements, and we assume that this fixed standard to be constant and unchanging. Thus, the length of the pole we measure, the size of the land we estimate, the weight of dough we determine, the course of direction we reckon—all these measurements are relative to some set scales for our convenient comparison. Therefore, units such as a yard, a gram, and a litre are all measuring scales we use for our day-to-day reference to do some useful work.

Thinking about this in another way, we may say that all the measuring standards we reckon are also relative to the other parameters, and thus, when we look at this feature a little further, we will notice that all our measuring standards are interrelated. For example, our current notion of *length* is in fact dependent on two other measuring standards, i.e. time and motion—we cannot calculate the length without referring to these parameters. Consider this: What is a *metre*? Historically, the length of a metre was defined as one-ten millionth of the distance from the equator to the North Pole, but this definition is now obsolete because of its inaccuracy. Modern science redefines a metre as the distance light travels through vacuum in one-$299{,}792{,}458^{\text{th}}$ of a second. But then what is a *second*? Gone are the days when we used to calculate a second by a certain fraction of the period of earth's orbit around the sun. Now a second is defined precisely as the basic unit of time needed for a caesium-133 atom to perform 9,192,631,770 complete oscillations (as measured by an atomic clock). Of course, the reader may see that 'oscillation' is a statement of position in space in a period of time, which again is a measurement based on length and time. In the same way, a *kilogram** is defined as the mass equal to the mass of 1.000028 cubic decimetres (dm³) of water at its maximum density at approximately 4 °C. By these definitions, the reader may notice that the units of length, time, and mass are all interrelated, and to measure one parameter, we use the reference of the other parameters. Thus, all our measurements in nature are relative to each other.

All our measurements in nature are relative to each other.

Is Motion Absolute or Relative? While all the measuring parameters are interdependent, the phenomenon of motion itself posed a different problem. The early scientists have realized that there is something intriguing about motion of objects. Consider another interesting thought experiment: an ocean liner was cruising along in the Pacific Ocean in a straight line with a speed of 100 km/hr, and a man on board is walking along in the direction of its motion at 6 km/hr. Now what is the *exact* speed with which the man is moving? We can see that with reference to the ocean liner, our man is walking at 6 km/hr, but with reference to the surrounding still waters, his speed can be said to be 100 + 6 = 106 km/hr. Thus, it can be readily seen that the motion of our man can differ by considering different *frames of reference.* But let us get a little curious and continue with our calculations a little further.

We can go on and take larger frames of reference and calculate the man's speed on his ocean liner. For example, an astronaut, Lucy, who is looking at the man on board from the moon, can say that he is moving at a speed of 1,674 + 106 = 1,780 km/hr (as our ocean liner is approaching Galapagos Islands, which are near the equator, and the earth rotates with a speed of 1,674 km/hr at the equator). You can consider further hypothetically: Lucy is now looking at our man from deep space in the solar system and say that he is moving

---

*   The 26th *General Conference on Weights and Measures* has voted to set a new definition for kilogram based on the 'Planck constant' of which we will learn in Ch 2-C.

at a speed of 108,000 + 1,674 + 106 km/hr (as the earth is hurtling round the sun with a speed of 108,000 km/hr). We will go a little further: our intrepid astronaut, as Lucy was, has now ventured into the deeper spaces of the Milky Way Galaxy (which is our home galaxy, Ch 4-A), and now she could calculate the speed of our man in the Pacific Ocean to be much faster because she can now see that our solar system itself is rotating and moving along with the Milky Way with a speed of 828,000 km/hr, taking along its stride the stars, the sun, the earth, and also our man with tremendous speeds. And furthermore, Lucy realizes that the Milky Way itself is moving in the gigantic *Local Group* (the name of our galactic cluster, Ch 4-A) and that the Local Group itself, as per the recent calculations, is moving at even greater speeds of about 244,792 km/s per megaparsec. Nobody knows as yet how far we could go on this way! In any case, our man's pleasant holiday trip in the Pacific has unwittingly become a whacking astronomical event with mind-boggling speeds!

Whatever that may be, it has now become clear that there could not be a 'fixed' standard reference to motion in the universe, and so a 'final' frame of reference which can be used universally does not exist! Hence, it appears that *the concept of motion is not absolute—it is 'relative'*. In other words, no object in nature is stationary but is in eternal relative motion.

There is no absolute motion in the universe – there is only relative motion.

The Real Problem: But then the matter is not so simple. If it were to be this straight, there was no need for all the scientific battles that have taken place across the history of science, and there was no need for the genius of Einstein to settle the issue with his ingenious theory of relativity! The real problem is that even though it seems to us that we have an obvious evidence for the relative behaviour of motion, scientists of the 'pre-relativity' period have considered motion to be *absolute*—of course with some justifications. This is because scientists have known that there are at least two sure-shot methods by which they could determine the *absolute motion* of an object in the universe, as we will see now.

The First Method: What if we have a *fixed frame of reference* throughout the universe which can be taken as a reference so that *absolute motion* of an object can be determined? In this situation, we can avoid comparing the motion of an object with the position of another object, as we have done in the example of the man in the Pacific Ocean. Scientists already have one such reference which was universally accepted as a standard (at that time), and this reference was called the *luminiferous ether* (Sec-B). If luminiferous ether is fixed and unmovable, as it was supposed to be, then an object's motion in relation to ether can be said to be absolute. In Sec-B, we will see how Einstein had dealt with ether and the speed of light in his *special theory of relativity* and solved the problem of absolute motion.

The Second Method: This is a little devious and complicated. Scientists have realized that the *inertial effects* exerted by the objects as they accelerate in space may also be employed to determine the absolute motion of objects, as we will see in Sec-D. Once again, we will see how Einstein had dealt with inertia and gravitation in his *general theory of relativity* and solved the problem of absolute motion once and for all.

Now we will see, step by step, how Einstein surprised the scientific world by resolving this issue in two folds—once in 1905 with his special relativity and again in 1916 with his general relativity.

Section B
Special Theory of Relativity

Theoretical Grounds for the Special Theory: More than 300 years ago, Galileo Galilei (1564–1642) conducted a thought experiment in which you shut yourself up in the windowless cabin of a ship which has a little bowl of water with water drops falling into it from above (Fig 1.1). Galileo said that as long as the ship is travelling in *uniform motion*,* the drops fall the same way as they do when the ship is

---

*   An object is said to be in uniform motion when it travels in a *straight line* and at a *constant speed*. Non-uniform motion occurs when the motion is *not* in a straight line (i.e. motion in a curved path) or when the motion becomes

stationary. Merely by looking at the falling drops, it is impossible for us to say whether the ship is moving or stationary without an outside view (i.e. without any outside frame of reference). And thus, the falling of water drops remains unaffected in both cases—the ship is moving at uniform speed or is in a stationary position—and there is no way of telling either way as long as the outside reference is cut off. Of course, when the ship takes a jolt or when it moves with *acceleration* (increasing or decreasing speeds), then the falling of water drops goes awry, moving either backwards or forwards, which would then indicate that the ship is moving in one direction.

It is also our common experience that if we are sitting in the cabin of a moving train and look out from the window at another train which is moving alongside of our train, we would be momentarily confused as to which train is moving—until our train gives us a jolt or starts accelerating. Of course, we will also be confident of our train's motion if another window shows the outside view of the stationary platform. Similarly, it's our usual experience that when we toss up a ball in a moving train (of course moving in uniform motion), it will come down straight into our hands as if the train is stationary, and the ball goes awry if the train suddenly accelerates. The conclusion is that as long as you are travelling in *uniform motion*, no experiment would be able to ascertain whether you are moving or stationary.

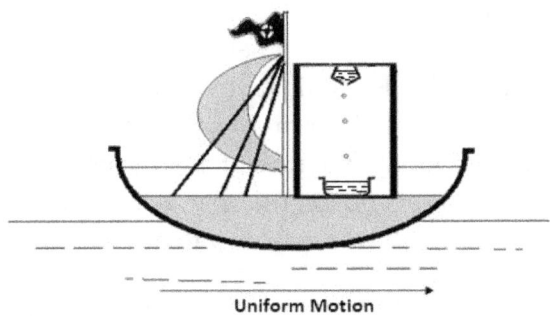

Uniform Motion

Fig 1.1: The Galileo thought experiment.

---

accelerated, i.e. either getting faster (when it is called *acceleration*) or slower (when it is called *deceleration*). See Ch 7-A for details.

There are also certain general features related to moving objects, which can be readily understood by us because they really form a part of our intuition and common sense. Consider the following situations: a policeman fires a rubber bullet from his gun to dispel a rioting mob. When he is standing on the ground, the rubber bullet travels, say, at a speed of 100 m/s, but if he fires his gun from an open-top police car which is speeding at 30 m/s, then the bullet acquires a total speed of 130 m/s. This means that the speed of the bullet depends on the source of the bullet. The speed of the source adds up to the speed of the moving object. You can also consider supersonic speeds: a bullet fired from the ground has a lesser speed than the bullet fired from the jet plane, which has a total speed of the bullet plus the jet. These are acceptably rational outcomes, and such calculations form a common part of our intuition.

But now let us take the example of *light*.[*] In the past, it was thought that light travels with an *infinite* speed, so the question of speed of light did not arise at all to the early scientists. However, in 1704, Sir Isaac Newton proposed that light is composed of tiny particles, which implied that the speed of light is finite, and subsequently, Ole Rømer demonstrated experimentally that light did travel by a *finite* speed. And modern research has precisely determined the speed of light as 299,792,458 m/s *in vacuum*.[†]

However, by the end of nineteenth century, an odd behaviour of light became apparent: the speed of light was found out to be independent of the movement of the source. In other words, light's speed does not vary with the movement of its source, as we have seen with the policeman's bullet. This fact was adequately documented by the early astronomical observations (and amply confirmed now by highly accurate modern experiments). This means that regardless of the motion of the light source, the speed of the light in vacuum is fixed at 299,792,458 m/s. It was contrary to our expectation that light

---

[*]   The term *light*, in a study like this, is used as a short form of 'electromagnetic radiation'. Thus, light not only means visible light but includes any of the electromagnetic radiation ranging from gamma rays to x-rays to radio waves (Ch 2: Table 2.1).

[†]   When we say *speed of light*, it means the speed of light specifically *in vacuum*. Light travels more slowly through other media such as air, water, and glass. The phenomenon of refraction of light is but the result of this property.

behaved this way, but scientists believed in this odd behaviour of light because this was supported by many experiments.

We will put this property of light in another way: for a light ray moving in space, the source of light can no longer be taken as a reference. But then in this perspective, another logical question would arise: if the speed of light through empty space is fixed and moving all on its own, then what was it moving in relation to? The answer to this seemingly simple question has really turned out to be of profound significance and has really added new dimensions to our understanding of the universe, as we will see now.

Ether and Michelson–Morley Experiment: Now we will examine the above question in another perspective. First we will come back to the problem of absolute motion. We have seen in Sec-A that one method to record the absolute motion of any celestial object such as the earth, moon, and sun is to use a universally fixed frame of reference. What could such a universal reference be? As we have already seen, there is a traditional fixed reference in nature called luminiferous ether (or simply ether), which was originally proposed by Aristotle more than 2,000 years ago. The existence of ether was well supported by the scientists both by intuition and by experiment. Way back in the seventeenth century, Otto von Guericke conducted a simple experiment: he emptied a glass bottle of its air and showed that sound waves from a bell hung in the bottle stopped passing through it because of the lack of conducting medium (i.e. air), but light still could pass through the bottle unhindered (as we could still see that bell!). This stood as an unimpeachable proof for the existence of ether – otherwise, how could 'some*thing*' like light pass through emptiness? Thus, luminiferous ether was thought to pervade the entire universe—present in every nook and corner of space in the universe covering all the vast empty spaces between the galaxies, amidst all the stars and planets in the universe, not only surrounding the earth but also covering every bit of space surrounding you and me. It was thought that ether literally bathes all the matter in the universe and suspends all the atoms in it. However, it was considered that this all-pervading ether can never be seen or felt or smelt or tasted or cannot be perceived by any means – it was simply believed that ether existed!

Of course, there was yet another evidence for the existence of ether. Since the time of Christiaan Huygens (in the seventeenth century) and Thomas Young (in the nineteenth century), it was known that light travelled by waves (Ch 2-B), and for the waves to propagate, there must be a medium, and the scientists have conveniently supposed that ether was the medium. Moreover, light from distant stars in galaxies had to travel through vast empty spaces between the galaxies before reaching the earth, so it was considered impossible for the light to travel in relation to stars, which were immeasurably far off each other, and so it was thought that light travels with reference to ether. So on all these accounts, the existence of a *fixed* and *motionless* ether was a natural hypothesis of the erstwhile scientists, and it could not be falsified or refuted in any way. And moreover, there was no need for the scientists to contest this view! And thus it was concluded that all celestial objects, including light, were supposed to move in relation to this fixed ether.

All objects in the universe, including light, travel in relation to ether.

But now we will see why the documentation of luminiferous ether became necessary for the scientists: the scientists at that time in the late nineteenth century had set out to measure the *absolute motion* of our earth in the vast, empty space. But then we can be sure of the existence of absolute motion of the earth in space only if we can somehow document its movement in relation to the fixed ether. As stated above, the logic was simple: instead of describing the motion of the earth in relation to the sun or other planets or stars (in which case it becomes a relative motion), we can determine the absolute motion of the earth by estimating its speed in relation to the fixed ether.

However, the crucial problem is this: how can we use ether as a reference experimentally if we cannot in any way detect it? But soon, the scientists have realized that the solution was the speed of light itself. Because light travels with a finite and fixed speed in ether without any reference to the source of light, the speed of light itself can be taken as a reference. Thus, to know the absolute motion of an object in space (e.g. the earth), we can measure the earth's motion in relation to the speed of light itself. And this ingenious concept was the logic behind the 'Michelson–Morley experiment' of which we will discuss below.

In 1887, Albert Michelson and Edward Morley had designed their famous experiment originally to calculate the absolute speed of the earth by using the above principle. However, the reader has to note here that because this experiment takes into account the movement of the earth (in ether) in relation to the speed of light (in ether), this experiment incidentally proves the existence of ether itself! With their ingeniously built equipment (the reader may get details of the experiment from other sources), they tried to measure the velocity of light in different directions of the earth's movement, which would then show the earth's absolute motion because the 'ether wind' is expected to drag upon the light when light moves in opposite direction but not when light moves in the same direction. But to their utter surprise, and despite their repeated and meticulous measurements, they could not find any difference in the speed of light – light travelled with the same speed in both the directions. Light behaved as though earth and ether did not exist! Michelson and Morley were at a loss to explain the negative result of their experiment. What could be the explanation of this odd behaviour of light? Why did not ether cause a drag on the light?

A strange explanation of this negative result of Michelson–Morley experiment was offered by Lorentz and FitzGerald in 1892. They presumed that ether puts pressure on a moving object in space (in this case the earth), causing it to *contract* in the direction of movement. Thus, the length of the object contracts more as the speed increases, and as a result, the speed of light itself remains unchanged (because now light has to travel for a shorter distance). Now the reader may get a legitimate doubt that in the case of shrinkage of length of an object, we can ascertain the contraction by measuring the length of shrinkage using a ruler. But it must be realized that along with the moving object, the measuring ruler (i.e. the measuring equipment) also contracts, thus keeping the measurement itself unaltered as if there was no contraction (remember the Poincaré thought experiment!). This phenomenon came to be known as *Lorentz–FitzGerald contraction*. This contraction, according to Lorentz and FitzGerald, is a *physical change* in the objects (later Einstein fine-tuned this idea, as we will see). However, this was considered an *ad hoc* hypothesis as it was just formulated to explain the odd result of an experiment, and there is no way we can either prove or disprove the hypothesis. Nevertheless, subsequent daintier experiments did not reveal any such changes, and thus, it was considered that this theory had failed to account for the negative result, but the problem of absolute motion itself had remained unanswered.

Einstein and Special Relativity: By this time, the story of relativity had already started in a nondescript corner of a small patent office in Bern. In 1901, as a young man, Albert Einstein, then a patent clerk, conducted a simple thought experiment: imagine that you are travelling in space along with a ray of light with a mirror in your hand (Fig 1.2). If you are moving at the speed of light and try to look into the mirror, what would happen? The light fails to leave your face and reach the mirror because you, your mirror, and the light are all moving at the speed of light, and consequently, you would not be able to see your reflection in the mirror. But recall Galileo's experiment (described above) – it concluded with a statement that as long as the observer is travelling in *uniform motion*, no experiment would be able to ascertain whether you are moving or stationary. However, Einstein's thought experiment could be able to say definitely that as soon as you attained uniform motion at the speed of light, your reflection in the mirror would vanish. This clearly points out that you can ascertain that you are moving with uniform motion at the speed of light by this 'optical' thought experiment! Thus, Einstein argued that either Galileo was correct or he was correct—but not both!

Fig 1.2: Einstein's thought experiment.

To resolve the issue, Einstein made an ingenious suggestion. His proposition was simple: light travelled in relation to the observer who makes the observation. Now reconsider the thought experiment with this assumption—you can see that even at the speed of light, the light rays still leave your face and reach the mirror because light now travels relative to you! Problem solved! With this assumption, you can still look at your face in the mirror even though you are travelling in uniform motion at the speed of light! Thus, with this thought experiment, both Galileo and Einstein could be adjudged correct!

However, much to the bewilderment of scientists at that time, the implication of this theoretical assumption was radical. Einstein's theoretical assumption has unwittingly proposed that *light did not travel relative to ether* or simply that *ether did not exist at all*. And so with this, the time-honoured concept of ether has to be discarded altogether! It is ironical to note here that just then, Michelson and Morley had conducted their elaborate experiment to arrive at the same conclusion which Einstein had concluded with his thought experiment (at that time, Einstein was not aware of the Michelson–Morley experiment, though famous the experiment was!). But then even though both the meticulous experiment and the ingenious hypothesis had pointed to the same result, the thinking pattern of an upstart such as Einstein and that of stalwarts such as Michelson and Morley was different – while Michelson and Morley had trouble with their preconceived notions on the ether dogma and preferred to stick with it, young Einstein was bold enough to jettison the concept of ether altogether. In fact, Einstein did not say that ether did not exist but only that ether has absolutely no effect whatsoever on the speed of light in space. But in effect, this nearly meant the same thing as ether did not exist. Nevertheless, whatever might have happened to the existence of ether, the big conclusion of Einstein's thought experiment was

The speed of light is constant and travels relative only to the observer.

This seemingly innocuous but radical assumption had startled the scientists of the time not only because it has some far-reaching implications on our understanding of the property of light, but it also plunges us into a queer new world of relativity. If the light is not travelling in relation to ether but is absolute all by itself, then light must behave in an absolutely fantastic way which must be quite contrary to our common experience, as we will see now.

Imagine an astronaut in a spaceship travelling in space at half the speed of light, and alongside, a beam of light passes in the same direction. If the astronaut measures the speed of this passing light beam, our common intuition says that the speed of light must be half its velocity because the astronaut is travelling with half of light's speed. But according to special relativity, the light beam is still passing by the astronaut at its full speed of 299,792,458 m/s because the light beam is now moving in relation to the astronaut himself/herself, who is making the observation (compare with our policeman's bullet above!). Now consider this: even when the astronaut travels towards the source of light with half the speed of light, the astronaut will not measure the light speed twice as fast, but he/she will still measure the speed of light at 299,792,458 m/s only. Thus, the speed of light has now emerged as a new absolute—it is absolutely and universally constant in reference to any object in the cosmos (or any observer making the observation)! This is the first cardinal principle of the theory of relativity.

Speed of light is absolute.

By the year 1905, Einstein has formulated the mathematics of the special theory of relativity, which has enshrined this cardinal principle. We will know why this theory came to be known as the 'special' theory when we deal with 'general' relativity. But for now, we will dig a little deeper into the affairs and see what happens to our world in special relativity.

Section C
Implications of Special Relativity

Our world is a predictable place. The lengths we measure, the time we record, and the pounds we weigh are all unchanging, and so all our measurements are expectably uniform at a given place and time. However, in the world of relativity, we will see that these 'standards' are merely our cherished notions, but the underlying theoretical truth is profound and deep, and consequently, it can be shown that all our concepts of measurement are not absolute but are only relative, the only one absolute entity in nature being the speed of light. We will now see how our notions change in the special theory.

Length and Time—Absolute or Relative? We will now see what happens to the common measurements such as length and time in special relativity. Hitherto, we have considered that the length and time we measure are standard and invariant (or unchanging) and could be relied upon for measuring a moving object in the universe. But now we will examine these measurements in the light of the new principles of special relativity.

The Lorentz–FitzGerald contraction theory had already suggested that when you measure the length of a moving object in relation to the speed of light, the length has to contract in order to keep the speed of light constant. Lorentz and FitzGerald had considered length at rest as absolute and thought that the length of an object contracts only with motion. Einstein went on the full way and said that not only in the case of a moving object but the length of the object at rest also cannot be taken as absolute any more, which meant that 'absolute length' by itself is a meaningless concept. Therefore, it can be said that there could not be any 'true' length in nature – all measurements are relative. In short, according to the special theory, the length of an object is not absolute but is dependent on the moving status of the observer.

Length of an object is relative to the observer.

Thus, it was shown in special relativity that the length of an object contracts *along the direction of its motion*. For instance, a 100-foot-long spaceship travelling at 99.99% the speed of light will appear one foot long to an outside observer, but the observer in the spaceship measures it as 100 feet only. Hence, there is nothing called *real change* in dimensions in nature, but every change in nature is relative and observer dependent (this is in contrast to the real physical change proposed by Lorentz and FitzGerald). In short, length (or as a matter of fact, any dimension) is no longer an absolute entity, and if we really want to define length precisely, its relative speed in space and time must also be incorporated in its measurement (see below)!

Time Dilation and the Case of Twin Paradox: Regarding the problem of absolute time, special relativity shows us that time is also relative and personal, as explained below. Before Einstein, scientists, in general, have agreed with Newton's *universal time*, which was thought to prevail throughout the universe. This universal time was considered as a gigantic clock ticking away eternally in a cosmic heartbeat with which all other clocks may be set. But with the advent of special theory, this concept of cosmic time is destroyed forever.

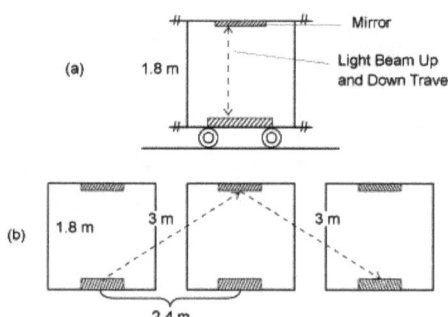

Fig 1.3: Einstein's 'light clock' experiment.

Einstein had shown, basing on his ingenious thought experiments, that time is flexible (not rigid), stretchable (not stagnant), and relative (not absolute). To measure changes in time, Einstein conducted a hypothetical 'light clock' experiment wherein he had used an imaginary contrivance which is far more precise and convenient for our thought experiment than a conventional clock. This imaginary clock measures the time taken for a pulse of light to bounce between two mirrors placed one above the other at a certain distance—each 'tick of the clock' is the time taken for a light pulse to go up and down once. Now if you measure time in a stationary train, the light ray moves up and down, tracing a path resembling 'I' (Fig 1.3a). Suppose the train is moving with a uniform velocity but at very high speeds nearing the speed of light. Then the observer in the train still finds that the tick takes the same time as if it is stationary because uniform motion and resting state make no difference to the observer in the train. However, for an observer who is looking at the clock from an outside platform, it measures a longer tick because now the light has to take a diagonal 'V' path to bounce between the two mirrors (Fig 1.3b). Because the speed of light is constant, it now takes a longer time to travel the longer distance—i.e. the same period of time for an observer on the train has become stretched (or expanded) for an observer outside the train. This means that time has 'dilated'. This phenomenon is called *time dilation*. Not only does the time change, but the physical measurements of the train, such as measurements of length, also contract along the direction of motion, and so you appear thinner to the observer on the moving train. Thus, greater the speed of your motion, slower the clock ticks, and more is the contraction in that direction.

The faster the motion, the slower the clock ticks,
and the more the object shrinks.

Thus, the special theory has shown us that time is relative and is observer dependent. There is no universal time, but there is only 'local' time or 'personal' time. This destruction of absolute time was an astounding accomplishment of special theory, which would rout our good old faith in the affairs of the world. The phenomena of length contraction and time dilation have many theoretical and practical implications, as we will discuss in Sec-F. But one weird implication of time dilation needs a special mention here—the so-called *twin paradox*. Consider this: if a person were to travel in space with speeds nearer to the speed of light, then the time for that person would tick slowly, and consequently, he/she would age slowly. Now suppose there were two twin sisters and one of them was an astronaut. If this astronaut sister, on one fine day, rides in her near-light-speed spaceship to some distant star and returns back to the earth after a few years, then she would find that many more years have elapsed on the earth. And this astronaut sister also finds that she has remained young, whereas her twin sister has aged significantly or has died a long time ago! This discrepancy in aging due to cosmic time travel is called the twin paradox. And if the astronaut sister were to travel by 85% of the speed of light, then her life expectancy would have extended from, say, her usual 70 to an extraordinary 700 years! And it can be said, by this analogy, that according to the special theory, if you were travelling on a beam of light, time would remain frozen forever, and you would never age at all!

Absolute Simultaneity: Because there is no universal time and because there is no absolute length, absolute simultaneity of any two events in the universe is a meaningless concept. For example, to measure the exact length of a moving train, you must know the precise location of its front and back ends at the *same* time—i.e. simultaneously. Because length of a moving object contracts along the direction of movement and as time dilates with movement, there is no precise way by which we can ascertain the absolute position of front and back ends of the train *at a given instant*. However, for the ordinary speeds with which our trains on the earth move, this difference is not accountable at all (and so luckily for us, we could board a moving train without a hitch!), but at speeds nearing the speed of light, the difference would become enormous.

Now consider this: the concept of instantaneity (i.e. the idea of *now*) is meaningful only for the very spot you occupy at a given moment, and you *cannot* assume that other events occur *simultaneously* with your 'now'. To sum up, we may say that no two events in the universe occur at the same time.

No two events in the universe occur at the same time.

Consequently, there is only a 'local time' for any moving object with a fixed reference which is relative to other local times, and there is nothing called a universal time (as Newton had believed). Hence, our earth has only a local time called 'Earth Time', which we would follow for our reference to set our clocks! What a conceptual destruction of our earthly notions that we have cherished all along!

Introducing $E = mc^2$—Is Mass Absolute or Relative? Now we will come to another crucial implication of special relativity by answering this question: is mass an absolute entity? To answer this question, we must first understand the concept of mass in the correct perspective.

Mass is a measure of amount of matter in a substance. A fluff of cotton of a mass of 1 gram has the same amount of 'fundamental matter' (electrons, protons, and neutrons, Ch 3) in it as that of a lead ball of 1 gram. We are taught in our high schools that the *weight* of an object varies from place to place, whereas the *mass* of an object does not vary and is the same at all places. Weight depends on the gravitational force acting on that object. For example, a stone boulder on the top of a mountain weighs a trifle less than what it weighs on the ground; or if you weigh 60 kilograms on earth, you weigh only 10 kilos on the moon but weigh 1,624 kilos on the sun (that is, if you could manage the blaze of the sun and try not to evaporate!). However, it is stated conventionally that the mass of an object is the same whether you measure it on the earth or over the moon.

But then really, the very idea of mass changes when we consider it in terms of special relativity—mass is no longer an absolute entity but is relative! However, mass shows an odder kind of variation in special relativity—the mass of an object changes with its motion because, as we have seen, the length and time of an object change with motion. In this perspective, it may be said that there are four general types of mass:

1. *Gravitational mass.* This is nothing but weight as we have already discussed above, and we have seen how weight of an object varies at different places.
2. *Inertial mass.* Mass of an object can be more precisely expressed in terms of the amount of force required to move it over a distance per unit of time. This is the inertial mass of the object. However, because movement over a distance is dependent on length and time, the idea of inertial mass is also a relative entity. We will see more of this in the general theory of relativity.
3. *Proper mass (or, rest mass).* Consider this: an astronaut carrying a steel ball in a spaceship has a mass which he measures the

same irrespective of its speed because the steel ball can be said to be at rest (or stationary) in relation to the astronaut in the spaceship whatever is its speed. This is the proper mass or the rest mass. This is also called the *intrinsic mass* of that object.

4. *Relativistic mass.* Now consider this. In the above situation, for an observer outside the spaceship, the mass of the steel ball changes as per the relative speed of the spaceship. This is the relativistic mass of that steel ball.

Now we will see how mass of an object changes in accordance with the special theory. We have already stated that the mass of an object changes with its motion because the length and time of an object change with motion, indicating that they have an interdependent relationship. For example, if the relative speed of two spaceships in space is *increased*, the observers in either spaceship will find that the *length* of the *other ship* is *shortened*, the *time* has *slowed down* (i.e. its clocks run slower), and the *mass* of the ship has *increased*. Now consider this: if the spaceships attain the speed of light, the observer in each ship will find that the other ship has shrunk to *zero* length, its clock's time has come to a *standstill* time, and the other ship has acquired an *infinite* mass. However, it must be noted that each observer in his/her own spaceship cannot find any change – as far as he/she is concerned, his/her ship's length, the measured time, and its mass remain the same.

Thus, for these theoretical reasons mentioned above, special relativity stipulates that at the speed of light, a moving object's *length disappears, time stops,* and *mass becomes infinite.* In other words, it may be said that the very existence of object in space ceases to exist once the object attains speed of light! And thus, it may now be summarily stated that no object in the universe can attain the speed of light – once the object attains speed of light, it ceases to have all physical dimensions and is no longer considered an object with mass (Ch 7-B). Therefore, the special theory has put a limit on the maximum speed attainable by any object in the universe to the speed of light (we will see below what happens to objects at speeds 'faster than light'). Thus, another cardinal principle of special relativity has now emerged:

No object in the universe can surpass the speed of light.

Also consider this interesting implication of the above discussion: when a spaceship has attained zero size with infinite mass at the speed of light, it exerts an infinite force on the propellers of the ship, and thus the spaceship comes to a standstill. But then the observer in the spaceship himself/herself would not find any change *except* that the cosmos outside the spaceship is hurtling backwards with speed of light and the galaxies and stars have flattened out into 2D discs with no measurements, and he would find that the cosmic time has come to a standstill. What a fanciful place that would be!

The Principle of Mass–Energy Equivalence: We will now look into *mass–energy relationship*: special relativity shows, in a queer way, that matter and energy are one and the same. To begin with, we will first see what exactly energy is. *Energy* can be defined as the capacity to do *work*, and work itself can be defined as a measure of *change* in a unit of *time* (Ch 6-A). This spatial change in a unit of time means that energy measurements are dependent on linear measurements. However, we have already concluded that all the three variables—length, time, and mass—are relative to the new absolute, i.e. the speed of light! Consequently, Einstein has mathematically shown that according to special relativity, *mass and energy are not only interrelated* but they are also *related to the speed of light*. In other words, we may say that mass and energy are *interchangeable*. This is the so-called *mass–energy equivalence* (not to be confused with 'equivalence principle', which we are going to study in general relativity, Sec-D). Einstein has finally formulated the principle of mass–energy equivalence in his celebrated formula:

$$E = mc^2$$

In this formula, m is the rest mass of an object, c is the speed of light, and E is the energy released when mass is annihilated (destroyed). Because of the large value of speed of light, it becomes apparent from this formula that a very small amount of matter, if annihilated, can release massive amounts of energy.

As a result of this grand principle of mass and energy, the century-old dictum of 'conservation of *mass* in the universe' has finally undergone a beautiful transmutation into the '*law of conservation of mass energy*'. In the later years, this principle would become crucial for the development of various advancements in science, and ironically, this principle had also led to the development of atom bomb, as we will see in Sec-F!

This mass–energy equivalence means that faster an object moves, greater becomes its relativistic mass, and consequently, more force is needed to accelerate the object. For example, it is shown that a small subatomic particle such as a muon (Ch 3-C) speeding in a particle accelerator at 99.9% of the speed of light attains a mass of about 22 times of its rest mass, and because of this gain in mass, it needs greater amounts of energy to propel the muon. But at 99.999% of the speed of light, the muon's mass will increase by a factor of 224, and at 99.99999999% of the speed of light, its mass would increase by a factor of more than 70,000! At these colossal masses, it needs a tremendous force to accelerate the muon. This is the reason why particle accelerators, such as Large Hadron Collider (Ch 3-C), are equipped with massive electromagnets to generate stronger and stronger magnetic fields to cope up with the acceleration of even the smallest subatomic particles. And finally, at the speed of light, it needs an infinite* amount of energy to push the muon!

An offbeat consequence of the special theory is worth a description here. Since the special theory showed that energy and mass are related and since heat is nothing but energy, it can consequently be construed that a cup of coffee would gain a trifle amount of more mass when heated and lose some mass as it cools. And the same is the case with a rechargeable battery, which, when fully charged, gains a tiny bit of more mass. But the problem is that we have no gadgets, as yet, sensitive enough to measure these infinitesimally small mass variations!

The Problem of Faster-than-Light Speeds: We have seen in the above sections that the special theory has put a limit on the speed

---

* In Ch 11-C, we will see that both *infinity* and *zero* can be treated on equivalent terms!

that can be attained by any object in the universe to the speed of light. However, right from the inception of the special theory, there has been some scepticism among the scientists on the concept of upper speed limit, which, of course, if proved wrong, would readily debunk the special theory. And in such a case, all the mighty physical principles we have learnt so far in the twentieth century would simply collapse, and consequently, we would go a century back, forcing us to revise the underlying working principles of our computers, GPS, atom bombs, and such others (Sec-F)! It is interesting to note that in 2011, a team of scientists working with the OPERA particle detector in Italy has made a scintillating claim that certain subatomic particles called neutrinos (Ch 3-C) could travel at faster-than-light speeds, but then this experiment was subsequently invalidated by the researchers as some observational error. And to this date, no experiment or observation has identified any such superluminal phenomenon.

However, we will now speculate theoretically what would happen to objects if ever they were to surpass the speed of light. Before going into this, the reader must understand the concept of faster-than-light speeds in the right perspective. The concept of the upper limit of the speed of light holds good only when the speed of light is measured in a *specific* inertial frame. Observations made outside this inertial frame can exceed the speed limit without desecrating the essence of the special theory. For example, imagine this situation: two spaceships near the earth travelling at three-fourths the speed of light pass each other in opposite directions – and if a third observer who is situated on the earth were to calculate their speed, the observer would measure the speed of the spaceships to be one-and-a-half times the speed of light. This is simply because this third observer is outside the inertial frame of the two spaceships and this observer on the earth has to really conduct two sets of experiments *simultaneously* in order to calculate the velocities of these two spaceships, which is impossible owing to the fact that absolute simultaneity is impossible, as already seen. Thus, when we say superluminal speeds, we are considering the events occurring in a single inertial frame. In the same way, the special relativity is concerned only with the moving *material objects* in the universe but has nothing to do with the movement of intervening *space*. Thus, at the time of the beginning of the universe (we will learn about the big bang in Ch 5), the space is speculated to expand in speeds far exceeding the speed of light, but this does not theoretically violate the special relativity principles. Likewise, it is now thought that the outer reaches of our universe is expanding at much faster rates than the speed of light (Ch 5-H). This is considered to be due to the expansion of space itself, and thus, it is not in contradiction with the special relativity. It must also be remembered that the special theory has nothing to say about what would happen when the speed of light is exceeded because the special theory simply forbids the defiance of this rule. However, we will only make hypothetical conjectures of what would be the upshot when an object supersedes this upper limit.

Now we will look into superluminal speeds. Mathematically, it can be said that if a spaceship is travelling at superluminal speeds, the cosmic time would then run backwards, and the stars would acquire 'negative mass', and the observer in the ship would perhaps find that the cosmos has turned inside out and become a mirror image of itself! However, the formulae of the special theory cannot be applied to these events (because the special theory has no such provisions), but then when applied, they would give out some illogical values to length, time, and mass that result in *imaginary numbers*, such as 'the square root of −1'. Now consider this: if you are a super sprinter and if you could start sprinting at faster-than-light speeds on your tracks, you would soon leave the light falling upon you and travel ahead of the light, and then you can see yourself racing towards you. Such a contingency means that an observer travelling at superluminal speeds can travel in the reverse direction—from future to past! In our present-day world, we are familiar with the cause-and-effect phenomenon, but in this weird world of superluminal speeds, the fundamental cause-and-effect are reversed—i.e. the effect may travel back to cause, or it would be like you may be dead *before* you are born! And in this faster-than-light setting, all such weird things would happen, which would make a fantastic sci-fi movie, like entering into some sort of world of H. G. Wells's *The Time Machine*. And then in this world of faster-than-light speed, you might be actually be watching the bygone dinosaurs rampaging across the forests, enjoying a meeting with your great-great grandfathers, or witnessing a dead man coming out alive like a zombie in a voodoo land!

With this, we have concluded our brief discussion on special theory. However, there are some other implications of special theory which would lead us to unravel certain profound universal truths of nature – they are dealt with in Ch 7-B. Now we will look into the essential principles of the general theory of relativity.

Section D
General Theory of Relativity

We have seen in the special theory that it is impossible to determine the absolute motion of an object in space as long as the object is in uniform motion. But consider this question: can we determine the absolute motion of an object in *non-uniform* (i.e. *accelerated*) motion? Most scientists in the early twentieth century have reconciled to the fact that absolute motion can still be detected in non-uniform motion. They argued that, after all, the effects of inertia due to gravity (see below) could be definitely employed to determine the absolute motion of the objects. However, Einstein had the uncanny intuition that if motion is relative in uniform motion, it must be relative in non-uniform motion also, but then the problem was how to prove it. Einstein obstinately persisted with this problem for eleven hard years after proposing his special relativity theory, and this hard thinking had almost driven him crazy as he had admitted himself once. But then he finally solved the problem of absolute motion once and for all. Einstein showed that all motion in *general*—either uniform or non-uniform—is only relative but not absolute. Hence, it is called the *general* theory of relativity (in contrast to the *special* theory, which is applied specially only to uniform motion). The reader may find that it may be a little hard to imagine the concepts of general relativity at the first shot, but once he/she understands its basic principle, he/she would be bewildered by its simplicity and its magnificence. Let us get down to the basics of general relativity.

Initial Musings: What made the scientists believe that absolute motion was so easily detectable in non-uniform motion? Studying objects in accelerated motion can be readily done by examining gravitational effects on moving objects. Newton had already established the mathematical principles of moving objects in his *law of gravitation*, which was simple:

Every object in the universe attracts every other object.

The mathematical formula of Newton's law of universal gravitation, $F = Gm_1m_2/r^2$, shows that the force with which objects attract each other increases as their masses increase and decreases as the distance between the objects increases (Ch 7-C). But then Newton realized that his law of gravity was incomplete and that there was something about gravity that defied a proper explanation. The following discussion would highlight the point.

Consider this historically famous experiment. In his *Leaning Tower of Pisa experiment*, which is perhaps only apocryphal, Galileo dropped two balls of different weights from the leaning tower and demonstrated that they both hit the ground at the same time. Assume, for example, that a heavy cannonball and a cork ball (assuming that this cork weighs a hundred times lesser than the cannonball) were dropped from a height—the mass of the cannonball being a hundred times more than that of the cork ball, our intuition says that the cannon ball would reach the ground before the cork ball, but astonishingly, they both land on to the ground at the same time (i.e. if we discount the negligible effects of air resistance).* This weird result had shocked the medieval scientists because it ran counter to the established thought, and Galileo himself, at that time, had no proper explanation for this weird fact. This oddity of gravity was properly explained many years later only when Newton proposed his laws of motion, and from then on, this feature became known as the *law of constant acceleration*.

We will look at *Newton's third law of motion* to explain this odd behaviour of objects under the influence of gravity (the reader may

---

* The physicist Brian Cox demonstrated this remarkable phenomenon as part of the BBC Two program *Human Universe* by dropping a bowling ball and a feather in the NASA's Space Power Facility at Ohio, the largest vacuum chamber in the world. The video is dramatically brilliant, very exciting, and worth watching.

find a comprehensive analysis of the Newton's first and second laws of motion in Ch 7-A and 7-B where they become relevant). The third law states that for every 'action', there is an equal and opposite 'reaction'. Here, *action* represents gravity, and *reaction* represents resistance to gravity, i.e. *inertia*. The third law means that the two forces acting upon an object are exactly equal and they act in the opposite direction. We will see how this law works out to explain the queer phenomenon of Galileo's falling objects. While the cannonball is being pulled down by the earth's gravity with a hundred times more force than that of the cork ball, the cannonball at the same time encounters inertia (i.e. resistance) in the opposite direction in exactly the same proportion. Similarly, the cork ball's gravity matches its inertia in the opposite direction exactly. In other words, the *force of gravity* on a falling object is always proportional to the object's *inertia* (Fig 1.4). This tug of war between the gravity and inertia is in full agreement with each other, a feature which was mysterious and unexplainable as if there was some sort of communication between gravity and inertia. Newton had realized about this remarkable correspondence between the two equal and opposite forces and was bewildered by their seeming coincidence, but he could not properly account for this phenomenon. And for centuries, this remained a mystery—until Einstein unravelled the secret with his general relativity.

A falling object's force of gravity is always equal to its inertia.

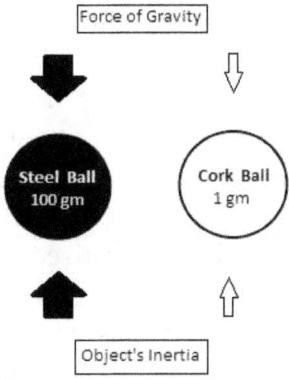

Fig 1.4: Equivalence of gravity and inertia.

Now we will come back to the actual problem of absolute motion: how are these inertial effects used to detect absolute motion in the universe? Here are three examples.

1. It's a common observation that when a bucket containing water is swung around a vertical axis, the centrifugal force will cause the surface of water in the bucket to become concave. You can say that the motion of the bucket is non-uniform because it does not move in a straight line rather it takes a curved course (recall that uniform motion always occurs in a straight line). Newton argued that this *physical effect* of concavity of water surface was a clear indication of the absolute motion of the bucket. It means that the observer who swings the bucket is stationary and the bucket itself is moving absolutely in space; otherwise, how is a physical effect such as concavity of surface of water possible? If movement of the bucket were to be relative to the observer, you are at perfect liberty to fix the bucket as stationary and the observer to be moving, in which case this concavity would not be possible! Thus, this inertial effect on the water in the bucket is an open-and-shut proof of its absolute motion, or so it was argued!

2. When a rocket takes off the ground vertically upwards, an astronaut in it is pressed against the seat, which is an obvious physical effect. This is because of the inertial effect (the body tends to be in a state of rest while the rocket accelerates upwards, Ch 7-B and 7-E). This effect offers a solid proof of rocket being in absolute motion and the astronaut remaining stationary.

3. The earth is not a perfect sphere but is very slightly oblong in shape with an *equatorial bulge* of about 42.7 km. This slight bulge is doubtlessly formed due to the centrifugal force, which is an inertial effect caused by the earth's rotation, and appears to be the result of absolute rotation of the earth in space. If the earth were to be in relative motion, then we can fix it as stationary with all the celestial objects around it moving in relative motion, in which case this equatorial

bulge could not be explained. Thus, this physical change in the earth's shape is also a clear example of absolute motion.

Thus, it may be concluded that the absolute motion of an object can be conveniently detected by using inertial effects. Scientists at that time had given up to argue and accept that special theory is special in the sense that it does not allow absolute motion to be detected in uniform motion and that absolute motion is still possible with non-uniform (accelerated) motion. However, as noted above, Einstein had a hunch that even movement of an object in accelerated motion is relative and went on to investigate further in this direction, which ultimately culminated in his ingenious theory of general relativity.

The Equivalence Principle: At the heart of general theory is the Einstein's famous *equivalence principle*. He stated that gravity and inertia are not different forces but are one and the same—two names given for a single force—and he went on to mathematically prove his claim through years of laborious toil and torment. Einstein said that we may call force in one direction as gravity and the other as inertia, but they operate in different directions. The following discussion makes the point clear.

Einstein explained the equivalence principle by another of his famous thought experiments. Imagine an observer in a lift (= 'elevator' in the US) being pulled up in space with a constantly increasing speed (i.e. accelerated motion). This makes the observer in the lift press upon the floor and experience the gravitational force. Consider this question: is the lift moving up and causing inertial effects simulating gravity, or is the observer moving down along with the universe and causing the gravitational effects? Einstein said that this is not a proper question – we can fix the frames of reference either way and arrive at the same conclusion—one force we call gravity and the other we call inertia. This means to say that according to the equivalence principle, both statements are true. This is simply because inertia and gravity are the same forces operating in opposite directions. For example, when we drop an apple on to the ground, we say that 'the apple has fallen to the ground'. But the fact of the matter is that the apple is falling to the earth as truly as the earth falling on to the apple (though the magnitude of fall of earth *towards* the apple is infinitesimally small!). However, by convention, we tend to think that the smaller object moves to the larger object. Thus, our perception differs only because of the difference in the magnitude of the force generated. In the case of Einstein's thought experiment, that of the lift being too small compared to the gravity generated by the universe, we tend to take the universe as our natural frame of reference and call this force as gravity and call the other force inertia. If we take the Galilean experiment, the cannonball is pulled by the attractive force of the massive earth which we call gravity, which is effectively countered by the attractive force created by the cannonball upon the earth, which we call inertia (Fig 1.4). But in fact, both these forces are two sides of the same coin! Thus, the big conclusion of equivalence principle is:

Gravity and inertia are one and the same.

Now consider this: if we take the sun as the fixed frame of reference, we may say that the earth is moving around the sun; alternately, we may also fix the frame of reference on the earth and say that the sun (along with the cosmos) is spinning around the stationary earth (i.e. as the geocentric theory of our ancients, Ch 5-A). However, according to the relativity theory, both of these statements are true, and it is only a matter of choice of reference. And on the same account, we may also state that the earth is falling towards the apple (as we have seen above) if we take the apple as our reference. But the human perception and logic are so designed that we always prefer to take the larger of the frames as our reference. It is like saying that 'a ball is lying over the table' instead of saying that 'a table is lying underneath the ball', though we are right either way scientifically. But then this automatic perception of larger reference by the human mind is so wonderfully utilitarian so that this perception allows us to get along with our lives on the earth in all perfection and efficiency!

Now consider another situation in the above thought experiment. When the lift is falling down in space, the downwards acceleration eliminates the gravity inside the lift, which results in a state of *zero gravity*. This state of 'free fall' of an object (i.e. moving freely under the influence of no other force except gravity, thus making the object weightless) causes the astronaut to experience a sense of weightlessness. This phenomenon is responsible for the free-floating astronauts orbiting in space (it is commonly and erroneously thought that the reason for their weightlessness is because of lack of gravity, the actual reason being that the gravity is effectively countered).

And finally, this equivalence principle has shown us that any object moving under the influence of gravitational force may also be shown to be moving in relation to other objects. Thus, according to general relativity, there is no absolute motion even in the case of non-uniform motion. So now the grand conclusion is that all motion in the universe is relative.

All motion—uniform or non-uniform—is relative.

These seminal concepts of general relativity have radically changed our understanding of the workings of the universe. The most important implication with the proposition of the equivalence principle is that the very definition of gravity has changed forever. Gravity has ceased to be reckoned as an 'attractive force' as was considered in classical Newtonian physics; rather it is now shown to be the result of 'spacetime warpage', as we will be discussing below.

SECTION E
Implications of General Relativity

What we have seen above is only half of the story of general relativity. The implications of general theory have some really far-reaching consequences in understanding the universe. Hereunder, we will briefly discuss some of the important implications of general relativity.

The Meaning of Spacetime: Einstein went on to explain the true nature of space, time, and gravity, and to understand these concepts, we need some simple discussion of geometry of objects in space. Imagine for a while about a single point in space—this geometrical point has no dimension. Now move the point in a straight line—this creates a line with one dimension (length) (Fig 1.5), then move the point at right angle to this line—this creates a plane of two dimensions (length and breadth). Move again at right angle to this—this creates a *space* of three dimensions (length, breadth, and depth). This is as far as we can go on with our imagination. We understand our world around us in a 3D setting. After all, we are creatures harbouring a 3D mind! However, mathematically, we may add another dimension, and this generates a *Euclidean space* of four dimensions. Of course, we may keep adding any number of dimensions theoretically (Ch 3-J).

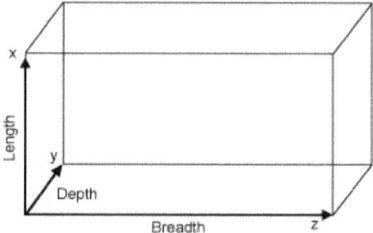

Fig 1.5: 3D coordinates.

We may now imagine the Euclidean line with one dimension on a flat surface. It has a *zero curvature* and covers an *infinite length*. It will never close on itself, and it goes on straight *eternally*. Now consider geometry on curved surfaces—the so-called *non-Euclidean geometry*. On a sphere (or a globe), the simple Euclidian geometry becomes contorted, and all the straight lines on the surface of a sphere have a curvilinear course (Fig 1.6). This curvature makes the surface bend back on itself and thus covers a *finite* area. Einstein worked on with non-Euclidean elliptical geometry to ascertain the real nature of moving objects, and this has ultimately explained the true nature of gravity itself.

Now let us consider this question: on a curved surface (as that of a globe), what is the *straightest possible line*? If you take the earth as an example, the largest circle (= equator) on it would also be the straightest possible line. In other words, earth's equator has the least curved path (or the straightest possible path), and all other latitudes have more and more curved paths. This great circle on any sphere is called a *geodesic*. And it can be said that it is impossible for two great circles on a sphere to be parallel (Fig 1.6).

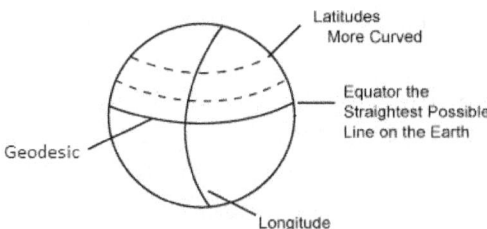

Fig 1.6: Geodesics.

Einstein realized that gravitational force would really be a function of the geometrical orientation of an object in space. But then we have already seen that special relativity has destroyed the very concept of absolute length. Hence, we may consider that all measurements specifying geometrical orientation are also relative, and consequently, it can be said that gravitational force is also relative (see below). Consequently, when we attempt to incorporate special relativity into gravity, it demands necessary corrections to the classical theory of gravity. To prove this mathematically, he has used a four-dimensional *Riemannian elliptical geometry*—but with an important difference: the three dimensions in Einstein's geometry are *space dimensions*, and the fourth dimension would be a *time dimension* (i.e. three spatial dimensions + one time dimension). In fact, it has been shown that the spatial dimensions may be freely exchanged with time dimension, and in this sense, it may be said that all the four dimensions may be treated as interchangeable. Thus, an object's spatial orientation can now be precisely ascertained not just by ascertaining its length, breadth, and depth but also by ascertaining its time dimension. This concept simply means that every moving object in the universe (or simply, every 'event' that takes place in the universe) can be described precisely as an event occurring in a *four-dimensional 'space-time'*. We can, in fact, say that space and time cannot be considered separately any longer in nature, but they can be regarded as a consolidated unit called the *spacetime*. This means that any moving object may be precisely located in space by describing its location in terms of these four dimensions—i.e. by describing it in spacetime but not by merely identifying it in space and time separately.

Any moving object in space may located by its spacetime.

Here, at this juncture, we have finally arrived at the most important conclusion of general theory of relativity. When we view an object in separate spatial dimensions (i.e. in 3D), we can say that one dimension of that object is relative to the other dimensions, and they become observer dependent (as we have seen above in the case of our spaceships and astronauts) – and this means that all measurements are relative to each other and are dependent on the observer's movement as long as we consider them in a 3D perspective. Or in simple terms, all dimensions in a 3D setting are relative.

However, when we describe the spatial orientation of an object in a four-dimensional spacetime, all the measurements become *invariant*—i.e. they are rendered the *same* for all observers irrespective of their relative movement. This means that the position of an object described in a 4D spacetime is absolute, and so this object appears the same to all observers irrespective of their own relative motion. Thus, in general relativity, we have arrived at yet another *new absolute*, and that is the invariance of *four-dimensional spacetime!* The reader may remember that we have already arrived at one absolute in the special theory—that is, the absoluteness of speed of light (Sec-B).

Four-dimensional spacetime is an absolute.

Thus, in the general relativity, 4D spacetime has become an absolute, and this has ultimately led to an important concept called the spacetime continuum, as we will see now.

Spacetime Continuum: Before proceeding further, it may be noted that the importance of the following ideas presented below may be appreciated better in Ch 7 – however, we will have an overview of spacetime here. In the above discussion, we have concluded that dimensions of an object in space are no longer relative when viewed in a 4D spacetime, and they become absolute and invariant without any relevance to the moving status of the observer. There are a few important implications which spring out of this concept. We may realize now that we no longer live in a 3D world, but rather, our world is a composite spacetime structure composed of four dimensions. All the objects (or matter particles) appear to us as composed of three spatial dimensions when we view time separately (for a clear analysis, see Ch 7-B), but they are, in fact, moving in 4D spacetime. And more importantly, it can be said that when viewed in this 4D perspective, all the objects in the universe travel at the speed of light. In consequence, no object in the universe can be considered at rest because even though it appears to be at rest with respect to the three space coordinates, it is still travelling through the dimension of time. This means that we can stop moving in 3D altogether and start moving with the speed of light! In other words, it can be said that when you are moving at the speed of light, your motion through 3D space is zero. This 4D feature is called the *spacetime continuum* (see discussion in Ch 7-B).

All objects in 4D spacetime are considered to travel at the speed of light.

However, it must be realized that spacetime continuum is a mathematical derivation, and we, the earthlings—being essentially 3D creatures—cannot 'visualize or imagine' the 4D world in our minds! The idea that time stops at the speed of light is an impossible concept to our 3D minds. For example, only in a 3D setting, we may claim that light has taken 13.77 billion years to travel from the big bang event to our present time (Ch 5-G), but for anybody who could have travelled along with light at the speed of light in four dimensions, it would appear that light has taken zero time to get from big bang to here!

Spacetime Curvature and Gravity: Thus, with these concepts of general relativity, spacetime has emerged as a dynamic entity consisting of interchangeable four dimensions. But then the time dimension can only be imagined in one's mind's eye and hence cannot be diagrammatically represented. However, conventionally, we will depict spacetime on a two-dimensional plane (e.g. the page of this book, as in Fig 1.7 a) as if it were some sort of stretched fabric with grid lines drawn across it – or, the reader may imagine it like a student's graph paper drawn on an elastic surface. This stretched fabric is undisturbed and flat when no object with mass is placed on it, but it would become warped when an object with a mass is placed on it (Fig 1.7 b).* This spacetime distortion or warpage gives the effects of gravity, as discussed below. The reader must understand that this spacetime fabric is all-pervading in the universe. It occupies the whole expanse of the cosmos, forming an infinite blanket of spacetime structure suspending all the matter of the universe in it. Thus, spacetime may be envisaged to exist as a continuous, flexible medium occupying the empty space that exists between all the objects and matter particles in the universe. In other words, the spacetime structure suspends all the galaxies and the stars and the planets of the universe within it, including you and me, and all the atoms and molecules that make the stars and planets and us!

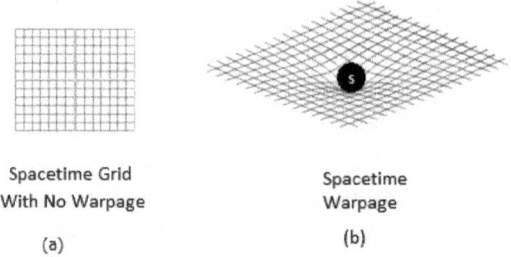

Spacetime Grid
With No Warpage

(a)

Spacetime
Warpage

(b)

Fig 1.7: Spacetime fabric and its distortion.

Einstein mathematically showed that all moving objects in nature tend to take the straightest possible paths. Now consider this: what is the straightest path in a spacetime curvature created by a massive object in the spacetime fabric? The straightest possible line on a curved surface is obviously its geodesic. Thus, for example, a planet can be said to be moving along the geodesic of its star's spacetime distortion (see below). In other words, it may be said that all the moving objects in space tend to take the *'path of least resistance'*, and that would be the geodesic of the spacetime distortion they are situated in.

---

* By saying that an object distorts spacetime, we mean, in reality, that the warping is three-dimensional, i.e. the warping happens in all directions around the object—up, down, sides, so on—much like an oil drop in the middle of a water medium. However, the 3D depiction of such a situation is impossible on a flat paper, so we draw a 2D picture and leave the rest to the reader's imagination.

A moving object tends to follow the geodesic of spacetime distortion.

Now consider this: when the sun is placed over this stretched fabric of spacetime, it warps the grid lines of the spacetime fabric and creates a dimple in the fabric (Fig 1.8a). And if a smaller mass such as the earth is placed in its vicinity, it will start rolling around the sun, travelling along the geodesic of the curvature of this depression (see below). This movement of the earth around the sun would create an impression that a 'force' exists between them (as was envisioned by Newton centuries ago!). Thus, the gravitational 'force' we see is merely an illusion, and the real mechanism of gravity is played by the geometrical distortion of the spacetime. Simply put, gravity is nothing but a spacetime warpage.

Gravity is a spacetime warpage.

Of course, the reader must realize that the earth also, as it revolves around the sun, creates a distortion in the fabric of spacetime—albeit of a smaller size compared to that of the sun—and when a smaller object like the moon comes in the vicinity of the earth's spacetime warpage, the moon will also start rolling across the spacetime curvature in its straightest possible path. However, the classical gravitational theory would say that the earth is under the influence of the gravity of the sun and the moon is under the gravitational pull of the earth, but in fact, they are just following the geodesics of the spacetime curvature created by heavier objects. In the same way, all objects in space (e.g. an apple falling from a tree or a stone thrown up into space) may be considered to create their own small dimples in the spacetime.

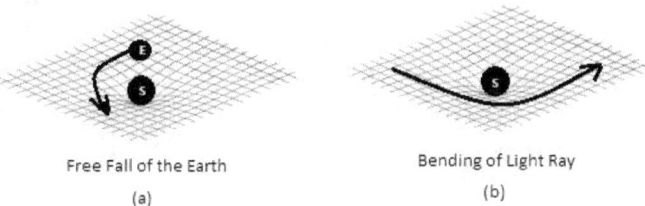

Free Fall of the Earth Bending of Light Ray
(a) (b)

Fig 1.8: Spacetime distortion changing the course of the earth (a) and light (b).

Newtonian gravity said that objects attract each other, but Newton did not understand how gravity is transmitted in space or how distant objects could influence the paths of other objects. The mechanism of gravity became known only with the advancement of general relativity—Einstein had shown that the true agent of gravity is the flexible fabric of spacetime.

Spacetime distortion is the mechanism of gravity.

In summary, we can say that Einstein's relativity has changed our understanding of the physical laws forever. In classical Newtonian physics, an object continues to move in a straight line in space with uniform velocity unless acted by a force. In general relativity also, an object moves in a straight line, but it travels in a composite framework of spacetime (instead of space). Gravity in relativity is not a force but is a spacetime warpage caused by a material object, and when another object falls in this spacetime distortion, it would follow the straightest possible path. Thus, it is not proper to say that the sun attracts and pulls upon the planets by its gravity, but it may really be stated that the sun creates a gravitational field by creating a depression in the spacetime around it, and when an object has fallen into this depression, it whirs (like a piece of cork in a whirlpool) and takes the straightest path to move around the sun. An object revolving round a massive object can be said to be in a state of free fall, but it is really travelling in the straightest possible line in space so that the trajectory of the object would appear to be curved towards the massive object, thus creating the impression of a free fall.

Not only objects but even light takes a bent course in the vicinity of gravity (Fig 1.8b). Newton, with his corpuscular (particulate) theory of light (Ch 2-B), also had predicted bending of light by gravity, but the concept of bending of light became clear with general relativity (see Eddington's experiment in Sec-F). Light appears to take a bent course in the vicinity of a strong gravitational field (such as a star's) not because it is actually bent but because the spacetime itself is bent, and light tends to take the longest path along the geodesic of this bent spacetime.

Einstein described this spacetime distortion in mathematical terms, but the mathematics of general relativity is bewilderingly complex. However, the equations of relativity may be understood simply as a relationship between the mass of the object and the amount of spacetime warpage it causes. And this is depicted in the final equation of general relativity, written as:

$$R_{\mu\upsilon} - \tfrac{1}{2}\,Rg_{\mu\upsilon} = 8\pi G/C^4.T_{\pi\upsilon}$$

The left side of the equation is called the Einstein tensor, which represents spacetime curvature, and the right side is stress-energy tensor, which represents the mass–energy distribution. In simple words, this formula depicts the extent of warpage of spacetime for a given unit of mass. John Wheeler, the eminent physicist, had superbly put the relationship between mass, spacetime, and curvature in an aphorism: 'Mass grips space-time by telling it how to curve, the space-time grips mass by telling it how to move.'

We will conclude our discussion on relativity by coming back to the original question of absolute motion in nature. General relativity showed us that objects accelerating under the influence of gravity are moving in a relative frame of reference, and thus, we may not be able to ascertain their absolute motion. Thus, all in all, the theory of relativity may finally be concluded this way: absolute motion in nature is impossible either in uniform or non-uniform motion. Or in simpler words

All motion in nature is relative.

The theory of relativity, as discussed above, is not merely a weird mathematical postulate, but it is a time-tested tool in the realm of modern science. Moreover, it has led to a number of technological advancements, as we will be seeing in the next section.

SECTION F
Proofs and Uses of Relativity

At the beginning of the twentieth century, there were many sceptics of the theory of relativity, and it was generally thought that a theory which is based on mere thought experiments could not have any serious meaning. Some scientists thought that the theory becomes valid *only* when an observer makes an observation of the events in the space because the theory was based on the subjective assumptions of a rational and thinking human being and could not have any objective significance. A general reader of this book, quite justifiably, may also arrive at somewhat similar conclusions after reading these weird concepts.

However, with each scientific discovery in the following decades, the objective veracity of the theory has been endorsed to the utmost degree of credibility. And gradually, owing to many observations and experiments in support of the relativity theory, many of its hardest critics of the theory have become its ardent followers. More importantly, scientists have even found many practical applications of the theory with great success, and in fact, some of the greatest scientific and technological leaps of the twentieth century were solely based on some of the concepts of relativity, such as mass–energy equivalence and time dilation. In fact, these new developments have a great practical bearing on our present-day lives, of course, for good or for bad, such as our use as GPS or our misuse as atom bomb, as we will see below. Not only this, but the equations of the relativity theory, disguised in some way, would creep into almost all aspects of theoretical physics of today, as we will be seeing throughout the book.

However, speaking from the other side, some of the phenomena of relativity, such as the twin paradox, are likely to remain as ghost theories for some more time to come, but their practical application would certainly become a reality when humans take up deep-space travel as a serious business. Here we will not go into the details of all these affairs, but we will look into some of the proofs of relativity.

Eddington's Solar Eclipse: One of the first proofs, and certainly the one which had made Einstein a science celebrity overnight, is the experiment conducted by Sir Arthur Eddington in the year 1919. First we will examine the basis of Eddington's experiment. Both general relativity and classical Newtonian gravity predicted that intense gravity has some influence on the path of light. Whereas the calculations of classical gravity have predicated a deflection of about 0.87 arcseconds, Einstein predicted, as per his general relativity calculations, that the deflection would be about 1.74 arcseconds. Shortly after the general theory was proposed, Eddington set out to calculate the deflection by measuring the shift in starlight as it passes near the sun during the solar eclipse (when the intense glare of the sun is shut off by the moon), and to the great surprise of many scientists, Eddington's calculations clearly showed that the bend of light is much more nearer to Einstein's prediction than the classical gravity. The results of this remarkable experiment attracted media of the time in a big way, which made a tremendous science sensation out of it. And this instantaneously made Einstein and his relativity theory popular household names across the world!

The relativity theory subsequently got additional theoretical support by correctly explaining the intriguing aberrations of the path of the planet Mercury—a slight anomaly of its path around the sun which could not be accounted for by using Newtonian gravity but was correctly resolved by using mathematics of general relativity.

Other Proofs of Relativity: There are many other proofs of relativity. Of course, the development of the atom bomb has stood, by itself, as a glaring proof of the special theory. Unfortunately, the success of building an atom bomb had really tarnished Einstein's reputation to some extent at that time, though we now know that his role in its development (as a part of the arms race during World War II) is very indirect and quite incidental.

The phenomenon of time dilation has also been proved correct. The average life of a fast-moving subatomic particle called meson (Ch 3-C) is experimentally shown to be longer than a slow-moving meson because meson's *proper time* runs more slowly as it moves faster. This stood as one of the solid proofs of the theory of relativity. Another curious feature validates the theory of relativity. If you look at the gigantic structure of the Large Hadron Collider (LHC) (Ch 3-C), which is used to accelerate the smallest of the subatomic particles such as a proton, it is surprising to notice that massive magnets are used to propel the smallest particles. But then really, why do we need such a massive collider to move so small a particle? The answer lies in the special theory – the faster the subatomic particles move, the greater is the force needed to accelerate them (because greater becomes their relativistic mass), and consequently, they attain supermassive momentum as they approach speeds nearer the speed of light (Sec-C). Thus, they need stronger and stronger magnetic fields to overcome their increasing mass.

GPS and Relativity: However, one of the most outstanding examples of relativity's practical significance is that of its role in our Global Positioning System (GPS). GPS satellites are situated at about 20,000 km above the ground with an orbital speed of about 14,000 km/h, and these satellites use atomic clocks that tick with an accuracy of 1 nanosecond (1 billionth of a second!). We have already seen that because of time dilation, the clocks on a moving object run slower (Sec-C), and calculations show that this time lag would be about 7,200 nanoseconds per day—a minuscule time to reckon by ordinary standards. But then consider this: the theory of relativity also says that gravitational force has an effect on the time – a clock runs *slower* as gravity becomes stronger. Consequently, because of weaker gravity at that height, the atomic clock in the satellite runs *faster*, and so the clock gains time by about 45,900 nanoseconds per day. Thus, the satellite's clock shows an overall increase of only about 38,700 nanoseconds per day (i.e. 45,900 – 7,200), and this increase would ultimately culminate in a staggering error in the distance measurements by the end of the day! And with such an error in your GPS system, if you ever want to navigate in an unknown city, for example, to reach out for your fiancé's rendezvous, you are more likely to land up in a cosy little bar, and you may guess the upshot! But then luckily for us, GPS is designed to take time dilation into account and adjust the atomic clocks to run slower (than when they were placed on the ground) to make our lives tolerably precise for our day-to-day navigation.

The theory of relativity also predicts several important outcomes (such as the speed of light, 4D invariance, and mass–energy equivalence) which have an important theoretical bearing on the development of a wide range of new concepts from the quantum theory to black holes to the big bang theory. The reader will come across these implications as she/he moves on to the subsequent chapters.

A Note on Gravitational Waves: Prediction of the existence of gravitational waves is certainly a staunch proof of general relativity. Gravitational waves are ripples in the fabric of spacetime created as an object accelerates in space. They are extremely weak disturbances and thus are hard to detect. Even though Einstein predicted gravitational waves in 1916 as a part of general relativity, scientists were able to detect them only after exactly a century. However, as the strength of these waves depend on the mass of the object, they are likely to be detected in some immensely violent cosmic events in the universe involving collision of some supermassive bodies such as two black holes collapsing into each other or two colliding neutron stars or huge supernova explosions (Ch 4-B). Gravitational waves are also called gravitational radiation, and they are thought to travel by the speed of light.

In 1974, astronomers at the Arecibo Radio Observatory, Puerto Rico, accurately predicted the rate at which two pulsars (Ch 4-B) attract and collapse into each other. General relativity predicts that such a binary system of stars (Ch 4-A) loses orbital velocity due to the generation of gravitational waves. This theoretical prediction of loss of orbital energy was actually detected in the above case, which stood not only as a validation to the theory of relativity but also as an indirect evidence for gravitational waves. There have been several attempts to directly detect gravitational waves ever since but went in vain largely because of their extremely weak nature. But then on 14 September 2015, the actual ripples of gravitational waves that came from deep space were finally detected by the advanced version of LIGO (Laser Interferometer Gravitational-Wave Observatory). These gravitational waves were found out to be generating from an extremely violent cosmic event of gravitational collapse of two giant black holes occurring at about 1.3 billion light years away from the earth!

Pitfalls of Relativity: So far, we have studied the magnificent achievements of the theory of relativity with many victories that have vindicated the theory across many decades. However, the reader has to realize that the theory is still far from complete, and it has left out some important lacunae in our understanding of nature. The important shortfall is its unaccountability in the world of smaller particles, i.e. the quantum realm – we will see in Ch 2 that quantum theory has its own rules which do not agree well with the principles of relativity. Moreover, any consummate theory in physics should bring about a unification of all the known four fundamental forces (gravitational, electromagnetic, strong and weak forces, Ch 2-A) into a single file of unified force (as discussed in Ch 3-F). The failure of the relativity theory to achieve this goal is another major drawback. And finally, the principles of relativity have no bearing on the singular events that have occurred at the beginning of the universe as envisaged in the big bang theory (as discussed in Ch 5).

All these pitfalls of the relativity theory bespeak of the need for certain refined modifications to the theory, which will surely be done by the future generation of scientists provided the young scientists of today would come out of the dogma of science and relinquish the authority of the establishment—as Einstein and several other stalwarts of science themselves have done in the past.

Having examined the theory of relativity in brief, we will now move on to the world of microcosm and study the quantum theory.

CHAPTER 2

The Microcosm
And the Bizarre Principles of Quantum Theory

Overview

In the preceding chapter, we have seen how the movements of objects in the macrocosm are governed by the principles of relativity, which are often concerned with massive objects with gravitational effects influencing each other's movements at very long distances. However, we know that gravitational force not only affects the movements of stars and planets but is also concerned with the movement of objects, such as apples on the trees and pebbles at the seashores. And we know that stars, planets, apples, and pebbles, in turn, are made up of atoms and subatomic particles, but then do the rules of relativity agree with the movements of these tiniest subatomic particles? The answer is a devious no – we have a separate set of laws presiding over the miniature world of atoms and matter particles! Let's now see what these peculiar principles are and how they govern the behaviour of tiny matter particles.

At the beginning of this chapter, we will look into an analysis of the intriguing wave–particle duality of light and see how this takes us into the world of quantum mechanics. Later, we will learn about many of the cardinal, yet weird, laws that dictate the movement of the smaller of the smallest particles. It is shown that in quantum theory too, as in the case of the theory of relativity, our intuition will not work – if the concepts of Einstein's relativity appeared uncanny to the reader, the laws governing the microworld would appear even more spooky. In fact, Feynman had remarked that 'if you think you understand quantum theory you have not understood it!' Thus, the reader may have to fly his/her imagination a little bit to get a grasp of the bizarre things that may happen in the ultramicroscopic quantum world. However, just as in the case of the relativity theory, even though the concepts of quantum theory appear utterly weird, they turn out to be really practical and workable in our daily lives, and these practical applications of the quantum theory are presented at the end of the chapter which may, at last, reconcile the reader to accord with its essential principles.

Now get ready for a few of the quantum surprises!

Section A
The Four Fundamental Forces

Before going into the actual study of quantum theory, a brief description of the forces of nature is undertaken, and this would introduce the reader to the essentials of quantum theory. First of all, we will see what exactly is 'force'. By a simple statement, we may say that force is an agent which causes a change in the motion of an object in space. As we have already seen, Newton's first law states that an object continues to be in a state of rest or in a state of constant motion as long as no force is applied on to it. Newton's second law states that with the application of force, the object changes its velocity, making it accelerate (Ch 7-A). Force can be mathematically defined as $F = ma$ (perhaps the second-most celebrated equation in science after $E = mc^2$), where F is the force, m is the mass of an object, and a is the acceleration. Therefore, it can be said that force exerted by an object is proportional to its mass and acceleration. For example, a tiny bullet moving at a great speed can cause the same amount of impact as that of a massive truck going at a low speed (a full discussion on force is undertaken in Ch 7-A). We will now move on a little further.

Force Fields: But then for a long time in the history of science, the exact nature of force remained obscure, and certain fundamental questions were left unanswered, such as these: How is force transmitted from one object to the other, and in what medium does it propagate? For example, how do magnets attract iron filings from a distance? Or how does the sun influence the movement of the earth from a distance of 150 million kilometres? Or in the modern parlance, how do microwaves heat up our pizzas in an oven without actually contacting them? Though we in the twenty-first century pride ourselves with some ready-made answers to these seemingly simple questions, the medieval scientists were nevertheless puzzled by the nature of the mysterious force and its transmission.

In 1831, Michael Faraday made an important contribution to our understanding of the forces of nature. He reasoned that some invisible force with some specified laws must be operating across the intervening empty space between the objects. Faraday imagined that these forces that fill up the space consist of some invisible tubes which do the pushing or pulling of objects, and he called them the *force fields*. The effects of these unseen force fields were readily demonstrated by placing a magnet underneath a paper with iron filings on it. The resulting regular lines of iron filings around the magnet would mark the lines of the force fields (Fig 2.1).

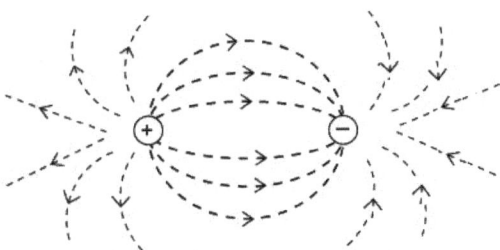

Fig 2.1: Lines of force in a magnetic field.

Since then, there were many advancements in various fields of science which led to a great improvement in our knowledge of these force fields, and some of these advancements are described in this book wherever they become relevant. In our current understanding, there are several types of forces that exist in nature ranging from the forces existing between large objects such as stars and planets to the forces guiding the smallest particles such as the atoms, electrons, and protons. We may categorize all the known forces that exist in the universe into four *fundamental forces of nature*. They are:

1. *Gravitational force.* Gravity is the first known force characterized in science and was discovered by Newton in 1687. It is the weakest of all forces, but it is a long-range force which works at very great distances. We will be dealing with this force throughout the book in some way or the other.

2. *Electromagnetic force.* The next force that was discovered was the electromagnetic force (EM radiation), which also goes by the general term *light.** We will be exclusively dealing with EM radiation in the following sections in this chapter, and as with gravity, we will also be discussing this force all through the book.

3. *Strong nuclear force.* This is a short-range force which is responsible for the binding of subatomic particles in an atomic nucleus.

4. *Weak nuclear force.* The weak interaction also works in short range within the confines of an atomic nucleus. The reader will learn more about strong and weak nuclear forces in Ch 3.

The reader may put a valid question here. There are so many varieties of forces we encounter in our day-to-day lives—mechanical force, heat, electricity, chemical energy, frictional force, pressure, vibration, sound energy, nuclear force, and so on. How do we account for them using this shortlist of four fundamental forces? It must be realized that any force we can think of—be it the force of weight of an apple falling to the ground, the revolutions of our ceiling fan, the vibrations of a loudspeaker, the energy we get from our food, or the power we use from a nuclear reactor—are merely a part of one of the four forces described above or, more commonly, a combination of these forces. For example, when we ride our bicycle along the

---

*   The term *light* in theoretical physics is used as a short form of *electromagnetic radiation*, and hence, light here does not mean only visible light but also includes any type of the radiation ranging from gamma rays to X-rays to radio waves (Table 2.1).

street, the tyres encounter a mechanical force called *friction* from the ground. This frictional force is not only the result of a combination of gravitational force exerted by the weight of the bicycle and the rider on the ground on a large scale but also the tiny electromagnetic forces that are generated on a small scale by the 'molecular rub' between the molecules of the synthetic rubber and the molecules of street tar! Anyway, we will discuss more about the fundamental nature of the various forces in Ch 7.

The reader must finally realize, however, that though we study these four fundamental forces separately, they appear to be somehow related theoretically to each other at some fundamental level, and it is the modern physicist's cherished dream to unify and understand all these forces as a single force (Ch 3-F)!

We will now begin the story of quantum theory by first understanding the nature of electromagnetic force (i.e. light), which, historically, has paved way for the development of quantum theory, as we will see in the subsequent sections.

Section B
The Wave–Particle Duality of Light

The present-day understanding of the quantum theory is the result of several decades of scientific research by generations of researchers. We will do better if we start at the beginning and deal with the historical debate of the wave–particle nature of light at the outset. However, the general characteristics of light (or electromagnetic radiation), as currently understood, are presented first so that the reader would appreciate the details of later discussion better. Subsequently, we will go into the actual wave–particle argument and then into the mires of quantum mechanics.

Light and Its Properties: Light (or EM radiation) is the most studied of the fundamental forces so far. It carries electromagnetic energy from place to place, the outstanding example being the energy carried from the sun to the earth in the form of sunlight. But then how does light, a fundamental force, propagate in empty space and transmit energy from place to place? We will have a brief account of the properties of light below.

All EM radiation, whatever is the type of radiation, has the following unique structure: light propagates by waves which are composed of two vibrating force fields—one electrical and the other magnetic; each field oscillates at right angles to the other, and the oscillations in turn occur perpendicular to the direction of wave propagation (hence, light is called a transverse wave). This propagation occurs in an undulating fashion with alternating *crests* and *troughs* in repeating cycles (Fig 2.2). EM waves have three characteristic features—wavelength, frequency, and amplitude. *Wavelength* is the distance between two successive crests or troughs, *frequency* is the number of waves that occur in a second, and *amplitude* is the maximum height (or depth) of a wave. Obviously, wavelength and frequency are inversely related. An EM radiation with more crests/troughs per second means that its frequency is high, and so its wavelength would be shorter, and *vice versa*.

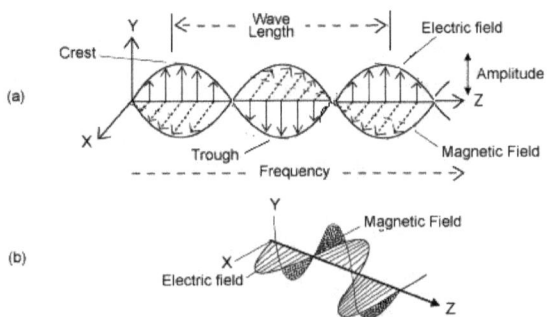

Fig 2.2: An electromagnetic wave.

There are many types of EM radiation depending on the frequency of its waves, and the term *light*, in the parlance of physics, encompasses a whole range of spectrum ranging from gamma rays (which have the highest frequency or shortest wavelength) to radio waves (which have the lowest frequency or longest wavelength). Our ubiquitous visible light* is situated in between these two extremes (Table 2.1).

---

*   The reader must note here that the wavelengths outside the red and violet

Table 2.1: Types of Electromagnetic Radiation

Wave Type	Wavelength (Meters)	Frequency (Hertz)	Energy (Electron Volts)	Ionizing Effects	Penetration	A Few Examples of Uses
Gamma Rays	10^{-12}	10^{20}	10^{7}	Yes	Pass through almost all materials for a short distance	Sterilization—kills bacteria and viruses in our consumables
X-Rays	10^{-10}	10^{18}	10^{5}	Yes	Pass though most materials, high interaction	In radiology and other medical uses
Ultraviolet Rays	10^{-8}	10^{16}	10^{2}	Yes (Less)	Low penetration, minimal interaction	Synthesizes vitamin D in skin, fluorescent lamps
Visible Light	10^{-7}	7.5×10^{14}	10^{1}	No	No penetration, no interaction	Perception in humans, photosynthesis, optical fibres
Infrared Rays	10^{-5}	4×10^{14}	10^{-1}	No	No interaction	Thermal imaging, night vision, burglar alarm
Microwaves	10^{-3}	10^{10}	10^{-4}	No	No interaction	Microwave ovens, information transfer
Radio Waves	10^{-3} to 10^{5}	10^{10} to 10^{4}	10^{-5}	No	No interaction at all	Communication over long distances, radio telescopes

All the properties of each kind of EM radiation, such as its energy, penetrating power, and ionization effects, are all dependent on its wavelength/frequency (Table 2.1). Waves with high frequency (short wavelength) carry more energy in them, and conversely,

range (such as radio waves, infrared rays, ultraviolet rays, X-rays, and gamma rays) cannot be perceived by the human eye directly.

low-frequency (long wavelength) waves carry less energy. For example, gamma rays which have the maximum frequency have the highest energy in them, and radio waves which have the least frequency have the least energy. In the case of the visible spectrum also, the colour of the visible light or the energy it carries is determined by their wavelength/frequency. Thus, violet light (wavelength about 0.0004 mm = 400 nm) has the highest frequency, so it carries maximum amount of energy in its waves, and red light (700 nm) has the least frequency and has minimum energy. But the violet light's penetrating power in a transparent medium is less than that of red light. This feature of light offers an interesting explanation as to why the evening sky should appear crimson red. During the midday (when the sun is directly overhead), the thickness of the atmosphere is less so that the short-wavelength (less penetrating) blue portion of visible light of sunrays could pass through the thin atmosphere directly from above, and so the noon sky appears blue. But as the day progresses, the sunrays have to pass tangentially across the evening sky, which makes the atmosphere progressively thicker. Now the blue light cannot pass through the whole thickness of the evening sky, whereas red light (which has more penetration) can pass through the atmosphere, and this makes the evening sky red. Of course, the same reason goes with the sun being reddish at sunrise!

EM radiation is a chargeless force. The reader may note here that though light is made up of charges (owing to the presence of electric/magnetic fields), it does not usually behave like a charged force because there are as many positive charges as there are negative charges so that they cancel each other out! As discussed already in Ch 1, all the EM waves in the electromagnetic spectrum (from gamma rays to radio waves) travel exactly at the same speed of 299,792,458 m/s through space.

Finally, there is one more important characteristic of light. Hitherto, we have discussed EM radiation as waves, but light has also been shown to propagate as particles called *photons*, and this intriguing feature of light would actually leave us at the doorsteps of quantum theory. Now we will see why!

Light—Wave or Particle? We have now arrived at the crux of the historical debate which has led to the development of quantum theory. Consider this: we have started the above discussion on light by stating that light travels by waves, and at the end, we have stated that light is composed of particles! These statements contradict each other—waves imply a rather 'diffuse' (or endless*) distribution of energy, whereas particles are defined (or discrete) points of matter. And, this will now bring up an austere question: is light composed of waves of energy or particles of matter? The story of this ambiguous concept of light is the actual starting point of our discussion on quantum theory.

Historically, the theory of propagation of light has a twisted course. At the beginning, in 1678, Christiaan Huygens proposed that light propagated by waves. He argued that the property of refraction, i.e. bending of light while passing from one medium to the other, is because of light's wave nature. Newton, in 1704, contested this view and said that light consisted of tiny particles – the property of birefringence and light travelling by straight lines had stood as support of this view, and this concept of Newton had become known as the *corpuscular theory*. And for about a hundred years, this 'particulate' theory of Newton had ruled the scientific thought!

However, in 1801, Thomas Young demonstrated a peculiar property of light called *interference*, which could be explained only by the wave property of light. The interesting phenomenon of interference needs a special description here. Young conducted his famous *double-slit experiment* which consisted of a light source that sends a beam of light on to a cardboard with two vertical slits in it and a photographic plate on the other side. When light was allowed to pass through a single slit in the cardboard, the resulting image would obviously be that of a single vertical bright band. But when

---

*   A wave has no beginning or ending. The reader may look at a stretched spring somewhere in its middle portion and try to mark a 'starting point' or an 'ending point', which turns out to be meaningless because there could not be any beginning or ending to each coil (or wave). Therefore, a wave is a continuous phenomenon. It is not a localized point, but it spreads endlessly (further insights in Ch 8-E).

both the slits were open, our intuition says that the result would be two vertical bands corresponding to the two slits. However, in this double-slit experiment, it was shown that there would be a series of alternating bands of bright and dark areas—the so-called interference pattern. Young explained this strange property of light by employing the wave nature of light (as an analogy, the reader may imagine waves in a water tank crossing two holes in a barrier showing the same interference pattern). Waves have two phases – crests (or peaks) are the topmost portions in a wave, and troughs are the bottom-most portions (Fig 2.2). Waves propagate by alternating series of crests and troughs. When light passes through two slits close to each other, the two light beams interact (or 'interfere') with each other, and when crests collide with crests, it reinforces the waves, creating a *constructive interference* (waves become more prominent). Troughs colliding with troughs also gives a constructive interference, but when the crests collide with the troughs, they cancel each other out, and this creates a *destructive interference* (waves become flat). The final result of this collision of waves would be an alternating pattern of bright and dark bands, i.e. the interference pattern. Thus, this property of light has proved beyond doubt that light travels by waves! Some of the subsequent experiments have also proved the wave theory of light to be correct, and thus, gradually, the dice turned over the scientific thought, and the wave theory came into vogue once again and stayed with the scientific community for some time.

The wave nature of light gained additional support from other scientific advancements. Consider these historical developments. In 1820, Hans Ørsted discovered that a compass needle deflects under the influence of a strong electric current, which denoted a relationship between electricity and magnetism. In fact, Ørsted was the first scientist to introduce the term *electromagnetism*. A little while later, Michael Faraday demonstrated that intense magnetism can also affect the nature of polarized light, and it was subsequently realized that electricity, magnetism, and light are all intricately related. But it was indeed the work of James Clerk Maxwell, in 1860, which has paved way for the modern scientific progress. Maxwell described electromagnetic force in a formal mathematical framework and showed that electromagnetic radiation propagates in waveforms through space with a fixed velocity (the speed of light was already determined nearly accurately by Fizeau). With these developments, light was thought to travel by waves beyond any doubt, or so it appeared to the scientists at that time.

But then it is an irony of history to note that by the early twentieth century, Einstein has once again revived the concept of the particulate nature of light, and this had actually stood as a bastion of the budding quantum theory at that time. Before trying to understand these affairs, we have to continue with the other scientific advancements for a while which have paved way for the development of the quantum theory.

Section C
Beginnings of Quantum Theory

In 1871, Ludwig Boltzmann theorized that the temperature of an object is due to the vibration of atoms in it—the faster the atoms vibrate, the hotter the object becomes, and *vice versa*. Subsequent theories by many scientists explained why heat flows from a hot body to a cold body and eventually equalizes, and these concepts had given birth to the field of *thermodynamics* (Ch 6). This theory of heat exchange has shown that all objects in the universe are in a state of constant exchange of energy, which means that all objects absorb energy from the surroundings and, at the same time, emit energy as radiation back into the surroundings on a constant basis.

Thus, it was long understood that every object at any temperature above the absolute zero emits radiation constantly to a variable extent. And because absolute zero is practically unattainable, it means that every object in the universe emits some sort of radiation into its surroundings at all times. The general rule is that the hotter the object, the shorter its emitted wavelength and more is the energy of radiation. For example, hot iron emits thermal radiation in the range of visible light at about 500 °C, making it 'red hot', and as the temperature of the metal increases, the wavelength of the emitted radiation decreases (i.e. its frequency increases), making it 'white hot'. We have no trouble in understanding this emission of heat in hot bodies. However, as noted above, it was shown that even cold objects do emit radiation in the infrared range into the surroundings at all times. We can, in fact, feel this infrared thermal radiation if it is intense enough (as, for example, in the form of warmth near a hot pan). And it must be realized that even objects with lesser and lesser temperatures (until they attain absolute-zero temperature) can exchange energy with the surroundings at wavelengths in the range of infrared regions (thus, even a cold slab of iron emits some amount of infrared radiation into the surroundings). It is interesting to note that all life forms (including human bodies) emit infrared radiation constantly into the surroundings. Nevertheless, this infrared thermal radiation cannot be detected by the human eyes but can be detected with the aid of certain special gadgets, such as infrared goggles—this is the basis of infrared photography!

Black-Body Radiation: The scientific understanding of heat transfer has subsequently led to the concept of *black-body radiation,* which had intrigued the later scientists of the nineteenth century. To understand this phenomenon, we will have to go to the basics. When a light ray falls upon matter, three things may happen—it may get *reflected*, it may get *refracted*, or it may get *absorbed*. Reflection gives the object its colour, refraction makes the object transparent, and absorption heats up the object. Usually, in most of the objects, all the above happen in varying proportions depending on the nature of the object's material and the type of radiation. Here we will concentrate only on absorption. In the case of absorption, the object gets heated up, and this object, in turn, emits back its absorbed energy into the surroundings in the form of radiation. This emitted radiation, theoretically speaking, has a wavelength distribution ranging from zero to infinite wavelength, but the only thing is that a particular object, though it emits radiation in all ranges, emits a maximum amount of emission only in a particular range of wavelength in the spectrum, and this is dependent on the temperature of the object. For example, a red-hot iron bar at 500 °C emits radiation at about 700 nm range, so it appears red. And as it gets heated to higher temperatures, it emits radiation at about 400 nm, so it appears blue. However, theoretically, the heated iron bar does emit radiation in all the possible ranges to some minimal extent!

Now consider this. A *black body* is an object which can absorb *all* the energy that has fallen on it without reflecting or refracting any of the light fallen on to it. However, black bodies are only theoretical objects, and in nature, there is no ideal black body which behaves in this perfect manner! Examples of near-ideal black-body objects are lampblack, radiators in furnace, carbon nanotubes, etc. And perhaps an ideal black body in nature would be a star like our sun, and conceivably, the absolutely perfect example of black body theoretically is a black hole (Ch 4-C)! A black body emits back radiation into the surroundings in the same manner described above—in all ranges but more in certain ranges. Thus, in practice, black bodies emit a specific wavelength range of radiation, depending on its temperature, and this is called the *black-body radiation*. But then theoretically speaking, when we consider electromagnetic radiation to be a wave and do mathematical calculations, an ideal black body should emit radiation maximally in the smaller wavelength range and minimally in the longer wavelength range. Thus, the scientists were puzzled by the peculiar relationship between the distribution of wavelengths of light emitted by the black body and its temperature.

However, when black-body radiation emitted in the ovens was experimentally studied (ovens and kilns are black-body radiators), it showed that the heat waves in an oven generated by its hot walls must have a whole number of waves but not fractions—the mathematical calculations simply would not allow any fractions. This is because in a lab setting, the total amount of energy in an oven is *finite* and fixed, and we cannot leave out any fractions. On the other hand, as we have discussed above, theoretical considerations have revealed that the possible number of wavelengths in an oven is *infinite*, and an infinite number of waves would translate into an infinite amount of energy! But surely, this is impossible—the amount of energy in an oven is but finite. This theory–experiment paradox shows that there is some defect in our understanding of this essential property of light!

To explain this intriguing phenomenon, Max Planck, in 1900, proposed a revolutionary concept that the energy carried by an electromagnetic wave does not come in a continuous waveform (as was supposed by Maxwell); rather, the radiation is emitted in lumps or packets. This means that electromagnetic force propagates in small fundamental 'packets of energy'—in discrete *quanta* (singular = *quantum*)*—and that these packets of energy can occur only in integers but never in divisible fractions. Planck also proposed that the energy contained in these packets is dependent on the wave frequency (or wavelength)—the larger the frequency (or shorter the wavelength), the more is the energy of each packet, and *vice versa* (Sec-B and Table 2.1). Thus, in an oven, only a finite number of waves (or more precisely, energy packets) can contribute to the finite amount of energy, thus the problem of infinite energy solved! Subsequent experiments showed that this ingenious theoretical insight agreed well with the experimental measurements. Further on, Max Planck showed that the proportionality factor between the frequency of radiation and the energy associated with the energy packet is a constant now known as the *Planck constant* (*h*), a constant which has profound implications in theoretical physics of the coming times and a constant which continues to puzzle even today's scientists (Sec-F). Planck constant is depicted in the formula: $E = hf$, where E is the energy content of a wave and f is its frequency. Modern estimates have put the Planck constant at $6.6260695729 \times 10^{-34}$ joules-sec—quite a tiny value for a great constant indeed! A modified Planck constant (called the *reduced Planck constant*, symbol $\hbar$, pronounced 'h-bar') is generally used to depict the quantization of the angular momentum (as, for example, in the case of depicting the energy of an electron moving around the nucleus), and $\hbar$ is obtained by dividing h by 2π.

Thus, Max Planck has shown that light behaves as a stream of particles but not as a continuous wave, and these particles (or wave packets) were later named *photons* (which are represented by the Greek letter *gamma*, γ). This is the first step in the development of the quantum theory. However, the full explanation of the quantum theory was due, and it needed the ingenuity of Einstein to come out

---

* A quantum can be an indivisible unit of anything. There are common examples of quanta: *staircases* are quantized with steps—you cannot have one and a half of a step—whereas a *ramp* is continuous. *Currency* is a quantized unit—you cannot have a fraction of a half-cent—whereas the concept of *money* is a continuous idea. Likewise, in the case of *particles* and *waves*, a quantum of energy is a particle, and it is a digital unit (photon)—you have only integers, but no fractions are allowed—whereas the concept of wave is a continuous idea.

with his explanation of another phenomenon called the *photoelectric effect*. We will now look into this intriguing phenomenon.

The Photoelectric Effect: Metals generally have plenty of electrons loosely bound in the outer shells of their atoms (Ch 3-A). These 'free' electrons can easily carry along the electrical energy from place to place across the metal wire, this, in fact, being the reason for a metal's good electrical conductivity. It is known for a long time that when energy (in the form of light) is thrown on certain metals, the metal atoms absorb energy and start ejecting fast-moving electrons from the metal surface. Our common intuition says that if we increase the intensity of light, the *speed* with which the electrons ejected increases. However, experimentally, it is shown that intensity of light does not affect the speed of the ejected electrons at all. For example, *infrared light* of any intensity shone on a particular metal does *not* emit electrons at all, whereas even a low-intensity *ultraviolet light* on that metal readily emits electrons.* In other words, what really causes electrons to be ejected from the metal surface is the frequency of light rather than intensity—the more the frequency of waves impinging on the metal, the more is the speed of electrons. In fact, as the frequency of light waves falls below a critical value, the metal stops emitting electrons altogether regardless of the intensity of the light shone. This phenomenon is called the *photoelectric effect*. This peculiar relationship between atoms and wavelengths was puzzling to scientists at that time, and before Einstein, it was not known why metals and light interacted in this way.

In 1905, Einstein explained the photoelectric effect by employing the quantum principle which was just then promulgated by Max Planck. Einstein proposed that the impinging light can actually be thought of as a stream of tiny packets of energy, i.e. quanta (or photons). He said that since the energy of each photon is proportional to the frequency of light, the more the frequency of the light wave (or lesser its wavelength), the more is the energy content of the photons.

---

*   The reader has to note here that the *number* of electrons emitted from the metal may increase with increasing intensities of that radiation, but their *speed* of travel in space is not affected by intensity. However, the speed of the ejected electrons is affected by the wavelength/frequency of the incident light.

With this knowledge, Einstein explained the photoelectric effect. Different metals require different ranges of energy for them to emit electrons from their outer orbital shells—i.e. each shell requires a minimum amount of energy to be supplied in order to release the electrons from its orbit. With this explanation of the photoelectric phenomenon, Einstein has proved that EM radiation carries energy in discrete lumps with specific amount of energy in each photon depending on the wavelength of light. The implication of this feature of light is that energy can be supplied only in whole integers but never in divisible fractions—that is, when photons with sufficient energy are supplied, the metal atoms absorb photons, thereby jostling the electrons out of their orbits. If the photons lack the minimum energy required to cause displacement, the metal simply does not emit any electrons. This is the 'all-or-none phenomenon' of light-metal interaction that gives us the photoelectric effect. This ingenious explanation by Einstein has indeed promoted the hitherto hypothetical quantum theory to its full theoretical form. And interestingly enough, this has brought Einstein his Nobel Prize—not for his famous theory of relativity, as is commonly thought.

Now look at this theoretical conflict: we have seen that the phenomenon of interference has proved unequivocally that light propagated by waves, whereas the photoelectric effect has also indisputably shown that light travelled in the form of quantum particles. Both of these theories appear contradictory, and so it's astonishing that light behaves like a wave at a given time as well as a particle at the same time! Hence we can conclude that:

Light behaves both like waves and particles at the same time.

Because of the solid evidence in support of both views, the scientists at that time have started to gradually reconcile to the fact that light behaves in this odd way, and light is now considered a special phenomenon in this way!

The Matter Waves: Until here, the scientists had no trouble in accepting the ambiguous nature of light, but from here on, the mind-bending concepts of quantum theory actually begin to appear, and the reader must now be ready for some of the most unintuitive ideas in the history of science. In 1923, Louis de Broglie made a theoretical suggestion that even matter particles should behave like waves when they travel, and subsequently, Davisson and Germer conducted a double-slit experiment using electron beams and proved that electrons indeed show interference pattern just like light waves, indicating that electrons also travelled in waves. Imagine for a moment a stream of electrons passing through the slits in a double-slit experiment. We would expect individual electrons to pass through one slit or the other at a time and reach the opposite side (like a football passing through one of the two goalposts situated side by side) and to cast two separate beams on the photographic plate. But surprisingly, electrons also behaved like waves of light and interfered with each other so as to give an interference pattern.

Since then, double-slit experiments were conducted using larger particles, such as atoms and molecules which also showed the interference pattern, which means that all matter particles in nature travel by waves. A thoughtful reader may immediately suppose that a *beam* of electrons may behave like waves, but an individual matter particle may retain the qualities of matter, thus behaving like a particle itself. It is worth noting here that even when experiments were repeated by using a single electron, it behaved much like a wave but not as a particle. The implication of these findings is that wave nature is not just a property of a beam of particles, but it is also the intrinsic property of an individual particle itself.

Matter wave is the intrinsic property of a matter particle.

It becomes an obvious theoretical implication that not only small subatomic particles but all objects (e.g. a tennis ball or a large boulder of an asteroid or a planet) also travel in waves, but then how do we account for the trajectories of larger objects being completely devoid of any wave-like movement? The mathematical equation of de Broglie's matter waves $\lambda = h/p$ (where λ is the wavelength, h is the Planck constant, and p is the particle's momentum) clearly shows that the wavelength of a moving particle is proportional to the Planck constant but is inversely proportional to the momentum, which, in essence, means that as the mass and velocity of a moving object increase, the wavelength decreases proportionately (both heavier masses and higher velocities shorten the matter waves simply because p, the divisor in the equation, is actually a product of mass and velocity ($p = mv$)). This makes the resultant wavelength of a large object so minuscule that it becomes totally undetectable!

However, it remained a theoretical challenge for the scientists to explain matter particles showing interference pattern—how did a single matter particle pass through both the slits at a time? Erwin Schrödinger had theoretically suggested that a single electron behaved like a 'smeared out' cloud with a wave pattern, and this has allowed the electron to pass through both the slits at the same time! But such an *ad hoc* explanation looked absolutely absurd and meaningless to the scientists at that time. And thus, this amazing property of matter particles stood unexplainable, and it needed a more conceptual understanding of nature to explain the phenomenon of matter behaving like waves!

Whereas this prankish behaviour of a particle has remained just as a mere conceptual mirage down the decades to the present, the wave nature of a particle itself was adequately explained by the later scientists. In 1926, Max Born made a queer suggestion to account for the wave nature of the electron. He suggested that the location of an electron in space must be interpreted more as a probability than as a certainty. He conceived waves more like packets of energy, or *wave packets*. Born mathematically showed that in a given field of wave packet, the place where we find the most momentum is the place where an electron has the highest probability of being found. Conversely, a place with least momentum is the place where an electron has the least chance of being found. If we carefully follow the above statement, it says that it is *only* a *chance* of finding an electron in one place or the other—i.e. the *probabilities* are simply either high or low—but we cannot be certain of the presence or absence of an electron in any given field whatever is its state of momentum. This essentially means that an electron in space may or may not exist physically at any given place!

Schrödinger, once again, went a step further and mathematically formulated an important equation (called the *Schrödinger equation*) which governed the shape and evolution of these probability waves, thus bringing Born's vague theoretical probabilities into accurate mathematical predictions. These probability waves later became known as *wave functions*. Thus, a wave packet is now envisaged as a combination of waves of different frequencies (Fig 2.3). They all crest together at the centre of a wave packet so that its energy is concentrated here, and at the periphery of the wave packet, they cancel each other out, resulting in diminishing energies. Hence, the probability of finding an electron (or any moving particle) at the centre of a wave packet is the greater than at the periphery.

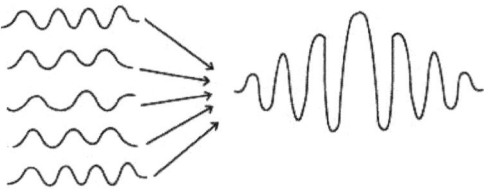

Fig 2.3: A wave function.

But then this probabilistic distribution of electron in a cloud of wave packet is something which did not go well with the physicists at that time. In the early twentieth century, classical Newtonian physics has still prevailed very strong in the minds of the scientists, and they were dealing with 'real' science where theories were no uncertainties but were 'concrete facts' with valid equations and completely dependable predictions. The science hitherto, throughout its journey, has taken a more deterministic course, and every event in nature was utterly predictable—a ball thrown up *did* come down to us dutifully under the influence of gravity, and a bullet shot from a gun *did* hit the target surely and certainly! But now with the advent of quantum theory, things have changed forever—the die-cast scientific concepts of the yore have now become obsolete in the quantum world, which is now replaced by a queer set of quite probabilistic or indeterministic laws. And these uncertainties would actually be dictating the events of the microscopic world of matter and energy hereafter with all their bizarre improbabilities!

The Problem of Observer's Effect: If the reader was not surprised by these strange implications of matter waves, then he/she would certainly be bewildered by another quantum phenomenon called the *observer's effect*, which is easily the most perplexing affair in whole of the realm of science! We will study this phenomenon briefly here.

Scientists made an intriguing observation while performing a double-slit experiment using electrons—if an observer makes an observation of the course of electrons (for example, by using a suitable particle detector) during the experiment, the electrons simply cease to show interference and pass through any one of the slits (which makes the interference pattern disappear on the screen)! And if the observer stops to observe (say, by switching off the particle detector), then the interference pattern reappears! This is completely weird, contrary to our intuition, and utterly diabolical! This impish behaviour of a matter particle behaving like a particle when observed and behaving like a wave when unobserved was totally confounding—nature going to the extent of playing pranks with us! This odd behaviour of matter particles is also seen in other quantum phenomena (such as quantum superposition and sum-over-paths, as we will see in Sec-D).

When observed, it is a particle! When not observed, it is a wave!

This ghostly phenomenon of quantum mechanics was explained by the so-called wave function collapse wherein, when we make an observation, several of the waves of a wave packet that represent a moving electron collapse into a single trajectory, allowing the electron to take a particular path through a particular slit, and the wave function dutifully reappears when we do not observe the event! However, the reader, on no account, must adjudge this observer's effect to be merely a perceptual phenomenon of the human mind. This bizarre phenomenon does show up certain physical effects in the quantum realm even with certain practical applications, as we will see in Sec-D.

For a long time, scientists had a great difficulty in accepting these counterintuitive findings in the quantum realm, but then these bizarre probability concepts gradually sank into the mainstream science of particle physics. The scientists have gradually reconciled to the fact that if that was the way the quantum world was designed to operate, and if that was the way it works, then it would work that way irrespective of human intuitions and inklings! However, this reconciliation of scientists has happened without any shred of understanding of the scientific rationale behind any of these bizarre quantum phenomena!

Section D
Heisenberg's Uncertainty Principle

The concepts of wave–particle duality and matter waves have several important implications in further developments in our understanding of nature, and one of the most important of them is the development of an ingenious mathematical principle called the *Heisenberg's uncertainty principle*, which has become the cardinal principle of quantum theory. We will first take a look at this uncertainty principle and then go into some of the weirdest phenomena of quantum theory that would actually dictate the exotic behaviour of the smaller matter particles in the quantum realm.

Historically, the uncertainty principle has replaced our understanding of nature by the century-old 'classical physics' (or 'Newtonian physics') with the new 'theoretical physics' with all its weird uncertainties and improbabilities. We will now briefly look into the essentials of this uncertainty principle. In 1927, Werner Heisenberg formulated the uncertainty principle, which described the theoretical inability to *simultaneously* determine both the position of a particle in space and its momentum with any precision. In other words, it means that if we could determine the precise location of a particle in space at a given time, we would not be able to determine its velocity with any more precision at the same time. The reader may visualize the Heisenberg's uncertainty principle in this way as it goes with a thought experiment: if we could capture a single electron and place it in a box and go on squeezing the size of the box to a smaller and smaller size in order to define its position in the box precisely, we should expect the electron to wriggle with even faster and faster velocities, thus making our simultaneous observation of its exact position and momentum null and void. Thus, the basic tenet of Heisenberg's uncertainty principle is, accuracy in position necessarily disturbs its concomitant motion, or simply

The more precisely the particle's position is defined,
the less precisely is its momentum known.

Heisenberg's uncertainty principle may be understood in the following way: the wave function of a matter wave may be said to contain all the information about the position and momentum of a particle. The position is given by the amplitude of the wave and its momentum given by the wavelength. First consider a sine wave. Because it has a single wavelength, this may enable us to determine the momentum precisely, but then, if we look at the position of a particle, it can be anywhere in the universe. Alternatively, if we depict a particle in a single wave, then its position may be accurately ascertained, but its momentum becomes imprecise because the wavelength has a probability distribution. Now look at the wave packet (Fig 2.3). We may see that the waves' amplitude waxes and wanes, and this renders the position to the particle in the wave packet uncertain, thus giving it only a probability distribution. But if we look at the wave packet again, we will see that it is made up of many different wavelengths, which puts the momentum of the particle in jeopardy. This is the essence of Heisenberg's principle!

To illustrate this uncertainty principle, let us see how scientists measure the exact location of an electron in space, and this would actually highlight the importance of this essential principle of quantum theory. Rays of light are used to probe the electrons to ascertain their location and velocity in space, and for doing this, the scientists make photons impinge on the electrons and study the bounced-off photons to know the electron's position and velocity. Scientists realized that the higher the frequency of light they use (i.e. shorter wavelengths), the higher is the precision with which an electron's location can be ascertained. This may be best understood by the following analogy: you draw a pencil sketch of a teddy bear about 5 cm in size on a paper and try to outline it with coloured beads; if you use large beads of about 1 cm to outline the sketch, the teddy bear soon loses all its shape, whereas if you use small 1-mm beads to outline the sketch, you can finely delineate the shape of your sketch. Large beads here represent longer wavelengths, and small beads represent shorter wavelengths—the shorter the wavelength (i.e. higher frequency), the better is the delineation of electron in space. But then here is a catch. If you use high-frequency photons, you may be able to precisely locate the electron; but high-frequency light waves carry higher energies, and they jostle the electrons more and more and increase their velocity—i.e. while the electron's location is precise, its accuracy of velocity is lost! On the other hand, if you use low-frequency photons (with less energy), the impact on the electron is minimal, so velocity can be studied accurately, but its location is not so precise (remember large beads!). Heisenberg said that this trade-off between the precision of position and velocity represents the innate property of nature itself, and thus, it has emerged as the fundamental principle of the microworld of quantum physics.

Heisenberg has mathematically formulated the uncertainty principle by telling us that a combination of *error in position* times *error in momentum* must always be equal to or greater than half of the reduced Planck constant:

$$\sigma_x \sigma_p \geq \hbar/2$$

where σ_x is the standard deviation of the position of a particle, σ_p is the standard deviation of its momentum, and $\hbar$ is the reduced Planck constant. The reader must note here that the uncertainty principle neither is observer dependent nor is the result of any technical fault in our measurement – it is just the way nature works in the background at the quantum level regardless of our observations and measurements. In fact, it is important to note that Heisenberg's uncertainty is not only limited to the 'position and velocity' of a particle but it is a general principle involving the products of any two conjugate pairs. For example, the precise measurement of the total amount of *energy* at a given precise *time* is also not possible. To illustrate this point, consider this – since a wave packet is a combination of different frequencies, the only way to know the total amount of energy of a wave packet at a given time is to study a wide range of frequencies (which represents the precise amount of energy of a particle), but in doing so, more time elapses so that the time determination becomes imprecise. This sort of conceptual understanding is very important and becomes useful in our study of the ambiguity of the physical principles in the subatomic world.

Therefore, this imprecision in the measurements, as noted above, is not due to limitations or faults in the measuring equipment or due to observer's parallax as a reader's wishful thinking might tend him/her to suppose. Rather, these uncertainties are inherent to the particles in the quantum realm, and this is the innate nature of a particle's existence in space. Thus, the basic tenet of the uncertainty principle may be read as:

> *The precise quantum state of a particle is the*
> *combination of both position and velocity.*

For a long time, a lot of scientists (including the prominent researchers who themselves have helped the quantum theory take roots, such as Einstein) were not convinced of how the nature could rely on such uncertain laws and thought that quantum theory covers only half the truth and that a finer theory is yet to come, which could resolve all the uncertainties once and for all and replace it with hard certainties. But experiments from time to time revealed that the predictions of the quantum theory were extremely accurate, and thus subsequently, many staunch critics of the quantum theory changed camps and became strong advocates of the very theory they desisted. After all, if nature intended to operate in a so unpredictable way, we must try to understand and analyze it as it is but not in the way we, the humans, wanted it to behave!

Thus, this conceptual understanding has ultimately brought *unpredictability* and *randomness* into the realm of mainstream science, and the weird concepts of the theory of relativity together with the wild improbabilities of quantum theory have occupied the centre stage of modern theoretical physics, and they have naturally become the 'theoretical tools' with which we study the universe both at its small and large scales (as we will see in the subsequent chapters). With this understanding, we will now look into some of the other major implications of the quantum theory.

SECTION E
The Weirdest Quantum Phenomena

In this section, we will look into some of the most intriguing quantum phenomena which are, in fact, offshoots of the quantum principle studied above.

Quantum Superposition and Schrödinger's Cat: When we consider a moving subatomic particle (e.g. electron) as a packet of energy, this energy packet can be shown, as per Maxwell's equations, to consist of an *infinite* number of waves. However, it has been theoretically suggested that not only do these waves result in constructive and destructive interference when they interfere with each other, but some of them may remain in the intermediate state for a transient period. These waves are called *superposed waves*.

Now consider this: when we reckon the subatomic particles as particulate, they are known to exist not only in several quantum states (quantum states are discussed in Ch 3-D) but some of them may exist in an intermediate state too. For example, an electron may have a spin-up state, or it may have a spin-down state (Ch 9-A)—or sometimes, as is suggested, it can exist in an intermediate state. This is called a state of *quantum superposition*. In fact, these superposed particles can theoretically exist at several places *simultaneously* in a quantum field of a packet of energy.

However, the reader must realize that this superposed state could not be actually observed, but we can see only its bizarre manifestations. Thus, as an observer makes an observation, all the possible intermediate states collapse into a single state of either one or the other state (see the observer's effect discussed in Sec-C) so that we can never be able to see an intermediate state of a particle or the phenomenon of a particle existing at several places in any of our experiments (because when we conduct an experiment, we will be observing the event, and as we observe the particle, it starts to exist only in either one of the two states!). In 1935, Erwin Schrödinger conducted a famous thought experiment to illustrate this observer's effect based on the *probabilistic* decay of a radioactive atom. It goes like this: a radioactive atom, which is placed in a closed box along with a live cat, has an equal probability of remaining intact or undergoing decay, and when the atom decays, it would kill the cat in the closed box by a certain mechanism. Now consider this: at a given point in time, an observer outside the box would not know whether the cat is alive or dead. Schrödinger said that at the quantum level, the unobserved cat in the closed box exists in both the states—dead *and* alive at the *same* time. Thus, the cat can now be said to exist in a 'superposed state'. But when we make an observation (i.e. open the box), the cat exists in only one state—*either* dead *or* alive!

Feynman's Sum-Over-Histories: In 1948, Richard Feynman has conceptually revised the double-slit experiment and matter waves and went a step further and challenged the basic assumption that an electron goes into *either* of the two slits; instead, he proposed a theoretical mechanism by which *each* electron goes though *both* the slits at the *same* time! Consider this: our common intuition says that an electron in a double-slit experiment which is about to enter the right slit would not have anything to do with the left slit – it would not just care about the existence of the other slit at all. But in the quantum realm, the electron does 'know' about the position of both the slits simultaneously. It behaves like a particle if there was only one slit, and it behaves like a wave if there were two slits. Here Feynman developed a mind-bending concept: he suggested that each electron making the journey from the source to the phosphorescent screen traverses *every* possible trajectory *simultaneously*—i.e. an unobserved electron would take an infinite number of trajectories from the source before reaching the target (Fig 2.4).

Feynman showed that we could mathematically predict the trajectories' combined average, which would be the same as the probability of a wave function (as discussed above), and the electron is likely to take this predicted pathway to make its journey to the destination. This phenomenon is called the *sum-over-histories* or *sum-over-paths*, a truly important contribution to the quantum theory by Richard Feynman! However, once again, it has been suggested that once we start to make an observation of the double-slit experiment, the infinite pathways collapse into one of the pathways (because of the observer's effect). But then statistically speaking, it most usually takes the predictable pathway!

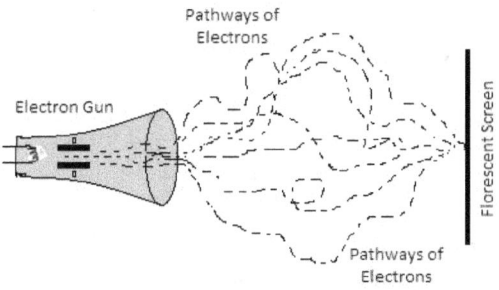

Fig 2.4: Sum-over-histories.

In fact, any particle travelling in space can be represented mathematically by its wave function. For that matter, any object— be it a small subatomic particle or a tennis ball or a boulder of rock or a planet—all moving objects can be represented by their wave functions. The trajectories of these waves can be mathematically shown to trespass anywhere between the source and destination. Theoretically the trajectories may take any curved or convoluted course and travel variable distances across the intermediary space between the source and destination (Fig 2.4), or the trajectory may even take a very long course and loop around some distant star before reaching the destination! In essence, it may even be said that an object can theoretically spread across the whole universe. However, the probability of finding a particle is higher at one place and lesser at the other places as directed by the sum-over-paths. Weirdly enough, on this account, we can claim that a grandma calmly relaxing in her easy chair can be mathematically shown, as she gets up and starts walking to the kitchen, to have travelled to the Andromeda Galaxy before reaching the kitchen (of course, stealthily, when not observed)! However, the infinite trajectories of the grandma took in her journey cancel each other out, finally leaving out a single predictable trajectory, which she takes. Fortunately for her, the predictable trajectory can be mathematically shown to follow the path as predicted by the Newton's laws of motion, and so our grandma could finally make her way to her cup of coffee in the kitchen! Thus we can say that, unobserved, she may be anywhere in the universe, but when observed, she is just hustling about in the kitchen. But then it can be mathematically shown that the sum-over-paths disappear rapidly as the mass of a moving object increases— so that our grandma could navigate across the house fairly safely and confidently. What a weird idea indeed!

Unobserved, a particle may exist anywhere –
observed, it exists in one state.

Therefore, to summarize, a quantum particle can be said to exist in all states at a given time, and consequently, it may be considered to spread across the whole universe at any given time! However, when we study them and make an observation, all the wave functions collapse, and the quantum particle presents itself in any one of its states as directed by the probability rules we discussed above.

Quantum Tunnelling: Now we will examine another intriguing concept which stems out of the quantum theory. By now the reader must have observed that the phenomenon of sum-over-paths may be considered more as a chance phenomenon of a moving particle. Consider this: a moving electron, though in all probability may follow the calculated average pathway and may reach its projected destination, has a very small but non-negligible theoretical probability that it may still take one of the devious courses and may not reach the calculated destination at all! Now consider this: normally, electrons do not pass through impassable barriers such as metals, but there is a certain non-zero probability that the electrons would pass through a thin metal plate because of the probability factor – and surprisingly, this phenomenon has been proved to be more of a fact than myth. This property is called the *quantum tunnelling*. Quantum tunnelling has really become more of a technological principle in the realm of electronics, and in fact, a *'tunnel diode'* works on this principle of quantum tunnelling. And the next time you listen to your favourite rhythm in your electronic gadget, remember that this exotic quantum principle is actually at play! Quantum tunnelling also plays a role in radioactivity (Ch 3-E) and in proton–proton fusion reactions in the sun (Ch 4-D) wherein this tunnelling property allows small subatomic particles to 'burrow' through formidable barriers, such as large nuclei.

Quantum Entanglement—'Spooky Action at a Distance': This is yet another extraordinary implication of the quantum theory. It is observed experimentally that photons in a laser beam, when they pass through certain types of crystals, split up into pairs of 'entangled photons'. It can be said theoretically that these two particles of a pair exist in a single quantum state so that they behave like a system as a whole, and these pairs become inextricably entangled in such a way that any change in one particle necessarily disturbs the other even if they are separated by infinite distances. This is because of a peculiar phenomenon called *quantum correlation* wherein the particles which have interacted at some point of time in the past retain their association and remain entangled in pairs for indefinite periods. Consequently, these entangled pairs show weird behaviour. If, for example, one particle makes a spin in one direction (Ch 3-C; Ch 9-A) the other particle would spin in an opposite direction, or if one particle rotates in a clockwise direction, the other would rotate in anticlockwise direction. Such a pair of particles are now said to be associated by *quantum entanglement*, and they could communicate almost instantaneously with each other even if they are separated by infinite distances across the universe! However, because any object or any piece of information in the universe travels with a speed limit of the speed of light, this sort of instantaneous quantum communication in the case of quantum entanglement is thought to be transmitted at speeds faster than the speed of light (though this, in principle, does not agree with the special theory)! Thus, the properties of these entangled particles, such as spin, momentum, and position, are said to be appropriately correlated between the pair particles even when they are located at infinite distances (perhaps, as is often said, even when located at the 'opposite ends of the universe'!). Einstein found this behaviour odd and unbelievable and called this phenomenon as 'spooky action at a distance'. But then one has to reconcile to the fact that this is an intrinsic property of the quantum particles, and it has to be understood as it is.

It is interesting to note here that recently, China has launched a 'quantum satellite' called *Micius* (named after the ancient Chinese philosopher), which would prospectively communicate with the earth using the principle of quantum entanglement. This quantum satellite, if it becomes really operational, would set new standards in the field of wireless communication and Internet security because any attempt to hack the entangled particles would, in principle, become virtually impossible, rendering this sort of wireless communication totally hack-proof!

Does God Play Dice with the Universe? Having narrated so many weird things about the quantum particles, the reader may be left behind with many questions: Why should particles behave in this odd way? Is quantum theory correct, or is there something wrong with our interpretation of nature? In other words, is the quantum theory a half-baked stuff (as some scientists may ask)? But these aren't valid questions any more to a learned scholar of quantum theory; these are all naive questions so commonplace only to a suspecting ignoramus. The simple fact in support of the validity of the quantum theory is that the theory is extremely accurate in predicting the events of the subatomic world – in fact, it is the most successful quantitative theory ever generated in the history of science! Moreover, quantum theory has become useful in the formation of various field theories, which, by themselves, are extremely successful theories (as we see in Ch 3). And also, these concepts have paved way for many important technological advancements of our century, as we will see in Sec-F. Feynman once wrote (quoted verbatim), 'Quantum mechanics describes the nature as absurd from the point of view of commonsense. And it fully agrees with experiment. So I hope you can accept nature as She is—absurd!'

All said and done, the quantum concepts may still appear illogical or irrational to the common man. But then this spooky sense of feeling is not for a commoner alone; it has also affected some of the greatest scientists too! Einstein (though he himself was the chief architect of quantum theory) was unhappy throughout his life about this probabilistic theory and could not reconcile with the glaring fact that nature went to such an extent to play pranks to conceal the truth (or reality) from an observer! And why should the fundamental nature of the matter and energy be founded so inexorably on the precarious uncertainty principle? However, in between the years 1925 and 1927, Niels Bohr and Werner Heisenberg have attempted to put an end to this ambiguity by interpreting quantum theory as a consummate theory wherein the subatomic world is chiefly governed by the uncertainty principle and by stating that the physical systems in the quantum realm would not have any definite properties but are ruled by the probabilities but not certainties. This set of 'theoretical commandments' has come to be known as the *Copenhagen interpretation*, so-called because these principles were laid out in a series of lectures held by Bohr and Heisenberg at the Institute of Theoretical Physics at Copenhagen (which was later named as Niels Bohr Institutet). This authoritative declaration of the validity of quantum concepts has really helped the quantum theory to gradually seep into the minds of the contemporary physicists.

But then recalcitrant as he was, Einstein had once remarked metaphorically, 'I, at any rate, am convinced that God does not play dice with the universe', meaning that the physical systems of nature could not gamble over the events, and so the quantum theory may still be theoretically incomplete. To this remark, Bohr had retorted with a sarcastic rejoinder: 'Einstein, don't tell God what to do'. Such historical altercations and theoretical debates only show that the pessimistic reader may be excused if she/he feels along with Einstein in these terms. Nevertheless, history of science teaches us to be somewhat tolerant of our impertinent scientists (or the so-called 'science outlaws') who continue to question the scientific establishment, and we must allow them to gracefully keep questioning. History of scientific progress, after all, is so replete with such out-of-bound questions (as we will see in the forthcoming chapters), and the answers to such questions had, time and again, shook the very foundations of science! Leaving this at that, we will now deal with some of the practical applications of the quantum theory.

Section F
Quantum Theory and its Applications

Though the quantum theory appears quite weird and unworldly to the reader, he/she must realize that this theory is indisputably responsible for many scientific and technological advancements of modern civilization. Moreover, the reader will notice as he/she reads on through the upcoming chapters that many of the recent scientific theories are protégés of quantum theory in the sense that the essence of quantum principles lies in the very core of all the later theories, such as the Standard Model of the atom to the big bang model of the universe and others as we will see in appropriate chapters.

However, the most outstanding and most direct human application of the quantum theory is in the realm of electronics. Quantum mechanics has revolutionized the concept of electronic gadgets and has led to the development of the fastest and the finest of the modern computers, including the so-called quantum computer, which may really become possible in the near future (see below). Hereunder, we will take a brief look into the theory behind the development of the modern computer, which highlights the importance of the quantum theory.

All *electrical* appliances (such as electric fan and electric heater) use *electric circuits* to get work done, the essential property of any electric circuit is to control the movement of electrons across it in a particular direction. Many electrical appliances use metals in their circuits. On the contrary, *electronic circuits* (which are the fundamental constituents of all electronic gadgets) use *'semiconductors'* (instead of metals) for circulating electric current in their circuits. But then what is a semiconductor, and what do they do in electronic circuits?

Before going into the basics, we will see how a metal wire conducts electricity. A good conductor of electricity has many loosely bound electrons in their outermost shells called the valence electrons, and these electrons can be technically considered to be situated along the wire in the form of a continuous sheet. This continuous array of 'static' electrons is called the *valence band*. When electric current passes through a metal wire, it excites these electrons, which will then jump from their existing energy level to a higher energy level.

And when they revert back to their original position, they release the energy, which is now transferred to another electron. This, in effect, carries electrical energy from atom to atom across the metal wire (Ch 8-F). This array of high-energy electrons which carry electrical energy from place to place is called the *conduction band*. Though electrons themselves have an intrinsic tiny electric charge on them (Ch 3-C), their contribution to the total amount of electricity that is being transferred along the conduction band is negligible, so the energy passed in an electric circuit is almost exclusively from that of an external source (e.g. a battery or a power station). Thus, for example, when we switch on our table lamp, the energy that is utilized to give us light is derived exclusively from that of the electricity source but not from the inherent electricity of electrons in the metal wire!

The story of transmission of electricity is a little different in the case of electronic circuits, which are made of materials called semiconductors. Now what are semiconductors, and how are they different from ordinary metals? Elements like *germanium* and *silicon* are insulators of electricity (i.e. bad conductors), but when they are mixed (or *'doped'*, as it is called in electronic parlance) with minute amounts of 'impurities' such as phosphorus, arsenic, boron, or gallium, they acquire a peculiar property of partially conducting electricity. In other words, they become semiconductors, and these semimetals are used in electronic circuits because of their peculiar behaviour. One interesting feature of semiconductors is that in an electronic circuit, the magnitude of the tiny electric current that is carried along the conduction band is also contributed by the intrinsic electricity present on the electrons in the valence band, which becomes appreciably large in a tiny electronic circuit (but not in the case of an ordinary electrical circuit, as explained above). Now consider this: as electricity passes through an electronic circuit, the electrons hop from valence band to conduction band, and in this process, the electrons effectively create some vacant spots in the valence band called the *electron holes*. In turn, these electron holes would be filled up by the electrons from the deeper shells of the atom, and this would create a flux of current across the circuit. In fact, these holes are the 'quantum wells' with certain discrete energy values which are bound by the rules governed by quantum mechanics. In the case of semiconductors, these tiny fluctuations of energy are unique and work wonders by demonstrating certain weird properties in the electrical resistivity in the electronic circuit (see below). Now we will see how semiconductors are used in the microcircuits of electronic gadgets.

Transistors: The heart of any electronic circuit is the *transistor*, which is made up of semiconductors. But then what are transistors, and how are they made? Semiconductors are of two types, depending upon whether they allow the electrons to be added to them when electricity is passed through them or allow the electrons to escape from them. In the positive type (*P-type*), the semiconductor absorbs electrons when current passes through it; conversely, in the negative type (*N-type*), the semiconductor gives out electrons to the adjacent field. Now place very thin sheets of N-type and P-type semiconductors side by side, and the result is a *diode* which would allow the current flow in only one direction. Transistors are created by using *three* layers rather than two (hence they are also called *triodes*), and they can be arranged in two sandwich patterns—NPN or PNP (Fig 2.5). When a small current is applied to the central unit (the *base*), one of the two things may happen depending upon which type (N or P) represents the base. It can either amplify the passage of current across the whole unit, or it can hinder the passage of current. Thus, each wave of current that passes through the transistor can be represented in two states—*on* and *off*. In other words, it can act like a *switch*. All the above discussion means that we have made tiny switches in the form of transistors which can be controlled dynamically within a circuit. And these tiny electronic switches (i.e. the transistors) would become the building blocks of 'logic gates' (see below). However, for the logic gates to become really operative, we need a suitable digital language. But then what is this 'digital language'?

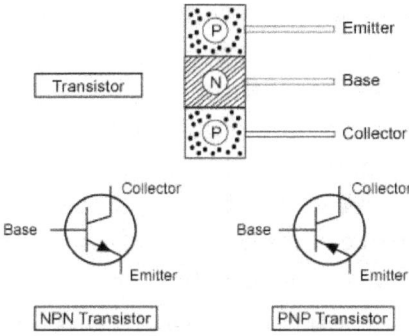

Fig 2.5: Transistors.

To understand this properly, we must know how *information* (Sec-E) can be processed and stored electronically. Any information in the form of sound, light, temperature, pressure, etc. can be converted into electromagnetic signals on suitable devices, and these signals can be converted back to their respective forms when necessary. For example, when we put in sound signals into an input electronic device (e.g. microphone), the sound signals are converted into electronic signals and stored on the computer chip (a single modern chip consists of billions of tiny transistors), and when these electronic signals are passed on to a suitable output device (such as speakers), they are again converted back to sound signals. Here's another example—light signals are stored as electronic signals in the chip of your digital camera, and when connected to a suitable display mechanism, they are converted back to light pulses, and this lets you have your picture back again. These electromagnetic signals can be electronically studied in two ways—as *analogue signals* or as *digital signals*. The difference between analogue and digital may be understood easily by studying your clock, which may indicate time in two ways—a continuous mode in the form of a dial with hands called an analogue clock or in discrete signals of numbers called a digital clock. The modern computers use digital signals for efficient communication.

The input of electronic signals needs a suitable *digital language,* and such a language was already developed by George Boole way back in 1854. What is this digital language? We know that *decimal system* (ten-digit system) of counting numbers uses ten digits—0 to 9—and it starts from ones, tens, hundreds, thousands, and so on. In

Boolean logic, the numbers are counted in a *binary system* (two-digit system) which has only two units—0 and 1—and all placeholders are multiples of 2. Any length of number can be represented by using binary language. All modern computers use binary language for processing information.

Now combine transistors with binary code. The transistors act as gates to the passage of electricity—switching them either 'on' or 'off', thus generating signals which can be made to pass with tremendous rapidity as the current flows. Thus, transistors can represent the current flow in two states—*on* (or 1) and *off* (or 0), and this is the feature that is utilized in representing the 'binary code'. This means that any input signal (like light and sound) can be converted into a binary code in an electronic circuit, which can now be studied electronically. This is the basic theory behind the functioning of a modern computer. Now the reader can see that all these magnificent computing features of our computers are fundamentally based on the quantum behaviour of electrons—more precisely, these digital properties may be said to be dependent on the most fundamental behaviour of electrons and electron holes in the semiconductors.

A brief history of the development of a modern computer chip is presented here so that the reader may appreciate the achievements of versatile human genius. Each *microprocessor* (called otherwise as *microchip* or simply as *chip*) in a computer consists of several transistors. Obviously, the number of transistors that are present in a chip represents the speed with which that device can perform. The first microprocessor released by Intel in 1971—the Intel 4004—contained 2,300 transistors on its circuit, whereas Intel's Core i7 (the quad) contains over 700 *million* transistors. And when you think of the latest version of chips containing over two *billion* transistors, you may imagine how far we have progressed in a matter of a few decades! And in the near future, transistors may measure about less than 100 nanometres. Compare this with the 'vacuum tube' of the first modern computer, the ENIAC, which measured a staggering three inches – each vacuum tube of yore can now accommodate about 1.5 billion-billion of our present-day transistors!

The reader may realize that the smaller the quantum size of circuits in a chip, the better is its performance – after all, the electrons have to travel for less quantum distances. Hence, the size of the transistor is all that matters—the smaller the smarter! The day of nanotechnology has come into the fore, and the yesterday's gossip of 'atomic-sized' transistors has almost become today's reality!

A Note on Quantum Computers: A quantum computer is a theoretical model of a computer which works on the principle of quantum superposition and quantum entanglement (Sec E). Whereas a traditional computer operates using the binary system, a quantum computer stores information in quantum states of superposition of 0 or 1—each bit called a *quantum bit* or *qubit*. Quantum computers can process information with such gigantic speeds that a 500-qubit computer can accomplish about 2^{500} computations in a single step, a number which far exceeds the total number of atoms in the universe! Some experimental quantum computers are indeed built in some of the institutions across the world, but a practical 'desktop' quantum computer is still a far cry. However, as an aside, it may be stated that such superlatively capable computers can theoretically unsettle today's corporate financial edifice because they can rip through their security encryptions, which are based on factoring large numbers that cannot be practically cracked open even by today's superfast computers. But then by the time quantum computers come into vogue, our encryption mechanism would also, hopefully, be reaching new dimensions!

The development of quantum computer underpins the validity of the quantum theory by employing the bizarre phenomena of quantum superposition and entanglement. Thus, we may state that quantum theory, though not completely understandable to the logical human mind, is a true theory that is operative in the background of everything we perceive in nature.

SECTION G
The Enigmatic Planck Units

In the theory of relativity, we have seen that our measurements of length, time, and mass are relative to each other and are not absolute by themselves. Our present-day understanding of the events of nature is chiefly based on this precept. But then we will now discuss an intriguing question: could there be a measurement in nature which is absolute, i.e. a measurement which is not relative to any other measurement? In other words, is there an absolutely fundamental unit of measurement of length/time/mass which constitutes the *intrinsic property* of matter itself and so cannot be changed? This question becomes practical when we attempt to measure the subatomic particles,[*] where, for example, we cannot accurately measure a subatomic particle using, for example, a nanometre—because a nanometre has a 'metre' in it, which itself is not an absolute entity (Ch 1-A). In the same way, we cannot weigh the mass of a subatomic particle without comparing it with an arbitrary standard, which is a relative entity by itself.

But then scientists have realized that if we take into account certain universal constants in nature, it appears that we can mathematically build an entirely trustworthy and absolutely fundamental unit of measurement. There are five such convenient constants which may come to our assistance, *viz.* the *speed of light* (c), the reduced *Planck constant* ($\hbar$), the *gravitational constant* (G), the *Coulomb's constant* (k_e) which governs electricity, and the *Boltzmann's constant* (k_b) which controls the statistical behaviour of large system of molecules. The speed of light (299,792,458 m/s) as an absolute entity is already discussed in Ch1, and we have also studied the Planck constant at the beginning of this chapter (Sec-C). To reiterate, Planck constant is a fundamental constant equal to the energy of a quantum of electromagnetic radiation divided by its frequency (derived from $E = hf$). It has the smallest value of $6.6260695729 \times 10^{-34}$ joules-sec. However, another modified but practical constant, called the reduced Planck constant, becomes useful while measuring the

---

[*]   What about 'quantum' as a measure? Is 'quantum' a measuring unit? Is it an absolute entity? We have seen the quantum unit of an EM wave as a pulse of energy carried as a packet of photon. A quantum of energy can be either small or large depending on the type of EM wave. A gamma wave has a short but intense pulse of energy carried within its packet with the dimension of the wave packet being about 10^{-12} meters (i.e. its wavelength), whereas a microwave has a long but weak pulse of energy in its quantum packet of about 10^5 meters (Table 2.1)! Therefore, the reader can realize that a quantum is not a unit of measurement by itself and so cannot be used to gauge the subatomic world. In short, quantum is not a constant unit.

mass of a matter particle, such as an electron revolving round the nucleus. This reduced Planck constant is considered the smallest angular momentum possible, and it is obtained by dividing the Planck constant by 2π.

The other constant which becomes operational in calculations of mass and energy is the *universal gravitational constant* (G), which characterizes the intrinsic strength of the gravitational force. This gravitational constant, the so-called *Big G*, appeared first in the famous Newton's law of gravitation, $F = Gm_1m_2/r^2$ – where F is the magnitude of force and m_1 and m_2 are the masses of two objects separated by a distance r. The value of G is estimated to be 6.673×10^{-11} N.m²/kg² which, fortunately for us, is so exceedingly small—so much small so that we do not attract and fall into each other when we sit side by side in a park (unless, of course, your partner is someone of Marilyn Monroe genre!).

A set of absolute and 'incomparable' measurements based on these fundamental constants have emerged, which are known as *Planck units*. In some measurements, Coulomb constant and Boltzmann constant are also included in calculating the Planck units, such as Planck temperature (see below). Planck units are now considered the absolute values in the universe.

Planck length (l_p) is a fundamental unit of length (derived from the formula $\sqrt{\hbar G/c^3}$), and its value is now estimated to be $1.616199(97) \times 10^{-35}$ meters, theoretically the smallest dimension possible (which is about 10^{-20} times the diameter of a proton). The significance of this number is that it is theoretically impossible to go in between two points less than the Planck length. *Planck time* (t_p) is the unit of time taken for light to travel in vacuum a distance of one Planck length (formula: $\sqrt{\hbar G/c^5}$), and its value is about $5.39106(32) \times 10^{-44}$ sec. Quite obviously, our day-to-day events are billion-billion-billion times bigger than these Planck measurements, and our scientists have not come anywhere near these measurements in our experiments using the current technology—the smallest recorded time interval being 12 *attoseconds* (10^{-18}), which is about 3.7×10^{26} times Planck time! However, the whole objective of these measurements is to make these units as objective as possible so that these Planck units become free of dependence not just from human observation or perception but also from a relativistic standpoint. In other words, at these extreme distances, the concepts of space, time, quantum rules, rules of relativity, and other physical entities cease to have any meaning. In short, they would become constants by themselves. But then there is not yet a real way by which we, the humans, can comprehend these absolute measurements by using our sensations and perceptions, which remain too gross for any such understanding!

Whereas Planck length is very short, *Planck mass* is very large (i.e. when compared) (formula: $\sqrt{\hbar c/G}$) – it is approximately 0.0217651 milligrams—commonly compared to the size of a flea's egg! This is owing to the fact that the denominator in this equation is G, which is extremely small (whereas, as the reader can guess, for l_p calculation, the denominator is c^3, which is extremely big). So then obviously, the subatomic particles such as electrons are lighter than the Planck mass by several magnitudes. It has been suggested that Planck mass defines the smallest possible black hole (Ch 4-C) in the universe. And it is also suggested that particles heavier than Planck mass do not possess an obvious wave function (as a tennis ball will not!) so that an interference pattern cannot be elicited in a double-slit experiment. The importance of *Planck temperature*, which reaches an incredibly hot number, is discussed in Ch 5-G in connection with the big bang theory.

However, the practical significance of Planck length, mass, and time is not yet clear. We do not know if they represent the *quantum gravity* (Ch 3-E), but it is speculated that at Planck length, the quantum gravity renders the spacetime a 'foamy structure' (Ch 3-H)—a vivid imagination indeed! Nevertheless, these fundamental Planck units are used in the development of the early universe (Ch 5-G) where it is supposed that the universe has started off at these infinitesimally small scales. The reader must realize that there is no currently available physical theory to describe such short measurements, and it is not clear in what sense the concepts of length and time become meaningful for values smaller than these units. And so we will put this possible question: can we say that any measurements falling below the values of Planck scales can be considered to denote 'vacuum' (or 'nothing')? But the reader may have to wait until Ch 3-H to decipher the exact meaning of 'nothing' in modern physics!

Thus far, we have discussed quantum theory in brief. We will now go into the world of atoms wherein we will study the various intricate features of matter particles and their interactions with forces of nature.

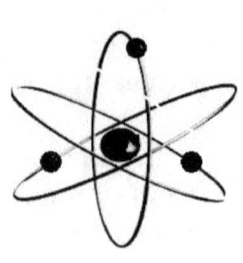
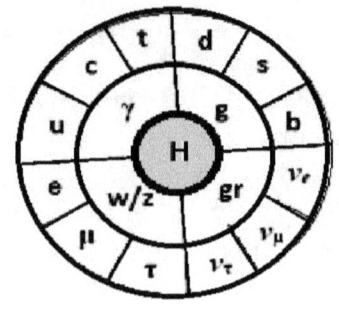

CHAPTER 3

The Standard Model
And the Structure of the Atom

Overview

I n this chapter, we will examine the microworld of atoms and look into a herd of smaller of the smallest subatomic particles that make up an atom. By the way, we will also see how various fundamental forces of nature participate in binding these matter particles together in an atom.

At the outset, we will learn about the older models of atom and see how scientific progress, especially in the field of quantum mechanics, revolutionized our understanding of the atomic structure and how this has unravelled some of its deepest secrets, which finally culminated in the development of the impeccable model of the atom called the Standard Model. In passing, we will come across an array of subatomic particles with all exotic names and behaviour which would certainly tax the reader's patience in remembering them. But then mercifully, we will see that we need to remember only a few of them, and the rest of the complex details needn't bother us to understand the essentials of the Standard Model.

Then we will study the four fundamental forces and their diversity, and having studied their diversity, the reader may realize that the modern scientists have been searching for a method by which all these four forces are unified, but in vain. Here we will review how this problem of grand unification is viewed today. In the subsequent sections, we will look into the fundamental problem of mass, which has intrigued scientists for a long time. We will then study how the researchers have solved this problem and defined the nature of the Higgs boson or the so-called God particle. The enigma of the antimatter is briefly discussed to show why matter (but not antimatter) dominates the universe. In the penultimate section of this chapter, we will plunge into the puzzling empty space to look into the world of 'nothingness'. And finally, to complete this discussion on the structure of matter, we will examine some of the alternative atomic models and discuss 'string theory' in brief, which is making rounds in the scientific circles of our times.

Now enter the variegated world of tiny subatomic particles!

Section A
The Atomic Models

Perhaps it was only a matter of philosophical intuition that Democritus (*circa* 460–370 BC) thought matter was made up of indivisible units called *atoms* (*a-tomos* = un-cuttable), but it took nearly 2,000 years for scientists to crack open the atom to proclaim that atoms are really 'cuttable' and that they have some smaller particles within them.

The science of atoms had actually begun in 1803 when John Dalton proposed his *atomic theory*, which stated that matter is made up of discrete atoms with characteristic chemical properties. It was the dedicating work of many scientists over the next several decades to show that each element was made of similar kind of atoms and that different kinds of atoms combine to form molecules. If the atoms were indivisible, then they must be fundamental, and if they were fundamental, then they must be similar to each other. But then why do atoms of different elements behave differently? What is that stuff which makes atoms different? In order to explain these intricate questions, science had to make its progress not only in the field of chemistry but in many other parallel fields, and by the dawn of the twentieth century, it was proved beyond doubt that atoms are not the fundamental units of matter but have some specific internal structure which dictates their chemical properties.

The First Atomic Models: Scientific progress in the field of electricity was certainly the prominent reason for the rapid development of the science of atoms. In the olden times, it was supposed, rather intuitively, that atoms themselves were electrically neutral. In 1897, J. J. Thomson studied cathode rays and theorized that they were actually composed of subatomic particles called electrons which were negatively charged. Thus, an electron ranks as the harbinger particle of modern physics.

With the discovery of electrons, scientists started speculating the atomic structure, and at first, it was logically conceived that an atom is made up of a positively charged 'dough' with negatively charged electrons randomly dispersed within it—the so called *plum-pudding model* (Fig 3.1 a). However, in 1911, Ernest Rutherford conducted his ingenious gold-foil experiment, in which a very thin gold foil was bombarded with positively charged alpha particles. He studied their scattering pattern, which showed us that the atoms do not have a random arrangement but have a rather well-organized structure. In this *Rutherford model,* an atom has a central tiny *positively charged nucleus* with a vast amount of empty space surrounding it with the *negatively charged electrons* hovering in this space. This model of an atom was descriptively called the *planetary model* (Fig 3.1 b)—electrons orbiting the nucleus as planets orbit the sun. The Rutherford model soon has enjoyed an immense success not only because of its simplicity but more importantly because it has explained the chemical behaviour of elements to a large extent.

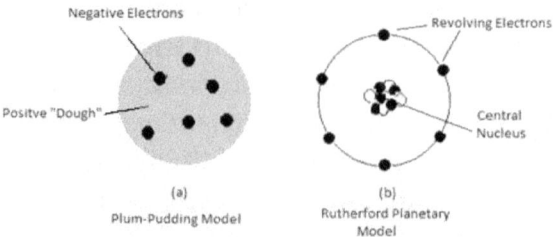

Fig 3.1: Early models of the atom.

But then what is the nucleus composed of? The gold-foil experiment had shown that the nucleus of an atom is a very tiny structure situated at its centre with a positive charge. But it remained an enigma for some time as to what it contained. Subsequent research, this again by Ernest Rutherford, showed that the nucleus itself was composed of tiny positively charged particles called *protons*. However, Rutherford's calculations also showed that the total mass of a nucleus could not be accounted for by the combined mass of protons alone. Some significant mass was found missing. He theoretically suggested that the nucleus contained some uncharged particles called *neutrons* that contributed to the mass of the nucleus. The existence of neutrons was experimentally confirmed only later by James Chadwick in 1932.

The Rutherford model of an atom is now complete with a compact nucleus consisting of a specific number of positive protons jiggling inside it along with a certain number of neutrons and with the same number of electrons orbiting round the nucleus (Fig 3.1b). This specific number of protons/electrons is called the *atomic number*, which varied from element to element. The reader may look at the *periodic table* and see that *hydrogen* ($_1$H) stands as the first element with a single electron orbiting its nucleus (which is simply a single proton). The next element is *helium* ($_2$He) with two electrons, two protons, and two neutrons. The next elements are *lithium* ($_3$Li) and *beryllium* ($_4$Be), so on and so forth. Each element has a specific atomic number, i.e. the number of protons decides the type of element. But the same element may have different number of neutrons in it so that it has a different *atomic mass*. Elements with the same atomic number but with different atomic mass are called *isotopes*. For example, hydrogen usually has one proton ($_1^1$H, when it is dutifully called *protium*), but it may also have an extra neutron ($_1^2$H, when it becomes an isotope of hydrogen called *deuterium*), or it may have two neutrons ($_1^3$H, when it is called *tritium*). In the same way, helium usually has two protons and two neutrons ($_2^4$He), or it may have two protons and but only one neutron ($_2^3$He). Gold atom, as another example, in general has 79 protons and 118 neutrons ($_{79}$Au-197), but its isotopes may have neutrons anywhere between 90 and 126 ($_{79}$Au-169-205). As a general rule, the nuclei of most of the isotopes are unstable and decay sooner or later into other stable elements (Sec-E).

There are ninety-eight elements that occur naturally on the earth. Some of the elements are abundant in the universe, some common, some rare, and some rarest. *Californium* (atomic number 98) is the last and the heaviest naturally occurring element in nature. Elements from atomic number 99 (*einsteinium*) are all synthetic, which means that they do not occur naturally on the earth but are synthesized in labs (however, they may be available naturally elsewhere in some of the stars and planets of the universe). So far, we could be able to synthesize atoms up to atomic number 119 (*ununennium*). It is worth mentioning here that even though we have a multitude of elements in nature, hydrogen is by far the most abundant element in the universe followed by helium (Ch 5-F).

However, as research progressed, scientists were able to study the properties of elements basing on the electron exchange during intricate chemical reactions, and they realized that the Rutherford model itself is not adequate to explain many peculiar properties of atoms. Rather, it became apparent now that electrons around the atoms do not revolve haphazardly but follow some definite patterns. Moreover, scientific advancements in many other related fields in physics have unwittingly prompted researchers to look for better ways of arranging electrons in an atom.

The Bohr Model: By this time, scientific progress had ramified into many other parallel fields of human knowledge accompanied by huge technological advances. Of special note here is the culminating work of Willard Gibbs which dealt with the dynamics of energy transfer leading to the development of an important branch of science called *thermodynamics* by the end of the nineteenth century (Ch 6). At about the same time, the theory of *electromagnetism* by James Clerk Maxwell showed that electricity and magnetism were but two sides of the same coin—an understanding which took scientific progress to new heights. But the pinnacle of advancement in atomic physics was made by Max Planck in 1900 when he postulated that electromagnetic energy can only be emitted in discrete packets of energy, which paved way for the development of the *quantum theory* (Ch 2-C). Gradually, all this new knowledge has crept into the science of atoms which has altered our understanding of their structure. Before going further into these advanced models, the reader may be advised to recollect the ideas presented in Ch 2 as this comes in handy here.

In 1912, Niels Bohr used the then newly proposed quantum principles to explain the characteristic atomic spectral patterns and postulated that electrons do not hover around the nucleus randomly but they are distributed in discrete layers around the nucleus called *electron shells* (Fig 3.2a). These electron shells were shown to be arranged in different levels around the nucleus. The innermost has the least energy and hence contains the fewest number of electrons, and the outermost shell has the maximum energy and hence contains the largest number of electrons (shells are further divided into subshells and orbitals, but we will not go into details here). This electron configuration has explained the chemical and physical behaviour of atoms of all elements with great precision. This means that we could now be able to predict the precise way by which electrons are transferred between atoms in the chemical reactions much in consonance with the energy transactions—a perfect example of evidence-based science indeed!

The Quantum Mechanical Models: Subsequent more advanced models of the atom are based on the developments in the quantum theory, and these models have successfully answered many intriguing questions which had hitherto bothered the scientists. Most importantly, the quantum theory has explained why electrons stay put in their prescribed orbits without getting attracted to the positive nucleus and ultimately collapsing into it. The explanation stands as follows: in 1926, Erwin Schrödinger proposed, basing on the work of Heisenberg and de Broglie, that electrons travel not only as particles but also in waves (Ch 2-C), and consequently, an electron's position around the nucleus cannot be precisely located but can best be determined only on the principle of probability. Thus, in quantum mechanical models, electrons orbiting the nucleus are considered as waves with specific wavelengths depending on their energy levels. These wavelengths are such that the crests of the waves fall on the crests of the succeeding waves so that the electrons keep orbiting the nucleus without losing energy. If the crests were to fall on to the troughs, then the electrons lose their energy, and their orbits collapse into the nucleus (Ch 2-B). This means that the energy that keeps the electrons in their orbits without falling into the nucleus is the energy of the electromagnetic wave. The more advanced quantum model is the *electron cloud model* (Fig 3.2b) wherein electrons are no longer considered as particles but are viewed as quantum fields, which may be depicted as 'smeared out' energy clouds.

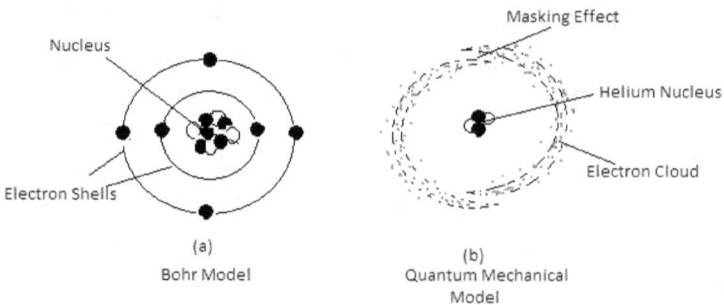

Fig3.2: Quantum models of the atom.

However, the reader may note here that the Bohr model is still the practical model which is widely used to understand and to routinely calculate the behaviour of electrons in our modern chemistry labs. But then the advanced quantum models are of great academic interest, and they help us in our theoretical understanding of the atomic structure in the quantum realm.

The Nucleus: Now we will see the happenings inside the nucleus. We know that atoms are infinitesimally small—smaller than one-billionth of a meter in diameter. In fact, *picometres* are used to measure the atomic lengths (a picometre is one-trillionth of a metre—i.e. 1/1000000000000 metres). The radius of a carbon atom, for example, is about 70 pm. While atoms are minuscule, the nucleus is even smaller and occupies a volume of just 0.0000000000001% of an entire atom! The vast empty space around the nucleus, in which electrons hover about, is calculated to be about 100,000 times larger than the nucleus. If the nucleus is of the size of a football, then the emptiness around it would be the size of the football ground itself! However, the nucleus is exceedingly dense. In fact, 99.97% of the atomic mass is contributed by the nucleus alone with all this mass squeezed into a small volume. What should finally impress the reader here is that the matter we so fondly look at and feel is composed mostly of nothing but a void of emptiness. Indeed, 99.999999999999% of an atom is just an empty void surrounding the tiniest of the nucleus, and almost all mass is concentrated inside the nucleus!

Now we will look into the amount of energy ('nuclear power') that is stored up inside this tiny nucleus. To illustrate this feature of the nucleus, we will take the differences in energy transactions that take place in chemical reactions and nuclear reactions. Atoms are not inert units of matter. Most of the atoms change their electronic configuration and participate in a variety of *chemical reactions*. Thus, all chemical reactions involve an exchange of electrons between atoms, which in turn exchange modest amounts of energy during these transactions. Chemical reactions are routinely observed by us everywhere in nature—an acid reacting with a base, a bar of iron getting rusted, starch burning into carbon dioxide and water, sugars giving us energy, and so on. In fact, our day-to-day lives are possible only because of a continuous exchange of electrons in the metabolic cascades of our cells (e.g. cellular respiration, Ch 6-D).

However, atoms may also participate in a set of far more complex and powerful interactions involving their nuclei. These are called the *nuclear reactions*. The reader should note that the chemical reactions, as discussed above, involve only paltry sums of energy transfer when compared to the gigantic amounts of energy that are transferred in the nuclear reactions. The reason for this tremendous amount of energy transfer in nuclear reactions is obvious. It is because of the simple fact that almost all the matter of the atom is concentrated in the nucleus, and during these nuclear reactions, some amount of matter is always converted into energy, thus releasing monstrous amounts of energies (in accordance with $E = mc^2$, Ch 1-C).

Having studied an atom briefly, our task now is to find out how matter inside the nucleus is arranged so that it can accommodate such massive amounts of matter and energy inside it. And for us to know these affairs, we need to look into the Standard Model.

Section B
An Overview of The Standard Model

In this section, we will have an overview of the currently accepted structure of the atom called the *Standard Model*, which has been tested time and again and proved to be an extremely accurate and predictable theory.

Up to the first quarter of twentieth century, scientists knew of only one *elementary particle*,* i.e. the electron. Since then, the science of 'particle physics' has progressed, and a herd of new subatomic particles were discovered in labs. This was possible by effectively breaking down protons, neutrons, and other matter particles in 'particle accelerators' and studying the resulting products. Though we had known many new subatomic particles by the middle of the twentieth century, the scientists had no conceptual understanding of how these particles were arranged inside the nucleus. In 1964, Murray Gell-Mann and George Zweig postulated the *quark model*, which gave us an idea of the arrangement of particles inside the nucleus. Since then, many significant improvements were made to this model, and it took nearly half a century to fully formulate the currently accepted Standard Model.

The essential idea of the Standard Model is that all matter is made up of 'elementary matter particles' which are bound together by 'force particles'. All elementary matter particles are divided into two categories—*leptons* and *quarks* (depending on their interaction with force particles, as we will see later). The common example of a lepton is the electron which is also the lightest, but there are several other kinds of heavier leptons. Quarks are elementary particles which make up the protons and neutrons in the nucleus. There are several kinds of quarks depending on their mass and charge variations (Fig 3.3 and Table 3.1).

---

* What are elementary particles? The smallest units of matter which cannot be divided further into smaller particles are called elementary (or fundamental) particles. Elementary particles have no internal structure. The electron is an elementary particle, whereas protons and neutrons are not (as they are known to be made up of smaller units of matter called quarks). Quarks are considered elementary as of today's understanding.

All these matter particles are bound down by any one of the three fundamental forces—EM force, strong or weak forces (the gravitational force, being much weaker, can almost be considered absent inside the atom, so we may cut it out from discussion here except in some special situations). In the Standard Model, all the forces are also considered as particles because, as we have learnt in Ch 2, force can be theoretically understood as packets of energy, and each quantum of energy may be considered as a 'particle' as well. In fact, these force particles are called *force carriers* because they literally carry force from one matter particle to the other. Thus, we have several kind of force carriers such as *photons* (for EM force), *gluons* (for strong force), *W* and *Z bosons* (for weak force), and *gravitons* (for gravity) (Fig 3.3). If we look at the atom in a gross way, it can be said that the electromagnetic (EM) force becomes operative chiefly outside the nucleus and binds electrons to the orbits, and inside the nucleus, the strong force is responsible not only in binding protons/neutrons together but also in binding the constituent quarks together as well.

So now we can see that there are grossly two types of particles for us to study—the matter particles which make up the mass of the nucleus and the force carriers which bind them together. Table 3.1 presents the chief particles of an atom, but there are far too many of these sub-subatomic particles, and discussing all of them would be somewhat confusing to the reader. And to add to the confusion, *all* the matter particles have their counterpart *antiparticles** (which are not depicted in the table). But then fortunately, it is not very essential for a reader to remember all the names of the particles to understand the Standard Model; hence, many minor details are omitted in the following discussion.

The Standard Model is a consummate theory because it efficiently answers many intriguing questions that plagued the scientists for a long time such as why positively charged protons do not repel each other inside the nucleus, how massive amounts of energy is stored inside the nucleus, why certain atoms become radioactive, and what confers mass to matter.

---

*   Antimatter particles are discussed in Sec-H. They are just similar to their counterpart matter particles and also behave exactly like them, except that they have an opposite charge. In other words, antiparticles are matter particles with *equal* but *opposite* charge.

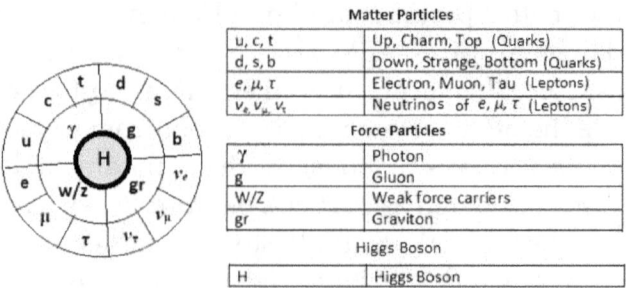

Matter Particles	
u, c, t	Up, Charm, Top (Quarks)
d, s, b	Down, Strange, Bottom (Quarks)
e, μ, τ	Electron, Muon, Tau (Leptons)
v_e, v_μ, v_τ	Neutrinos of e, μ, τ (Leptons)
Force Particles	
γ	Photon
g	Gluon
W/Z	Weak force carriers
gr	Graviton
Higgs Boson	
H	Higgs Boson

Fig 3.3: Particles and forces of an atom.

The reader may also note, as mentioned above, that numerous particles with exotic names are described in the Standard Model, and it is a futile attempt to remember all of them. However, the reader is advised to ease himself/herself with the innumerable names we may come across in this intricate model but to grasp its essence. We will first discuss matter particles and later the force particles.

SECTION C
Particles of Matter

Before commencing a discussion on these mysterious particles, let us first study the prototype elementary particle of matter, the electron, which is the simplest of all particles, and it has the most basic qualities of an elementary particle. In fact, all other matter particles in particle physics are studied in comparison with electrons.

Electron: An electron (symbol *e*) is a lepton (see below for details on leptons). It is the most easily accessible and hence the most extensively studied particle in particle physics. Electrons are responsible for many of our day-to-day activities—for example, the static electricity we commonly experience on a balloon is due to electrons hovering over its surface. The electricity that passes across our metal wires is because of the electrons carrying charge from one end to the other, the picture we see on our TVs is owing to the electrons bombarding the fluorescent screen, the chemical reactions in our labs are due to transfer of electrons between atoms, all biological processes (hence life itself) function because of exchange of electrons between organic molecules, and so on – the list runs endlessly!

The mass of an electron is estimated to be about 9×10^{-31} kilograms, which can be depicted as 0.000,000,000,000,000,000,000 ,000,000,000,9 kilograms! This rest mass of the electron is taken as a 'standard', and all other particles are conventionally weighed in multiples of it. For example, a proton weighs 1,837 times the mass of an electron, and tau (see below) weighs 3,520 times of an electron. The energy of an electron is estimated to be 0.511 MeV (mega-electronvolts), and in accordance with the mass–energy equivalence ($E = mc^2$), the mass of an electron may also be expressed in terms of energy as 0.511 MeV/c^2.

Electrons carry electrical charge on them, and this charge is called negative. The size of the electric charge on an electron is 1.6×10^{-19} coulombs, or putting it in another way, we can say that 1 coulomb is equal to the charge of 6 million-million-million (6×10^{18}) electrons. Once again, this unit of charge is taken as *one unit* of charge, and all other particle's charge is compared to it. Thus, the electron is considered to have a charge of 1. A proton also has a charge of 1 because though the proton weighs heavier, it carries an equivalent charge same as an electron but with opposite sign.

And what could be the size of an electron? Now this is a tricky question because an electron cannot be considered a localized particle in the truest sense of the word; rather, it is considered more as a cloud of energy (Sec-A). But by convention, its size is estimated to be about 10^{-18} metres. Electrons behave like tiny magnets because of their electrical charge and momentum, which causes them to *spin*. However, 'spin' in quantum mechanics does not really mean that the particles physically spin like a top, but it is more of a mathematical expression. We will deal with this property in Ch 9-A, suffice is to state here that an electron has the smallest spin of 1/2.

The antimatter particle of electron is the *anti-electron*, but it is more commonly called *positron* (e^+). It has all the properties of an electron but with an equal and opposite charge. Having learnt briefly about electron, we will now peep into the nucleus and do a general survey of the other particles associated with the Standard Model.

Quarks (Fig 3.3): These are elementary subatomic particles constituting protons and neutrons. Quarks (symbol: q) exist in six varieties, and they are arranged in three pairs (Table 3.1)—up/down, charm/strange, and top/bottom (all these names do not carry any literal meaning but are only fanciful labels—*up* is no upper, *charm* is no charmer, *top* is no topper, and so on). Of course, all these particles have their corresponding *antiquarks* ($\bar{q}$). Quarks are known to form different combinations, giving rise to several complex particles. Thus, a *hadron* (as you see in the name 'Large Hadron Collider') is a collective name of all particles composed of quarks. Hadrons are usually of two types: *baryons*, which have three quarks in them (protons and neutrons are the leading examples, though there are several other baryons); and *mesons*, which have one quark and one antiquark in them (e.g. *pion*, π).

Protons and neutrons put together are called *nucleons* (= residents of nucleus), which make up the nuclei of all atoms in existence. Because these nucleons belong to the class of particles called baryons, we may call all our matter technically as *baryonic matter*. It is worth repeating here that all the matter contained in our known universe—all the planets, stars, and galaxies in the known universe, all the mountains and rivers, all the men and beasts—are made up exclusively of baryonic matter (we have no evidence of existence of any other type of matter in nature)! Each proton has three quarks— two up-quarks and one down-quark (= uud); each neutron has one up-quark and two down-quarks (= udd). Thus, we may conclude that all our atomic nuclei are made up of a combination of only up-quarks and down-quarks (and these are the lightest of all quarks, see Table 3.2).

Mesons are highly unstable because they have in them a quark (= matter particle) and an antiquark (= antimatter particle) which annihilate (= destroy themselves by mutual attraction) readily into a great flash of energy. Thus, mesons do not exist in nature but are generated in particle accelerators for very brief periods in the process of breaking bigger particles.

Table 3.1: Elementary Particles of an Atom

Matter Particles	6 quarks	Up, down, charm, strange, top, bottom
	6 antiquarks	—Same as above—
	6 leptons	Electron, muon, tau, 3 neutrinos
	6 antileptons	—Same as above—
Force Carriers	Electromagnetic force	Photons
	Strong force	Gluons
	Weak force	W and Z particles
	Gravitational force	Gravitons (not present in atoms)

Quarks have varying masses and are divided into three generations depending on their masses (Table 3.2). Whereas the up- and down-quarks are the lightest, top-quark is the most massive (about 40,200 times that of an electron). The reader may see, as noted above, that all the baryonic matter is composed exclusively of only the lightest of the quarks—i.e. the up and down quarks. The heavy quarks do not participate in the atomic structure, but they appear only transiently during nuclear reactions. However, the exact mass of an individual quark is not to be taken too seriously because we cannot isolate a quark and measure its mass directly. But then it is generally accepted that up-quarks have an average mass of 1.7 to 3.3 MeV/c^2; down-quark, 4.1 to 5.8 MeV/c^2; charm-quark, 1270 MeV/c^2; strange-quark, 101 MeV/c^2; top-quark, 172 GeV/c^2; and bottom-quark, 4.2 GeV/c^2 (Table 3.2).

The quarks have electromagnetic charges on them, but they are peculiar in the sense that they do not occur in integers (as in electrons) but are fractionated. If you look at Table 3.2, you will see that by adding fractions of charges of valence quarks in a proton (uud), it is (+2/3) + (+2/3) + (−1/3) = +1 (i.e. proton has a positive charge of +1). For a neutron, if you add up three valence quarks (udd), (+ 2/3) + (−1/3) + (−1/3) = 0 (i.e. a neutron has no charge).

Table 3.2: Classification of Quarks

Particles (Mass in eV/c²)	1st Generation (Least massive)	2nd Generation (More massive)	3rd Generation (Most massive)	Charge
Quarks	Up (1.7 to 3.3 MeV/c²)	Charm (1270 MeV/c²)	Top (172 GeV/c²)	+ 2/3
	Down (4.1 to 5.8 MeV/c²)	Strange (101 MeV/c²)	Bottom (4.2 GeV/c²)	–1/3
Leptons	Electron (e) (0.511 MeV/c²)	Muon (μ) (105 MeV/c²)	Tau (τ) (1776 MeV/c²)	–1
	Neutrino-e (massless)	Neutrino-μ (massless)	Neutrino-τ (massless)	0

Quarks are unique in the sense that, in addition to electrical charge, they have some other kind of charge on them called the *colour charge* (the term *colour* denotes nothing of our ordinary colours but is only an imaginary term to differentiate it from electromagnetic charge). The colour charge exists in three colours: red, blue, and green (once again, these colours carry no meaning in their true sense). And colour charges have their anticolours: antired, antiblue, and antigreen. Colour charges are mediated by the strong force (see below). Protons and neutrons have all the three colours, and thus, they remain *colour-neutral,* which means that all the three colours cancel each other out. Colour charges are known to be continuously exchanged between quarks in a hadron. Since protons and neutrons are colour-neutral, all the matter we see around us in everyday life is also colour-neutral and stable. The colour charge has no influence on leptons such as electrons. This is the reason why strong force (however strong it may be) is limited to the confines of the nucleus and will not spread out and affect the electrons spinning around it. Quarks never exist in isolation but always exist in combinations because they are only stable when they combine and become colour-neutral. Quarks are the only particles in nature which are known to be influenced by all the four forces—electromagnetic, weak, strong, and gravitational (though the last being negligible). Like electrons, quarks are also half-integer-spin (i.e. spin-1/2) particles.

Leptons (Fig 3.3): Leptons are another class of elementary particles upon which a strong force has no influence at all. An electron is the leading example of a lepton. There are six types of leptons—three are charged, and three are neutral (Table 3.1). A charged lepton may be an *electron* (*e*), *muon* (*μ*), or *tau* (*τ*), of which tau is the heaviest particle (weighs about 3,520 times an electron). The neutral leptons are called *neutrinos* (see below). All leptons have antileptons (e.g. electron has positron, muons have antimuons, and so on). Leptons are also spin-1/2 particles.

Muons and taus are generally not found in ordinary matter as they are highly unstable and they decay spontaneously into lighter particles. Leptons are influenced by all fundamental forces except a strong force because they have no colour.

Neutrinos: A neutrino is an uncharged lepton (symbol: Greek letter *v* for '*nu*') and is the second most widely occurring particle in the universe (next only to photons). Each charged lepton is associated with one type of neutrino – consequently, they are named as electron-neutrino (v_e), muon-neutrino (v_μ), and tau-neutrino (v_τ). All the neutrinos are spin-1/2 particles. Neutrinos also have antiparticles; whereas neutrinos have a left-handed spin, antineutrinos have a right-handed spin. Neutrinos bear neither electromagnetic charge nor colour charge. It was originally assumed that neutrinos are massless. However, in 2002, Takaaki Kajita and Arthur McDonald have shown that they do have very small masses of their own (some have estimated their mass to be about 10^{-37} kg, and that would be about 17 billion times lighter than a neutron!). Neutrinos are predicted to travel with nearly the speed of light; the claim of their faster-than-light travel has consistently been disproved by many recent experiments (Ch 1-C)!

Way back in 1930, Wolfgang Pauli had predicted the existence of neutrinos theoretically basing on an interesting observation: isolated neutrons (those which are not associated with protons inside a nucleus) are highly unstable, and they decay soon into a proton and an electron. However, as per the law of conservation of momentum, the decay process should also release another fast-moving particle to account for the disappearing mass. These hypothetical particles were named neutrinos (= a 'little neutron' in Italian), though their existence was confirmed experimentally only much later.

We know of two chief sources of neutrinos in nature. Our sun is known to generate neutrinos in large numbers during the nuclear fusion reactions (Ch 4-D), which would sweep across the space and swarm the earth in such great abundance that an estimated 60 billion solar neutrinos pass through every square centimetre of our bodies every second! However, owing to their near-massless state and having no charge, they would not interact with absolutely anything, and thus, they pass through all the objects they come across unhindered and unnoticed! Consequently, neutrinos not only pass through the total span of the earth as if it does not exist, but they will also pass through the entire interstellar space unhindered by stars and galaxies! And fears set aside, they are not likely to cause any harmful effects on human physiology whatsoever because they will not participate in any reactions!

Neutrinos are also theorized to be generated in great numbers at the time of the big bang by the decay of neutrons (Ch 5-G), and these early neutrinos are speculated to have travelled for great distances across the universe from the time of their birth 13.7 billion years ago, in transit passing through billions of stars and planets, all of which, of course, could never be able to stop their exodus through the infinite space and time.

In summary, it is interesting to note here that even though hundreds of subatomic particles are described in the Standard Model, all the simple and stable atoms around us are composed of only five types of fundamental particles—three matter particles (electrons, up-quarks, and down-quarks) and two force particles (photons and gluons). The rest of the matter particles are unstable and so exist only transiently during nuclear reactions during the conversion of massive amounts of nuclear energy into matter particles, and so they will not participate in the formation of atoms around us.

A Note on Particle Accelerators: How do the physicists study these smaller of the smallest particles? No microscopes (including the so-called electron microscopes) would work because visible light has much too longer wavelengths, which would 'smudge' the details of these particles (see explanation in Ch 2-D). But then the subatomic particles are studied by indirect methods such as by 'breaking' open atoms in huge laboratories called particle accelerators—also called atom smashers—and studying the showers of smaller particles that are ejected out of them. We will now briefly learn about the particle accelerators.

Particle accelerators are mammoth devices located in very big facilities containing gigantic magnets designed to accelerate small electrically charged particles at tremendous speeds (sometimes speeds nearing the speed of light) and using them to bombard larger particles, such as small atoms and protons, to split them up into showers of smaller sub-subatomic particles which are then studied. The magnets used in the accelerators should be massive enough because particles, as they gain speeds nearer to the speed of light, gain superlative momentum (Ch 1-C)—for example, a proton, when accelerated to speeds 99.9999991% the speed of light, attains an energy of 7 TeV (teraelectronvolts, or the energy of a trillion electrons)!

Particle accelerators are one of the most important tools to decipher the structure of an atom and to discover new particles. The chief purpose of a particle accelerator is to drive enough energy into a very small volume of space in order to create a matter particle with certain mass—the more energy injected into the space, the more massive the particle created. Hence, they must really be called particle creators. The present-day accelerators are quite efficient so much so that they may create very massive particles, such as the Higgs particle (Sec-F) with a mass of 126 GeV (giga-electronvolt).

The *Large Hadron Collider* (LHC) built by CERN near Geneva, Switzerland, is by far the largest particle accelerator in the world, which weighs a staggering 38,000 tons and runs for 27 kilometres in length, situated 175 metres deep inside the earth. In this expansive circuit, a particle travelling at the speed of light would make 11,000 rounds in a single second! There are other important accelerators in the world such as Fermilab's *Tevatron* in Batavia, Illinois; *KEKB* in Japan, and several others.

SECTION D
The Nuclear Forces

Having described the matter particles, we will now see how the fundamental forces operate within the nucleus to bind matter particles together. Here we will concentrate only on two fundamental forces, the strong and the weak interactions, put together called the *nuclear forces* because they operate chiefly within the bounds of the nucleus.

The Strong Interactions: Strong force is the most powerful force in nature—estimated to be more than100 times stronger than the EM force and about a million times stronger than the weak force. Strong force not only prevents the nucleus from breaking apart but also keeps the quarks together to form stable protons and neutrons.

As mentioned above, strong force in the nucleus is represented by colour charge; consequently, particles without colour such as leptons (e.g. electrons and neutrinos) are not influenced by it. The force carrier of the strong force by which it shows its effect on particles is called the *gluon* (= 'glue particle'). Gluons, like photons, are massless particles with an integer spin of 1. However, gluons (unlike photons which are not charged) have colour charges on themselves and are shown to come in eight types (the significance of which is yet to be known).

The force field of strong force has an interesting feature: in 1970, David Gross, Frank Wilczek, and David Politzer discovered that the strong force between two quarks becomes weaker as they come nearer and becomes stronger as they move apart. In other words, it can be said that as the distance between quarks *increases*, the attractive force between them becomes *stronger*—something like a rubber band holding two boxes together which, when pulled apart, increases the strength of attraction. Consequently, when the distance between quarks *decreases*, the attraction between them *diminishes*. This phenomenon is known as *asymptotic freedom*. Now the reader may see that at shorter distances, the strong force is weak so that the quarks can move freely in relation to each other and behave more like independent particles. This feature of strong force, in fact, allows gluons to ply between protons and neutrons within the nucleus, and because of this exchange of gluons, the quarks change their colour constantly!

Yet another important implication of asymptotic freedom is worth our dutiful consideration. Because the strong force becomes too great when quarks are pulled apart, it becomes thermodynamically advantageous for the quarks to snap away from each other after separating for a critical distance (as a rubber band snaps away if you pull it for greater distances); by this way, the energy of strong force is conserved. However, the conserved energy in this event is converted into a mass particle in the form of an antiquark, and this quark and antiquark pair up together to form a meson which soon annihilates, emitting a great flash of energy. This is the reason why breaking up a proton in a nuclear reaction releases tremendous amounts of energy (Ch 4-D)! This is also the reason why quarks cannot have an independent existence.

Gluons are short-range particles. They can travel only for very short distances, about one femtometre (10^{-15} meters) at a stretch (unlike photons which can travel for infinite distances to the end of the universe!). Thus, the strong force is really strong within the confines of the nucleus but can barely extend its influence beyond the boundaries of the nucleus.

The strong force which binds the quarks together is thought to 'leak' out of the nucleons in the form of short-range nuclear force; this is called the *residual strong force*. This is a weaker force than strong gluon force, and this becomes progressively weaker and dies down exponentially over short nuclear distances. However, it is strong enough to hold the protons and neutrons of a nucleus together.

The Weak Interactions: As the name implies, this is a weak force operating within the nucleus, but it is much stronger than gravity. Weak force mediates many nuclear decay reactions and hence is involved in the phenomenon of radioactivity (Sec-E). In this sense, weak force is generally considered a 'disruptive' force, causing disintegration of nucleus (while the strong force usually keeps the nucleus intact). The force-carrier particles for weak force are two charged particles, W^+ and W^-, and one neutral Z particle. When weak force is exchanged between two elementary particles, these particles undergo certain change in their properties. In other words, weak forces modify elementary particles. Here is an important point for the reader to consider: we must not suppose that elementary particles are indestructible or unchangeable. Elementary particles can either undergo complete disintegration into energy (called annihilation) or change into some other particles. When such a process of transmutation occurs under the influence of a weak force, then a particle is said to have changed its *flavour* (once again, this has nothing to do with our culinary flavours!). For example, when a weak force carrier is exchanged, an up-quark can change flavour and become a down-quark, or a muon can change its flavour and become an electron. The process of such transmutation of quark has been the most crucial step in the fusion reactions in the sun by which solar energy is liberated, as we will see in Ch 4-D. Weak force may interact with all matter particles, including neutrinos. The reader will shortly see that both weak force and electromagnetic force are unified into a single force called electroweak force, which carries a tremendous theoretical importance (Sec-F).

Before we go further into other details of nuclear events, we have to become acquainted with the concept of 'bosons' and 'fermions' which, in fact, sheds light on the fundamental nature of matter and energy.

Fermions and Bosons: All the particles in the universe studied above—both matter particles and force particles—can be functionally classified into two groups, the *fermions* and the *bosons*, depending on their behaviour in the subatomic world.

Before attempting to understand the meaning and significance of fermions and bosons, we need to know about the *Pauli exclusion principle*. In 1924, Wolfgang Pauli made a remarkable observation that no two electrons could exist in the *same quantum state simultaneously* (the reader would know the significance of this as he/she reads on). Pauli's observations were based on the spectral emission of the atoms. The atoms, when heated, were known to give out a specific pattern of spectral emissions depending on the temperature. These specific spectra could only be explained by assuming that electrons were following certain basic rules to fill the orbits around the nucleus. As we have seen, an electron not only carries electric charge but also has another property called spin. In addition to spin, an electron was also known to exist in the orbit in *three quantum states* depending on its special geometrical state. Each one of these states is assigned a quantum number—*viz.* a principal quantum number (*n*), an orbital quantum number (*l*), and a magnetic quantum number (*m*). Thus, along with spin, these three quantum numbers define an electron's state in an orbit. Pauli showed that no two electrons can exist in the same quantum state in an atom, which means that no two electrons can have exactly identical properties (indicating that no two particles can occupy a single position in space around the nucleus). Thus, the process of filling up of electrons in an atom has a limitation as to which orbit they have to occupy first and which next, the final aim being that all electrons are in different quantum states. It was also realized that the orbits around the atoms represent energy levels. The nearest orbit has the least energy and so contains the least number of electrons, and the farthest orbit has the highest energy and so contains the most electrons. Consequently, the general rule of filling up with electrons is that the atoms tend to have the least energy levels filled up first before filling up the next energy level. This ergonomic requirement carried a tremendous theoretical as well as practical significance because it simply explained the chemical behaviour of the elements with great precision (simply because all chemistry is based on movement of electrons in the atomic orbitals). It was subsequently shown that this fundamental principle of quantum states is not peculiar to the electrons in orbits but applies to all particles in general in the subatomic world.

But then every rule has an exception! And as it turns out, shortly after this discovery, Satyendra Nath Bose mathematically showed that *not all particles obey the exclusion principle* but certain particles would exist in the same quantum state at a given place and time. This astounded the physicists at that time. It simply means that an infinite number of particles could occupy the same place at the same time, and this was an impossible derivation. However, Einstein realized the importance of this idea and worked together with Bose to create *Bose–Einstein statistics,* which laid down the foundation to an important concept called boson.

Now we will make the nomenclature clear: all particles which obey the Pauli exclusion principle are called *fermions* (named after Enrico Fermi); all particles which do not obey this principle are called *bosons* (named after Satyendra Nath Bose). A general description of these two groups is presented hereunder.

Fermions are described euphemistically as 'antisocial', which means that they 'fight' for independent positions in the available space in an atom—i.e. no two fermions can occupy the same position in space. All fermions have a fractional spin (1/2, 3/2, 5/2, and so on) and so have an *innate asymmetry* in their existence. Quarks, leptons, protons, neutrons (as a matter of fact, almost all matter particles), and even most atoms are examples of fermions. Because of the antisocial behaviour of fermions, it is impossible to squeeze them into a single confined point. If not for this property, all the matter of the earth would have collapsed on to a single point in space within no time and would have ultimately disappeared into nothing (see Ch 9-A for further discussion). For this reason, fermions are considered the building blocks of matter. Consequently, it has been said that had the world not been ruled by the exclusion principle, the subatomic particles would *not* have organized into separate clusters as protons, neutrons, electrons, etc. and would not have stayed in their confined locations; and consequently, well-defined atoms were not possible. And all the matter would then be in the state of an infinitely dense soup of undifferentiated particles. Then in that case, neither you nor I would have existed!

Bosons, on the other hand, can be described as 'social', which means that any number of bosons can occupy the same position in space at a given time. Bosons are shown to have an integer spin (0, 1, 2, and so on). All force-carrier particles are bosons (e.g. gluons, photons). But here is a caveat: composite particles with even number of fermions also behave like bosons. For example, a meson with one quark and one antiquark behaves like a boson (but a proton with three quarks behaves like a fermion). Interestingly, helium nucleus is considered bosonic as it contains an even number of nucleons.

Hence, helium is thought to exhibit certain weird properties at near absolute-zero temperatures. It is *never* known to crystallize at any temperature, it *never* turns solid, and it remains as a 'superfluid' with strange qualities like zero viscosity and no surface tension.

To summarize, we will make a rough blanket statement of fermions and bosons here so that further discussion becomes simple and practical:

In general, all matter particles are fermionic; all force carriers are bosonic.

Bose–Einstein Condensate (BEC): The theoretical upshot of the concept of the boson is the existence of Bose–Einstein condensate. First consider this to understand BEC: matter in the universe exists in four states. In a *solid* state, the matter particles in a substance exist in closest approximation to each other with least movement. In the *liquid* state, the matter particles gain more momentum and move wider apart, and in the *gaseous* state, the particles gain most momentum and move faster and faster. There is a fourth state of matter called the *plasma* state where, at very high temperatures, the particles cease to exist as organized atoms but are ripped off their structure and exist in high energy states, forming a soup of protons, electrons, etc. Plasma state of matter exists in the core of the stars where nuclear reactions occur (Ch 4-D), but it can also be seen on our earth in minuscule forms at certain superheated places such as inside the fluorescent lamps, neon signs, and thunder lightning. We can see that the temperature of a material object has to increase tremendously in order to pass from the solid state to plasma state.

Now we will go in the reverse direction and see what happens. When the matter is *supercooled* to near-*absolute-zero* temperatures (0 Kelvin or −273.15 °C), what would be the fate of the atoms? Could the atoms stop their movement absolutely? Would the particles become indistinguishable and superimpose on one another at one place? Bose and Einstein worked out mathematically and stated that such a state was indeed possible, and this was called *Bose–Einstein condensate* (BEC). However, this was only a theoretical argument, and it was unknown if such a state could exist. The technology then simply did not allow labs to reach such supercooled states, and thus, the existence of BEC was viewed with derision by some scientists. However, in 1995, Eric Cornell and Carl Wieman of the University of Colorado cooled rubidium atoms to near absolute-zero temperatures, thus creating a state of BEC, which vindicated the theory, proving Bose and Einstein correct.

Section E
Essentials of Radioactivity

Having read about matter particles and force interactions, we are now ready to learn about an intriguing phenomenon called radioactivity. The discovery of radioactivity has been an important landmark in the history of science, and a major portion of progress of human civilization in the last century may be attributed to radioactivity. This weird phenomenon of powerful radiation emitted by atoms was thoroughly investigated by a stream of eminent scientists, including Wilhelm Röntgen, Henri Becquerel, Marie Curie, and others; and by the turn of the twentieth century, the science of radioactivity has not only become an important technological tool in all fields of science, technology, and industry but has become an indispensable tool in many medical investigations and therapeutics! We will briefly look into this phenomenon.

The Basic Principles: Radioactive elements are unstable by virtue of either an imbalance between the number of protons and neutrons in their nuclei or owing simply to the very large and redundant sizes of their nuclei. These unstable nuclei have a tendency to become stable either by correcting the imbalance in the proton/neutron ratio by various methods (which we will see shortly) or by discarding the redundant nucleons (= protons/neutrons) from the nucleus. In the process of becoming stable, an unstable radioactive nucleus emits a lot of energy in the form of intense radiation and/or various small subatomic matter particles, as described below. Before going further, it's relevant for us to first see what makes a nucleus stable.

Atomic nuclei are bound down by certain amount of energy called the *nuclear binding energy*. The term *binding energy* is misleading, and so it needs some explanation. The binding energy of a nucleus actually signifies *the amount of energy needed to break up* all its constituent nucleons into individual protons and neutrons (for an easy understanding, the reader may look into Ch 6-D for a similar binding energy in chemical reactions between molecules and atoms, the essential principle in both the cases being the same). Speaking the other way, this means that when protons and neutrons combine together, some amount of energy is liberated, and if you want to separate them again, then the same amount of energy needs to be supplied. For example, when a slow-moving proton combines with a neutron (forming a hydrogen-2 nucleus), about 2.23 million electronvolts (MeV) of energy is liberated in the form of gamma rays. And when they are separated, the same 2.23 MeV of energy has to be supplied, and this energy needed to break them apart is called the *bond energy* of $^2_1$H nucleus. Because energy is equivalent to mass, we can say, in other words, that the total mass of an atomic nucleus is always less than the combined mass of its constituent nucleons (nucleons = protons + neutrons).

The total mass of a nucleus is less than the combined mass of its nucleons.

Now consider this very important point: an atomic nucleus with *lesser* nuclear binding energy is *unstable* because this nucleus is 'loaded' with more energy, and consequently, only a small amount of energy is needed to break it up. Such unstable nuclei tend to become stable by disintegrating into smaller nuclei with greater binding energy (nuclei with less energy). And in this process of disintegration called *decay*, the excess of binding energy is liberated as either energy or matter or both. Thus, large atomic nuclei with lesser binding energies tend to decay into smaller ones with greater binding energies, or a radioactive 'parent' nucleus (e.g. uranium-235) with lesser binding energy decays into smaller 'daughter' nuclei with greater binding energies (krypton-89 and barium-144) (Fig 3.4).

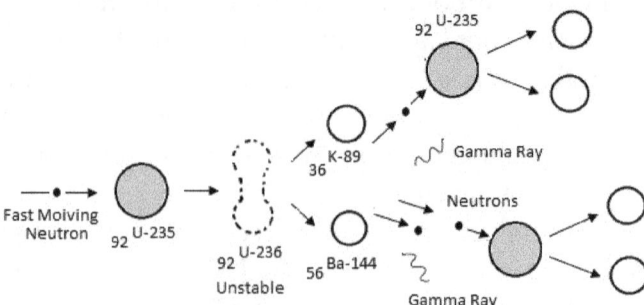

Fig 3.4: Uranium-235 chain reaction.

Sometimes, a very small nucleus may also become unstable because of the discrepancy in the proton/neutron ratio. Generally, for elements of lower atomic numbers, equal number of protons and neutrons make the nucleus stable, but as the atomic number increases, the nucleus achieves stability by increasing the number of neutrons. For example, the most stable nucleus of helium has equal number of protons and neutrons. For stable nuclei of other elements such as lithium, beryllium, and boron, a single extra neutron is needed; for heavier nuclei such as iron, copper, silver, and gold, many more extra neutrons are needed to make their nuclei stable (the reader may look into the periodic table to conduct an exciting survey of this phenomenon!). Hence, nuclei lose their stability if this balance between nucleons is lost, and such nuclei tend to change their configuration by changing the ratio of their nucleons so that nuclear stability is restored.

Thus, we may say that an unstable radioactive nucleus may become stable by changing its nuclear configuration by one of the following four ways: alpha decay, beta decay, electron capture, and spontaneous nuclear fission. We will briefly look into these nuclear phenomena.

Alpha Decay: In alpha decay, an *alpha particle* (= two protons, two neutrons) is ejected from the nucleus of a radioactive element, and as can be expected, this happens in very large nuclei with many number of nucleons. Of course, after alpha decay, the parent element changes to some other element because the atomic number changes (e.g. unstable polonium-210 releases alpha particle and changes to stable lead-206). This sort of disintegration results in the release of energy owing to the fact that the parent nucleus always weighs more than the combined mass of the daughter nucleus and alpha particle (the missing mass accounts for the released energy). This *disintegration energy* is converted into kinetic energy, which is shared equally between the original parent nucleus, the daughter nucleus, and the alpha particle (because of the conservation of momentum) – however, the emitted alpha particle zips away faster because of its smaller mass.

Beta Decay: In beta decay, *beta particles* are emitted. A beta particle could be either an electron or a positron. In *beta minus decay*,* a neutron decays into a proton, thereby increasing the element's atomic number by 1. In this process, an electron and an antineutrino are released (as per the laws of conservation, see below). In *beta plus decay*, a proton decays into a neutron, thereby decreasing the atomic number by 1. In this process, a positron and a neutrino are released (again, as per the laws of conservation). And it is needless to say that the daughter nuclei are relatively more stable because of a change in atomic number/mass after the decay. The details of beta decay and the mediation of weak force is studied in Ch 4-D and 9-C.

Another mode of decay is the *electron capture*. In this process, the proton of an unstable nucleus captures an electron from one of the nearby shells, thereby becoming a neutron, and this reduces the atomic number by 1 (e.g. the proton-rich $^7_4$beryllium decaying into stable $^7_3$lithium). In this process, a neutrino is released. Electron capture is also accompanied by the release of an X-ray as result of an electron jumping into the lower orbital to replace the absorbed electron.

---

* An example of beta decay: carbon-12 has six protons and six neutrons in its nucleus which is stable, and carbon-13 has one neutron extra in its nucleus which is also stable. But carbon-14 has two additional neutrons, and thus, it becomes unstable and tends to lose one electron (by beta minus decay) and becomes nitrogen-14 within a specific period of time (called half-life, see below). This phenomenon, in fact, is utilized in the technique of *carbon dating* wherein the estimation of the ratio between C-14 and N-14 in a given sample of geological interest is taken to determine its geological age.

Nuclear Fission and Nuclear Reactors: Lastly, a very large nucleus with many redundant nucleons may break down spontaneously into two roughly equal parts (which represent two relatively lighter elements). This process is called *spontaneous nuclear fission* (not to be confused to *'fusion'*, Ch 4-D). Spontaneous fission is generally accompanied by the release of tremendous amounts energy in the form of gamma rays and the release of a number of neutrons (see below) and other fast-moving particles. There is another form of fission called induced nuclear fission, which is of tremendous interest to us – we will discuss this in some detail.

Induced nuclear fission is commonly employed in our nuclear reactors and atom bombs. Here, large nuclei are bombarded with neutrons to induce radioactive decay in them. The radioactive decay, in turn, releases more neutrons, and these released neutrons bombard other large nuclei in the vicinity, thus initiating decay in them. This continuous array of decay reactions is called a *nuclear chain reaction* (Fig 3.4). We will illustrate this process of chain reaction with a typical example used commonly in our nuclear reactors: $_{92}$uranium-235 is one of the most common radioactive elements used in nuclear reactors because of its 'fissile' property and its high energy yield. U-235 is a stable atom with a natural half-life of over 700 million years. But when a neutron is added to the nucleus, it becomes highly unstable and starts breaking down into smaller nuclei. Thus, the chain reaction starts when a slow-moving neutron bombards the $_{92}$U-235 nucleus (Fig 3.4). This collision results in the formation of a highly unstable compound nucleus ($_{92}$U-236) which soon breaks down into $_{36}$krypton-89 and $_{56}$barium-144, and in this process, the extra three neutrons are released out (89 + 144 = 233; 236 – 233 = 3, thus leaving three free neutrons), which is accompanied by the liberation of a huge amount of energy to the tune of about 215 MeV in the form of gamma rays (which we use to generate electricity). The released neutrons have tremendous kinetic energy, and thus, they collide with other uranium nuclei which, in turn, undergo fission, releasing more neutrons—this keeps up a chain reaction.

In a nuclear reactor, the chain reaction is controlled by using heavy metals (called filters) which can absorb neutrons so that energy is released slowly and so can be harnessed by us to generate power when necessary. In contrast, in a *nuclear bomb* (which also works on the same basic principle), these chain reactions go unchecked, and so they occur in an uncontrolled fashion with exceptional rapidity, resulting in the release of explosive amounts of energy in a nuclear blast!

General Characteristics of Radioactivity: It is interesting to note that in any nuclear reaction, all the important conservation laws of nature are protected. *Conservation of mass/energy and momentum* is achieved throughout the decay process in the form of mass of the emitted electrons/protons, neutrinos/antineutrinos, and/or in the form of liberated gamma rays, and/or in the form of kinetic energy and the momentum of the fast-moving particles. *Conservation of electric charge* (i.e. electric charge can neither be created nor destroyed) is achieved, which means that the algebraic sum of all the electric charges in each reaction remains constant. In the same way, *conservation of lepton number* is also achieved—'lepton number' for each proton/neutron is zero. But for electrons and neutrinos, it is +1, and for positrons and antineutrinos, it is –1. Thus, the reader may observe that in beta decay, if an electron is ejected out of the nucleus, an antineutrino also accompanies it. And when a positron is ejected, a neutrino accompanies it – these combinations tally the lepton number.

All radioactive elements have a 'half-life'. But then what is half-life? Radioactivity is intriguing in the sense that whereas the probability of radioactive decay of a nucleus of any single atom at a given time cannot be predicted, the rate of decay of a large group of nuclei in a given sample of radioactive substance can be statistically predicted. Thus, the lifespan of a radioisotope element is determined by measuring the length of time that is required for the decay of half of the nuclei in a given sample of that element—that is the so-called *half-life* of an isotope. The half-life of each radioisotope element differs greatly, ranging from a fraction of seconds (hydrogen-7 with the lowest recorded half-life of 2×10^{-23} seconds) to billions of years (tellurium-128 with a maximum measured half-life of 2×10^{24} years, which is many more times the age of the universe!).

In conclusion, we may say that radioisotope elements decay by one of the four decay processes described in order to achieve stability to their nuclei. The formation of radioactive elements in the universe is discussed in Ch 4-B. We will now look into some of the most important topics in particle physics.

SECTION F
Unification of Forces

The Standard Model, as presented above, shows how matter particles and forces interact with each other in building up the material universe. But the reader can see that each force has its own distinct manifestation and appears as a separate entity. However, even before the time of Einstein, the 'indomitable unifier' as he was, there have been attempts by several scientists to unify all the four forces of nature into a single force, but much in vain. It was impossible to understand them all on a common footing. But then theoretical science has changed its face with decades of changing technological progress, and with this, the concept of unification has changed. In this section, we will see how far our modern scientists have progressed to reach the goal of unification. First we will outline the problem of unification.

It's of a great historical insight to note that the earliest known unification of forces was already done way back in 1864 by James Clerk Maxwell when he unified the electrical and magnetic fields into electromagnetism. It was also the earliest example of employing *gauge symmetry*—a mathematical tool to unify the variable parameters of a system. Since then, immense knowledge has accumulated in the field of theoretical physics which, ironically, has actually deepened the chasm in between the four fundamental forces and made their unification theoretically rather more challenging. But then now we will see that the more diverse these forces appear in nature, the more unity they demonstrate!

We have seen that both the theory of relativity and quantum mechanics essentially deal with the fundamental forces in one way or the other. Roughly speaking, gravity and electromagnetic forces are chiefly focused in the relativity theory, whereas strong and weak nuclear forces are mainly dealt with in quantum theory. However, it is an obvious fact that the rules of the theory of relativity do not agree well with that of quantum theory, and so it was soon realized that the first step of unification is to incorporate relativity into quantum mechanics. The theory of relativity has showed us the rules that govern the moving objects in the universe, but what are the rules that govern a moving particle in the quantum world? In 1927, Paul Dirac formulated a mathematical understanding of the motion of matter particles in force fields, which has led to the development of an important branch in physics called *quantum field theory* (QFT). However, a major hitch in the quantum theory is that the quantum world is chiefly ruled by the uncertainty principle, which would tend to make rules arbitrary and render judgement inconclusive. Moreover, when we set out to study the moving subatomic particles, they appeared to take an infinite number of pathways to travel from place to place (see sum-over-histories, Ch 2-E), which would theoretically give the matter particles an infinite mass and energy. But then we know that all subatomic particles have a finite mass! In order to surmount this infinity problem, scientists had to do some tweaking of the mathematics of quantum field theory, and this has led to a really versatile mathematical exercise called *renormalization*. Renormalization would now effectively take care of the infinity problem. In fact, renormalization has become an effective tool in theoretical physics to cure the infinities and has become a standard procedure which was subsequently applied to many other field theories as well.

By the middle of the twentieth century, stalwarts such as Wolfgang Pauli, Richard Feynman, Julian Schwinger, Freeman Dyson, Sin-Itiro Tomonaga, and others have effectively used renormalization to show how electromagnetism could be incorporated into the quantum theory, and this has led to the birth of *quantum electrodynamics* (QED). The essential concept of QED is that the photons are the smallest possible bundles of energy, and matter particles share photons in order to pass on the force. Consequently, Feynman showed that electromagnetically charged particles (e.g. electrons, positrons, muons), after exchanging the force carriers, change their pathways as depicted in his famous *Feynman diagrams* (Fig 3.5). But then here, the exchanged photons exist as *virtual photons* (not as real particles), and so they cannot be detected in any way by any experiment. However, these virtual particles should not be thought of as merely mathematical entities, but they may be considered to 'physically' exist in the quantum world albeit for very transient periods so as to cause their detection impossible. This consideration has turned out to be a valid concept because when the field equations are applied to matter particle interactions, these virtual particles are needed to save the law of conservation of matter–energy. Hence, these virtual particles are reckoned to exist in reality but for a certain very small and ephemeral periods of time (see Sec-H for details). Whatever it is, the big conclusion of QED is that the EM force could finally be brought into the quantum realm.

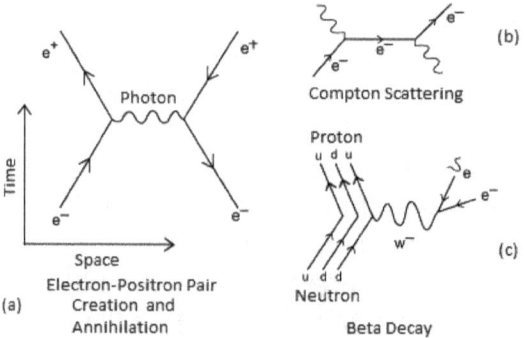

Fig 3.5: Feynman diagrams.

Electroweak Interactions: Having brought EM radiation into the quantum fold, the next major leap of scientific progress that happened was the unification of EM force with weak force. In 1967, Sheldon Glashow, Abdus Salam, and Steven Weinberg effectively clubbed the electromagnetic force with the weak force, giving us the *electroweak interaction*. They showed that both EM radiation and weak force are actually one and the same and behave like similar forces at superlatively high temperatures (such as at the time of the big bang, Ch 5-G). Technically speaking, it can be said that *symmetry* exists between photons and weak force-carrier particles at these very high temperatures. Their calculations showed that the EM force and weak force have equal strengths at very short distances (10^{-18} meters) and at very high temperatures. As the universe expanded and cooled down (Ch 5-G), the symmetry is broken down so that these two appear as different forces with photons being massless and weak force carriers being very massive. Whereas it was already known that photons are massless, the existence of the massive weak force carriers was only as a theoretical prediction at that time, and for a long time, they were not demonstrated experimentally. Many years later, however, they were actually observed in the particle accelerator at CERN, which gave the electroweak theory much credence. And now the electroweak force is incorporated into the mainstream theoretical physics.

Though EM radiation and weak force were incorporated into the quantum realm, the reader can see that the strong force was clearly left out of unification. The discovery of a unique charge, in 1973, called colour charge in the nucleus (Sec-D) has led to the development of yet another quantum unification called *quantum chromodynamics* (QCD) (chromo = colour, representing the colour charge). As can be guessed, renormalization remained an essential tool here also to bring strong force into the quantum fold and to develop QCD. In 1979, the existence of massless force-carrier particles, called *gluons*, was discovered with an integral spin of 1, which gave credence to the QCD theory. One peculiar feature in chromodynamics is that the gluons themselves carry colour charge (whereas in QED, the photons themselves are electrically neutral). However, the significance of this feature is not known.

Thus, all the three non-gravitational forces (strong, weak, and EM radiation) have been accommodated in the quantum realm. But then the crucial riddle in this unification of forces remained unanswered. Consider this: we have seen that at high temperatures, the EM and weak forces unite into a unified electroweak force, but unifying strong force with electroweak force at these high temperatures is impossible because the strong force in these conditions becomes weak and dissociates from the rest. However, scientists have been attempting to build a *grand unified theory* (GUT) on the premise that at *even* higher energies (may be, at the time, shortly after the big bang, Ch 5-G), all the three non-gravitational forces may have the same strength and behave similarly, and so they can still be unified. It has been speculated that these events could take place at energies in the range of about a thousand million million GeV.

At present, no experiment can prove this theory because the best particle accelerators at present could only reach energies up to about 100 GeV, and in the future, we may be able to build some very-high-energy accelerators. But to build a machine that can move particles at millions of GeV, the machine has to be immensely massive—perhaps as massive as the mass of our solar system! GUT theory also predicts that the protons (which are one of the most stable particles in the universe, Ch 9-C) would spontaneously decay into lighter particles because of their internal quantum uncertainty, but till date, no such decay has ever been observed. These theoretical impossibilities have made GUT more of a relic than reality. Today, the scientists have gone a step forward of unifying *all* the *four* fundamental forces *and* matter particles in one single consummate theory—ambitiously called the *theory of everything* (TOE). However, till date, this unified, all-encompassing, universal TOE has eluded the scientists in many ways, and it has remained only as a theoretical dream!

A Note on Quantum Gravity: The one outstanding pitfall of the Standard Model, as already noted, is that it does not incorporate gravitational force in any of its field theories. Gravity is so extremely weak that its influence on the subatomic particles is almost negligible, and consequently, there is no accepted way to weld gravity to quantum mechanics. However, several theories have come up in the recent years, but all of these are partial theories, and some of them are more speculative than being scientific. But we must be wary of judging speculations in science. Some speculations have really brought about significant leaps in scientific progress in the past as we have learnt in the previous chapters, and we will be seeing such successful speculations in the subsequent chapters too! We will see here how this quantum gravity becomes important in unification.

The most important implication of gravity in the quantum realm is that of the issue of gravity's role at the time of the big bang. Soon after the big bang, the rate of expansion of space is chiefly dependent on the slackening of the pull of gravity, and thus, gravity must also exist in its most miniature form at the time of the big bang—i.e. this form of gravity at that time should have acted at a very short range. However, these small-distance gravitational forces could be understood more in terms of the quantum rules than by the laws of relativity. Hence, a theory of 'quantum gravity' was proposed which is supposed to govern the earliest events of the universe which sheds light on our full understanding of events at the time of the big bang (Ch 5-G). However, as the reader could readily see, in these quantum settings, Heisenberg's uncertainty principle presides over the scenario along with its attendant probabilities, and this would menacingly come in the way of a ready unification.

Nevertheless, one inevitable prerequisite of quantum gravity is the existence of a force-carrier particle because the essential principle of quantum theory is that all force fields exist in quanta, or packets, of energy, and so gravity is also predicted to exist in quanta called *gravitons*. Gravitons are predicted to be massless, chargeless particles which would travel at the speed of light with a spin of 2 (whereas photons and gluons are spin-1 particles). They are speculated to have no antiparticles. However, no such gravitons are detected so far either in nature or in particle experiments!

SECTION G
The Enigmatic Mass and The Higgs Boson

The problem of mass in our ordinary world is one thing, but the phenomenon of mass at the quantum level is quite another thing. Something weird happens to the concept of mass as we go down into the quantum realm. We know that energy and mass are interconvertible, but how exactly is mass created in nature at the fundamental level? Decades of painstaking research and observation went in to answer this question to a great extent, as we will see below.

First of all, we will see what intrigued scientists when they have dealt with the problem of mass at the quantum level. We have seen that almost all mass of any object is contributed by the nuclei of its constituent atoms—the contribution of electrons to mass is next to nil (which means that an object's mass is what the nuclei weigh!). We have also seen that an atomic nucleus is made up of hadrons (protons and neutrons), and these particles, in turn, consist of quarks, which are bound by gluons. However, a very queer thing about the mass of the proton/neutron is that only a very, very tiny part of the hadron mass is contributed by quarks in it. Consider this: a proton has a mass of 938 MeV/c²,* out of which the quarks account for only about 5 to 15 MeV/c², which is merely about 1% of the mass of a proton. This means that about 99% of the amount of mass contained in a proton is *not* conferred by quarks themselves and so could not be properly accounted for the total mass of nucleons. Gluons, being massless, do not contribute to the mass of the nucleus either.

Then how is this missing mass accounted for? The answer is straight—the energy that is trapped inside the nucleus itself contributes to its total mass. Mass–energy equivalence once again comes to our rescue and explains the mystery of this missing mass. While the gluons transfer their energy to the quarks, the quarks, 'energized' by the gluons, move with great speeds within the confines of the nucleus, thus imparting relativistic mass to the quarks. Thus, a major portion of mass of a nucleon is contributed by the *kinetic energy* of its resident quarks. Thus, it is the kinetic energy (i.e. the relativistic mass) that we are calculating when we weigh the nucleus, not the proper mass of quarks itself. This also explains why in proton decay the product particles zip away rapidly into the surroundings. The product particles gain kinetic energy from the missing mass.[†]

But then there is a yet another fundamental question related to the mass of the matter particles. We have learnt about a host of particles in the Standard Model—some essentially massless, some light, some heavy, and some very heavy. For example, while the up-quarks in a proton have a mass of around 5 MeV/c², the charm-quark's mass is 1.35 GeV/c² (GeV = giga-electronvolt), and the top-quark is a whopping 175 GeV/c². Equally intriguingly, while the photon is essentially massless, the W and Z particles register a mass of 80 and 91 GeV, respectively, although these force-carrier particles fundamentally belong to a single class (i.e. electroweak force). Now the big question is, what is that which confers these

---

*   MeV/c² denotes mega-electronvolt, which is a unit of mass as per the special theory of relativity because mass and energy are interconvertible.

†   It is experimentally observed that the products of decay of a massive quark or lepton *always* have *lesser* mass than the parent particle itself—meaning that there would always be some missing mass in these interactions (Ch 9-C), which is converted into kinetic energy.

particles their mass? There must be a mechanism by which some particles acquire mass and some do not. Scientists have proposed a theoretical mechanism, called the Higgs mechanism, by which the fundamental particles obtain their mass. Below is a brief account of the current understanding.

Higgs Mechanism: We will have a better understanding of this mechanism if we start from the beginning. We have seen in the double-slit experiment (Ch 2-B) that light demonstrated interference pattern, which meant that a massless photon would pass through *both* the slits *simultaneously* to cause interference pattern. Further on, it was shown that not only photons but also mass particles, such as electrons, also demonstrate interference pattern. The result of this experiment meant that the trajectories of particles that have travelled from the source to the end could have taken an *infinite* number of paths with several possible loops and kinks, as we have seen in sum-over-histories (Ch 2-D; Fig 2.4).

But consider this: the simplest and most direct path possible for a particle to take would be a straight line—the one without any kinks. Feynman's mathematical rules predicted that a particle with zero mass (such as a photon) would possibly take only this perfect straight path—no curves, no loops, or no kinks. In other words, the probability of a zigzag course for a massless particle would be very low for a massless particle. However, it was shown that the propagation of a particle with a mass (such as an electron) would possibly take a somewhat zigzag course. Furthermore, the mathematical rules have indicated that the more massive the particle is, the more would be the zigzag pattern and that the lesser the mass, the lesser the zigzag course. Speaking the other way, it can be said that the mass of a moving particle is determined by the zigzag pattern. But then a basic question remained: what has caused this zigzag pattern?

It was natural for the scientists to predict that there is some kind of a 'field force' that is operating in the intervening empty space in which the particles travelled. And this field force would have interacted with different particles in different ways so as to give them their zigzag courses and consequently their masses. This field force is now thought to be all-pervading—i.e. it would exist right amidst the space between the smaller of the smallest particles of the universe, and also it would exist in the whole of the vast empty voids of the universe! This means that the 'emptiness' of the whole of the universe is not really empty (Ch 2-G) but is occupied by some weird force which would interact with matter and impart them their mass. It is thought that this field mechanism would not interact at all with certain particles, making them massless, but interact strongly with certain others, giving them their masses and implying that this 'ethereal' field force must have some peculiar properties to act in this way.

In 1964, a consortium of theoretical physicists led by Peter Higgs, including Robert Brout, François Englert, Gerald Guralnik, Carl Hagen, and Thomas Kibble, described a mathematical model by which the force field would interact with the fundamental particles and give them their mass. This field became known as *Higgs field*, and the mechanism by which the particles could acquire mass is called the *Higgs mechanism*. Higgs also predicted that a particle is associated with this new field, and it was called the *Higgs particle*. Mathematical calculations suggested that this particle will have an intrinsic spin of 0. Because integral spin is the characteristic feature of boson (Sec-D), Higgs particle is supposed to be bosonic in nature. Hence, it is also called the *Higgs boson*. Furthermore, it was also predicted that Higgs boson itself would be a very massive particle—ranging from 115 GeV/c² to anywhere up to 600 GeV/c²—but with a very short life, much lesser than a nanosecond.

Thus, in summary, it is proposed that a massless particle acquires mass when it enters and interacts with the Higgs field (Fig 3.6), and conversely, a particle which does not interact with the Higgs field remains massless. Therefore, a photon or a gluon which do not interact with Higgs field would remain massless, and an electron which interacts with the Higgs field would acquire mass. Now the reader can see that this Higgs field is the most crucial particle in the Standard Model, without which all other fundamental particles remain massless. Higgs particle is also thought to have an implication in the early big bang. It is speculated that just after the big bang, the Higgs field was zero, but as the universe cooled and the temperature fell below a critical value, the field grew spontaneously so that any particle interacting with it acquired mass.

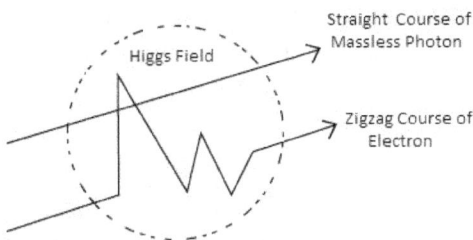

Fig 3.6: Higgs field.

On account of this crucial nature of Higgs particle, some people have dubbed it as *God's particle* (though Higgs himself desisted such a scintillating title!). The reader can now see that Higgs particle is the key to the Standard Model, and the veracity of this model depended upon the discovery of this particle. Moreover, it is predicted that this discovery may herald the beginning of a 'new physics' (which perhaps could also explain the enigma of 'dark matter', Ch 5-H). Naturally, a massive hunt was set forth in search of this versatile particle—at the cost of billions of dollars—in almost all the particle accelerators across the world. At first, there were only tantalizing glimpses of the Higgs boson. A few of the suspicious particles discovered bearing the characteristics of the Higgs boson were soon found out to be only decoy particles. However, finally, in 2012, a new particle with a mass of 125 GeV/c^2 was discovered which has a zero spin and an 'even parity' (see below)—features which are considered to be the fundamental attributes of a Higgs boson. In 2018, the LHC, in its ATLAS experiment, has studied this particle and demonstrated Higgs interactions with massive particles, which finally confirmed the discovery of Higgs particle. And with this discovery, the Standard Model, which stands out as a hallmark of human genius, can be said to be *almost* complete!

Section H
Matter–Antimatter Dilemma

We will briefly go into another problem that has pestered the scientists for a long time but appears to be somewhat resolved now. This is the dilemma of excess matter over antimatter in the universe. First we will understand the problem.

We have seen in the Standard Model that all matter particles in nature have their corresponding antiparticles (Table 3.1) which look and behave exactly *equal* in all respects—except that they have an *opposite* charge. For example, each lepton has an antilepton, electron has positron, muon has antimuon, and neutrinos have antineutrinos (though they are chargeless, their spin is opposite); likewise, each quark has an antiquark – up, down, charm, strange, top, bottom—all have antiquarks. Even composite particles like protons have *antiprotons* which are composed of their corresponding antiquarks. A neutron, though chargeless, also has an *antineutron* which is composed of corresponding antiquarks. Mesons also have *antimesons*—for example, pion is made up of one up-quark and one down-antiquark, whereas *anti-pion* is made up of one down-quark and one up-antiquark. Interestingly, we have even an *'anti-atom'* in existence. We can envision its structure: it would have a nucleus consisting of antiprotons and antineutrons with positrons revolving around it. The simplest possible anti-atom in nature would be a hydrogen anti-atom (i.e. *antihydrogen*) with a single antiproton and a positron. Indeed, a few of such antihydrogen atoms were created by the CERN scientists after some really laborious efforts.

Antiparticles are conventionally depicted by a bar placed over the symbol of matter particle—e.g. u for up-quark; $\bar{u}$ for up-antiquark. The theoretical existence of antimatter was first predicted by Paul Dirac in 1928, and they were discovered subsequently in particle accelerators in 1932. Gravity affects matter and antimatter indistinguishably because gravity is not a changed force.

However, all antiparticles are highly unstable because when they come into contact with their particles, they both annihilate each other, resulting in the release of massive amounts of energy. It's a theoretical dream for a physicist to synthesize antimatter in some meaningful quantities so that this annihilation energy can be harnessed to power the spaceships to explore the distant stars and planets of this endless universe, but we must be wary of such excesses because this might also lure us to build new 'anti-atom atom bombs' which may become capable of searing away our habitat on the earth in a jiff (remember Dan Brown's book *Angels and Demons!*).

Now we will discuss one big conceptual problem with antimatter: it is known that there is no natural antimatter in existence in the universe. The universe we know is exclusively composed of only 'normal' matter. All the objects in the observable universe—the planets, the stars, the distant galaxies, and virtually anything we know of—are made up of only normal atoms. But then we will see in the forthcoming chapter (Ch 5) that at the time of the big bang, both matter and antimatter were created in equal amounts, and there was no special preference of one over the other. Thus, it is conceivable that an equal proportion of antimatter is distributed across the universe. But then why this preponderance of matter over antimatter in our present-day universe? Why is normal matter dominating?

CP Violation: The answer for this preponderance of matter over antimatter appears to lie in the phenomenon of *CP violation*. Before we go into CP violation, we must know something of *CPT-symmetry* (charge–parity–time reversal symmetry). C-symmetry stands for *charge conjugation symmetry*, which means that two particles of opposite charges with similar mass and other properties, like spin, behave exactly the same in all ways except that their charges are opposite. For example, an electron and a positron behave exactly the same in all ways except that their charges are different. In short, C-symmetry is the same for particles and antiparticles. P-symmetry stands for *parity symmetry*—the laws governing two similar particles with opposite mirror-image properties (right-handed/left-handed or clockwise/anticlockwise) are similar to each other. T-symmetry stands for *time reversal symmetry*—the time elapsed for two similar particles moving in opposite directions agrees totally even if the direction of their movement is reversed. For example, if time is reversed for a forward-moving particle, it moves back in exactly the same way backward. This CPT-symmetry was considered so invincible to the scientists for a long time that any violation of this looked absurd.

However, as science progressed, more precise experiments were conducted using daintier gadgets, and it became clear that the citadel of CPT-symmetry is after all not as strong as it was presumed. In 1956, Tsung-Dao Lee and Chen-Ning Yang observed that, under certain conditions, the weak force particles do not obey

C-symmetry and P-symmetry but still their combined symmetry (called CP-symmetry) was not violated. In 1964, James Cronin and Val Fitch discovered that even CP-symmetry is not obeyed by certain particles (e.g. K-mesons) under certain conditions. The essential meaning of this astounding discovery is that the laws governing the particles and antiparticles for their mirror images may behave slightly differently if their direction of travel is reversed. This non-adherence to time reversal by the CP-symmetry is famously called *CP violation.*

Now we can explain the matter–antimatter dilemma. This CP violation has been shown to have some cosmological implications: as the universe started expanding at the time of the big bang, it has taken the time forward, but because of CP violation, it is speculated that the early universe has not obeyed the T-symmetry. Thus, the CP-symmetry is broken, and this has caused a slight imbalance in the annihilation of matter and antimatter, leaving behind a slight excess of the ordinary matter. This 'slight' excess of matter constitutes all the stars we see, all the planets that are in existence, all the matter we are made up of. In short, this excess matter constitutes the entire universe that we have today. CP violation also has some important thermodynamic implications in the universe of which we will study in Ch 6-E.

In conclusion, we may say that the information presented in the sections above is the essence of the Standard Model. But then is this the only model of an atom we have? Or do we have any other models in our armamentarium to define matter in another way? We will examine an alternate model, called string theory, in Sec-I, which is under intense scrutiny by many researchers at present. But before going into the string theory, we will examine the concept of *nothing* in the realm of theoretical physics.

SECTION I
Nothing is Something

Hitherto, we have studied various subatomic particles that constitute matter in the universe, but we are left out with a question of what is that which exists in between these particles? The reader may assume that there exists a space of stark nothingness in these empty voids, which is devoid of any activity whatsoever. However, the concept of 'emptiness' or 'nothingness' has changed radically ever since quantum theory has taken hold of our understanding of nature, and it is now clear that the empty space is not absolutely empty but is bubbling constantly with energy fluctuations and proxy particles. In fact, it is now known that the empty space is really a potential arena for energy exchanges, which can even show some physical effects in the real world, as we will see below.

The idea of empty space not being really empty came out initially as a theoretical consequence of Heisenberg's uncertainty principle. Consider this: for empty space to exist, the value of a field force at a given place and its rate of change must be exactly zero. If that is not the case, the space would not be truly empty. However, the uncertainty principle states that the total amount of energy of a field and its elapsed time cannot be determined simultaneously with any precision (Ch 2-D). In other words, there is a fundamental limit on how accurately we can measure the energy of a field at a given place and time. Consequently, it can be said that nature does not allow the existence of empty space. Rather, the empty space must exist with a state of minimum energy at all times. And the energy of this empty space is called the *vacuum energy*.

Vacuum energy is the minimum energy that is contained in empty space.

Now consider this: we have also seen in the special theory of relativity that time appears to slow down as a particle moves faster. Eventually, time stops ticking and comes to a standstill when the particle reaches the speed of light (Ch 1-C and Ch 7-B). But then it can be theoretically assumed that when a particle moves faster than light, then that particle would actually appear to move backwards in time. Since the rules of Heisenberg's uncertainty preside over the events in the quantum arena, the particles are allowed to act as if they move faster than the speed of light even though we cannot precisely measure the speed of particles for very short periods. Thus, in effect, we can say that such particles are moving backwards in time (Fig 3.7). In this scenario of briefest periods of time, there would be particles which flip forwards, backwards, and again forwards as they jiggle along in the quantum space. And if we draw a Feynman diagram of the movement of a *single* particle in such an event, then we are likely to encounter *three* particles at a given time (Fig 3.7). Of course, they are not real particles, but they *do* exist for very brief periods of time in the quantum realm. These ephemeral particles of the quantum world are termed the *virtual particles*.

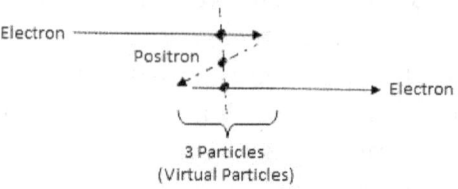

Fig 3.7: Virtual particles.

Therefore, it is possible for virtual particles to spontaneously pop out of empty space, borrowing energy from nothing—like gas bubbles popping up from 'nowhere' in the middle of a just-opened soda bottle. But then these virtual particles may take the form of either a particle or an antiparticle (because nature does not have any preference of particles over antiparticles, Sec-G). Consequently, these particle–antiparticle pairs can coexist for the briefest period of time before they annihilate and return the energy debt paid back to the vacuum. This is the so-called *zero-point energy*. Nevertheless, it must be stated here that things that are very strict laws in our macroscopic world, such as the laws of conservation of mass and energy, may not be strictly adhered to in the quantum world—but with the caveat that these laws are broken only for very small intervals of time (perhaps in the range of Planck seconds!). Consequently, in the case of virtual particles also, there is a temporary violation of the law of conservation of energy so that one of the lighter particles may become a pair of heavier particles, but these heavy particles would quickly annihilate and rejoin the original particle as if they were never there!

But then here, one major objection remains to be addressed. Mathematical calculations show that the virtual particle pairs that arise out of a unit of vacuum could be infinite in number, and thus, it is theoretically possible that they may generate an infinite amount of energy. And as per the general relativity, this infinite energy would tend to curl up the empty space into an infinitely small size, which, however, is contrary to the observation. Of course, scientists have come up with a few newer theories now which may circumvent such hurdles (Sec-J), but they have not been completely successful in their explanations.

For many years, these concepts were considered to be merely hypothetical with no practical value. But then as research progressed, these virtual particles did show their telltale signs of existence in the experiments. Consider this phenomenon: a hydrogen atom has a proton as nucleus with an electron revolving around it, and when energy is supplied, the electron jumps from a lower-energy shell to a higher-energy shell, and when it loses energy, the electron shifts back to the ground state. This shift causes the electron to vibrate at different frequencies, which shows up as a characteristic spectral pattern. The Schrödinger equation (Ch 2-C) has allowed us to calculate the predicted frequencies *almost* exactly—but *not* exactly. The spectrum would still have a few complicated waves which could not be properly accounted for. In 1928, Paul Dirac proposed an improvement to these equations, but they also failed to predict the precise behaviour of an electron in the hydrogen atom. However, a few years later, it was realized that the Dirac equation could actually give us an immaculate prediction only when we take the existence of virtual particles into consideration. This peculiar fact could be understood by the following simple theoretical explanation: an electron in the hydrogen atom really behaves like an electron–positron pair of *virtual* particles (a theoretical vagary which is conveniently allowed in quantum mechanics), and when the particle behaves like a negatively charged electron, it tends to move towards the proton. But when the particle behaves like a positively charged positron, it tends to move away from the proton, causing quantum fluctuations. These infinitesimally minute fluctuations of the electron, when considered in calculations, could finally explain the characteristic spectral shift of hydrogen, precisely lending a solid proof of the existence of virtual particles.

Further evidence of virtual particles was supplied by other experiments. In 1947, Hendrik Casimir proposed that the popping up of energy fluctuations in empty space could actually be demonstrated. He had conceptualized that when two thin uncharged metal plates in vacuum were placed adjacent to each other (separated by nanodistances), it could be construed theoretically that more virtual particles exist around the plates on their outer surfaces than in between the metal plates (because the wave patterns of the virtual particles in the vacuum tend to cross the thickness of the thin plates to reach the outer surfaces), and this differential in energy from outside to inside would generate a seemingly mysterious force that would tend to push the metal plates together. This is called the *Casimir effect*. Ever since this effect was proposed, many experiments were conducted which have conclusively documented the existence of vacuum energy and virtual particles. More recently, Pasi Lähteenmäki measured Casimir effect in between two moving mirror plates (i.e. *dynamical Casimir effect*) and showed that flashes of light energy could indeed be generated from vacuum!

Quantum Foam: As science progressed, it gave birth to newer insights which led to some alternative concepts of vacuum energy. John Wheeler, in 1955, proposed that spacetime occupying the empty space need not be smooth but that it can show infinitely tiny fluctuations (in the range of Planck scales). The uncertainty principle will not allow these fluctuations to keep pattern, and this feature makes these entities indefinite or irregular, and hence, they are speculated to behave like ever-changing regions of space and time. He suggested that this feature of empty space would render the spacetime a 'foamy character'—the so-called *quantum foam*. However, as per the calculations, this hypothesis suggests that photons at certain fields must travel at speeds *lesser* than the speed of light (which also essentially contradicts Einstein's theory of special relativity—the speed of light in vacuum is constant, neither less nor more), and as of today, no experiment has succeeded in documenting such a radical finding. Nevertheless, the concept of quantum foam has become very popular among many scientists, and they vouch for its existence, hoping for some experimental evidence to turn in its favour!

Section J
The String Theory

The success of the Standard Model in explaining the structure of matter has been stupendous especially with the discovery of the final piece in the puzzle, the Higgs boson. However, having said this, we may also admit that the Standard Model by itself is not a consummate theory, one of the most obstinate problems that it is facing at present being the problem of gravity. The non-inclusion of gravitational force into its fold has still remained the Achilles heel of the Standard Model. However, scientists have realized that the only way to fill in the gap between gravity and the Standard Model would be to weld quantum theory with the relativity theory—i.e. to understand gravity in terms of 'quantum gravity'. Although the concept of quantum gravity was proposed long ago by the theorists, it has not been a very successful theory. However, there have been some speculative theories in the offing, such as GUT and ToE, which would attempt to unify all the forces of nature in a meaningful way, but they are still half-baked and not universally accepted as it were. But then with more conceptual advancements, new models of matter particles and forces have started to emerge in the scientific scenario, of course all of them with some hypothetical guesstimates but with no concrete factual evidence. Even so, here we will briefly look into the evolution of one of the leading models which gained some varying degrees of acceptance among the scientists.

The string theory is one such prominent candidate that is thought to fit in the slot of unification. It is the first ever concrete attempt to do the job of unification to any significant extent. We will study the string theory briefly. Way back in 1926, Theodor Kaluza and Oskar Klein made attempts to incorporate gravity into electromagnetism. To understand the concept behind this unification, consider this. Euclid had said more than two millennia ago that a *line* has a single dimension—length; a *plane* has two dimensions—length and breadth; a *solid* has three dimensions—length, breadth, and depth; and *nothing has four dimensions*! But then Einstein has shown us that we can take Riemannian geometry and mathematically construct a four-dimensional spacetime structure with three space dimensions and one time dimension (Ch 1-E). However, human beings, being essentially 3D creatures, could not imagine objects in a 4D spacetime in their mind's eye, though we, by erudition, know that all objects in the universe can be represented in a 4D form (Ch 1-E and Ch 7-B).

Kaluza and Klein suggested that gravity and EM force can be combined and envisioned as one entity only with the help of another dimension—the *fifth dimension*. Ever since this proposition, mathematicians have worked out the existence of more dimensions, and some initial mathematical calculations have suggested the existence of a staggering number of twenty-six dimensions. However, with some later insights, these dimensions were proved to be theoretically redundant, and with the introduction of yet another mathematical model called a *Calabi–Yau space*, the number of dimensions was curtailed to a manageable six-dimensional manifold. This concept of extra dimensions was thrown into oblivion for a few decades, but in 1984, Michael Greene and John Schwarz showed a renewed interest in them and worked up a concrete concept called the *superstring theory* (or simply the *string theory*), which gained some popularity among researchers in general at present.

The string theory views the basic building blocks of matter not as pointlike elementary particles (as we have viewed them all along in our Standard Model). Rather, in the string theory, the fundamental matter particles are viewed as infinitesimally small *one-dimensional vibrating strings* of energy. These minuscule strings are thought to have only length as its dimension (with no other dimension!), and the size of these strings is hypothetically proposed to be on the Planck scales—i.e. unimaginably small not only for the current scientific experimentation but even for the human intuition. All matter particles and force carriers are represented as vibrating strings, and these vibrations are supposed to occur in a ten-dimensional setting with six dimensions wrapped up (or curled up) on infinitesimally small Planck scales and the remaining four dimensions are exposed to us (which form our familiar four dimensions). A string is proposed to be either *open* or *closed* (loop-like), each with different properties. Whereas the strength of the vibration of these strings represents the mass of the matter particle, the pattern of vibration is supposed to differentiate the various other properties of particles, such as charge and spin.

The underlying principle of string theory is called the *supersymmetry*, which is a mathematical concept wherein an entity remains invariant (unchanged) even after several transformation operations. The string theory assumes that matter particles and force particles obey supersymmetry, by which they may be unified. Consequently, it is postulated that every fundamental matter particle has an associated massive force-carrier partner, and every force-carrier particle has an associated massive matter particle partner, thus representing supersymmetry partnership of both matter and force particles. For example, a quark is supposed to be associated with a massive force carrier called *squark*. Likewise, a photon is associated with a fermionic partner particle called *photino*, and *glutino* is the name given to a fermionic partner of bosonic gluon, and so on. The concept of *supergravity* is also proposed to weld supersymmetry to general relativity.

There are many versions of the string theory, of which one theory, called the *M-theory*, proposed by Edward Witten in 1995 has gained some popularity over the years. Witten brought various forms of the string theory under one heading and formulated another theory where the basic building blocks of matter are not stringlike but looked more like *membranes* (or called simply *branes*).

Though the string theory has some wide acceptance among researchers, it has several pitfalls apart from being largely incomplete. Basically, the string theorists have no convincing explanation as to why the six dimensions are so infinitesimally curled up in space as to render their detection impossible. Moreover, even though the principle of supersymmetry is the stronghold of this theory, none of the exotic partner particles proposed by the string scientists are identified so far in any of the particle accelerators. Moreover, even though the string theory has been theoretically applied to many of the existing problems in physics, such as the origin of the universe, black holes, and quantum gravity, it has been met with only limited success in explaining these phenomena. However, it is interesting to note that these extra dimensions and exotic possibilities of matter have also prompted some scientists to speculate over the existence of other 'parallel worlds' (i.e. multiple universes apart from our own), which possibly have certain different physical laws operating in them. We will discuss parallel worlds in Ch 4-E.

Having discussed the world of small particles in this chapter, we will go on to the next chapter which describes a bigger world out there, encompassing countless number of massive stars and galaxies spread over vast distances in the cosmos.

CHAPTER 4

The Magnificent Universe
And the Life Cycle of Stars and Black Holes

Overview

Having examined the microworld of atoms in the preceding chapter, we will now venture out into the never-ending skies and study the big and bigger world of celestial objects out there. In the first section, we will conduct a general survey of the universe, see how big and infinite it is, and then try to locate our own earth in this vast cosmos. In here, we will also come across some of the salient features of the universe.

Next we will probe into the life cycle of a typical star by learning how stars take birth in the universe, how they grow big, and how they eventually die out. We will also track down the lives of certain giant stars and understand why they are destined to end their lives as exploding supernovae or interminable black holes. In our story of the stars, we will also see how various elements are synthesized in their cores and how they are disseminated into the vast spaces across the universe. This study shows us why we, the humans, are cosmologically paraphrased as 'stardust'! Over the next section, we will explore into the innards of our sun and briefly learn about its intricate anatomy, and we will then see how the sun burns its fuel in its nuclear furnace not only to emit immense amounts of energy but also to eject a countless number of exotic matter particles from its surface.

In the final section, we will put forth the question of the existence of universes other than our own in this infinite cosmos and see what some of our leading scientists think about parallel universes. And we will also discuss 'wormholes' as a possible solution to time travel in the future. And at the end, we will hunt for the existence of extraterrestrial life lurking somewhere in this gargantuan universe!

We will now take off to an exciting expedition to the depths of the cosmos!

Section A
The Extent of The Universe

One of the ancient and perhaps the most intriguing questions posed by mankind is the question of the extent of the universe—questions such as how big the universe is or where it begins or ends. But then we do not know if the universe is finite and bounded or if it is infinite and everlasting, and we do not know if we have some other universes like ours coexisting in this vast expanse. There are no definite answers in the present day, and it is inconceivable that there would be a definite answer in the near future. Nevertheless, we have some amazing facts and figures related to the *observable universe* as presented below.

We will first look into the general arrangement of stars and other celestial objects in this expansive universe. Stars are the chief constituents of the universe. They are the huge luminous bodies of matter in the cosmos with nuclear reactions occurring deep within them generating tremendous amounts of energies which make them shine. A majority of stars have planets revolving around them.* Most of the stars in the universe do not exist in isolation but are grouped together, and these groups of stars are called *galaxies*. Some are small galaxies which may contain about a few thousands of stars, but most

---

*   The planets of other stars (i.e. other than that of our sun) are called *exoplanets*, and thousands of such exoplanets have already been discovered—with new ones being added to our list almost every day.

of the galaxies have some millions of stars in them. And a few of the very large galaxies have even hundreds of billions of stars in them. Galaxies may take many shapes and forms—spiral, elliptical, irregular, some bright, some dim, and some emitting not visible light but light of other forms such as radio waves (e.g. radio galaxies). Many of the galaxies are thought to contain giant black holes (Sec-C) at their centres. In between these galaxies, there exists some gigantic emptiness of space called the *voids*, which dominate the vast expanse of the universe (Sec-B). However, some of the galaxies are closely bunched up to form large complex groups called the *galactic clusters*. Some galactic clusters may, in turn, group together in huge assortments called the *superclusters*. The above description is the general arrangement of stars in our known universe.

Now we will look into our local system—i.e. where we, the humans, reside—and see how stars and planets are arranged around us. Our star system, where we live in, is the *solar system*, and our star is the sun. The solar system has eight major planets (Mercury, Venus, Earth, Mars, Jupiter, Saturn, Uranus, and Neptune) revolving round the sun, and many dwarf planets, such as Pluto, Ceres, and Eris. The solar system itself is located in a medium-sized galaxy called the *Milky Way Galaxy*, which is estimated to lodge about 10^{11} stars. The Milky Way is a spiral galaxy, and our sun is located in the outer reaches of one of its arms. The Milky Way is known to be rotating, and it is estimated that the sun takes about 250 million years to go round the galaxy once. Our galaxy has a huge black hole at its centre named *Sagittarius A*. As our neighbouring galaxy, we have the *Andromeda Galaxy*, which is at least two times bigger than our galaxy, whereas another adjacent galaxy called the *Large Magellanic Cloud* (LMC) is smaller. The Milky Way, LMC, Andromeda, and many other galaxies are all located in a jumbo galactic cluster called *the Local Group*. The Local Group, along with many other such galactic clusters, is located in a galactic supercluster called *Virgo Supercluster*. Now the reader may see that this is our correct 'address' in our universe. So, folks, next time, when you are prompted to give your 'complete' postal address in your job applications, do not forget to include all the above celestial details!

The extent of the observable universe has expanded vastly with scientific progress. Consider this: it has been estimated that we could be able to count about 3,000 stars with our naked eye on a clear night sky, whereas a light telescope could detect about 100,000 stars. Now with sophisticated telescopes, we could detect even the faintest light from the farthest of the far-off stars, and now we have estimates of as many as 100 billion stars in our Milky Way Galaxy alone. But obviously, the universe is far larger. According to the recent estimates from the Nottingham University, there are a dizzying 2 trillion galaxies in the observable universe, and the total number of stars could be about 10 billion trillion, i.e. 10^{22} (some observations even put this figure up to 10^{24}). And the span of the observable universe is equally overwhelming; it is speculated to be about 92 billion light years across!*

The universe has about 2 trillion galaxies containing 10 billion trillion stars!

None of the galaxies are stationary, but they are in a constant state of relative movement, and sometimes they encroach upon each other, leading to gigantic celestial collisions. For example, the vastly bigger Andromeda is getting closer to the Milky Way at a rate of about 300 km/sec and may eventually gobble up our galaxy in the 'near' future, but that cosmic near future is not too sooner, so do not panic!

Is the Universe Uniform? By a simple observation of the night sky, one could say that the universe is uneven in its distribution with all the stars distributed in irregular groups and with spaces of empty voids in between them. This disorderly distribution has, in fact, given the constellations of stars some really bizarre appearances which had prompted our ancients to give them some imaginary and exotic names of zodiac. We may also see that the solar system

---

* Here the reader may take a brief interlude and ponder over human existence in the universe. Our 'address' in the universe (as noted above) becomes begrudgingly insignificant if we take into account the sheer magnitude of the universe. Moreover, if we have any inclination towards considering our human life as anything special, as was the cherished notion of the ancient theologians, then such a hope would be thwarted if we consider the monotony of the universe. The place where we live appears no different on a large scale from the rest of the universe!

itself is variable in its arrangement, the sun being a massive body (contributing nearly 99% of the matter of the solar system) located at its eccentric centre (Ch 5-A) surrounded by miniature planets all in different sizes with varied mineral composition and with eccentric orbital movements. In short, we find our universe highly variegated in its architecture.

However, astronomers have a different perspective about the distribution of matter in the universe. They realized that though the universe appears irregular on a 'short scale' (our earth, the solar system, and even our galactic clusters may be studied in 'short scales' when compared to the exceedingly vast stretch of the universe!), the average density of matter in the universe is uniform throughout its expanse when taken from a sufficiently large viewpoint. This feature of uniformity of the universe is called the *cosmological principle* (not to be confused with 'cosmological constant' described in Ch 5-A). The cosmological principle encompasses two essential features of the universe—that it is *homogeneous* and *isotropic*. Homogeneity means the universe appears the same at all places, and isotropy means the universe appears the same in all directions. This means that if you sample a sufficiently big chunk of the universe, there would be no difference in the density of matter from one place to the other and that there is no preferred direction in the universe.

The universe is homogeneous and isotropic.

Einstein had realized that the notion of homogeneous distribution of matter in the universe is a mathematical necessity in modern science because it stands as a practical tool for us to study the universe as a whole. If we were to consider the universe to be irregular from place to place, then we would not be able to make any reasonable generalizations, and so no mathematics or science would work out for us to study the universe at large! But even so, we now have some solid proof of this cosmological uniformity from observations of the CMB radiation, as we will be seeing in Ch 5-F.

The Olbers' Paradox: Here is an intriguing puzzle. We have seen that there are countless stars distributed throughout the universe, and we may presume that they have existed for some infinitely long periods of time. But consider this: if such is the case, the night sky on the earth must be as bright as the day because the starlight from these innumerable stars in an infinite universe is likely to cover the entire night sky by at least one star in any given direction, and this starlight collectively would shine the dark side of the earth (just like the day by the sun) (Fig 5-2, Ch 5). However, paradoxically, we find that the nights on the earth are dark and cool, cosy, and congenial for our comfort and existence. This feature of the night sky has intrigued scientists. We will discuss about this issue in Ch 5-F and see how the scientists have attempted to resolve the paradox with the help of the big bang theory.

We will now take up various interesting structures and events that highlight our universe for our study. We will first look into a typical star in the universe and see how it takes birth in space, becomes senescent, and eventually dies.

Section B
Life Cycle of A Star

The universe is essentially made up of matter and energy*—matter arranged in the form of various atoms with various forms of energy interacting with matter. Atomic matter is distributed across the universe not only in various visible forms, such as stars or planets, but also in the form of diffuse molecules spread across the vast 'empty' spaces that exist in between stars and galaxies. In fact, the reader must note that the amount of this matter in the voids of space is far greater than that which makes stars (Ch 5-H; Fig 5.5)!

From the standpoint of human beings, scientists have named this expansive space that exists around the earth as the *outer space*, which, by definition, starts from about 100 km above us. And it is

---

*   The following description is all about our knowledge of the 'known' matter of the universe, but scientists have postulated that there exists some unknown matter and energy which dominate the universe (called *dark matter* and *dark energy*). This mysterious form of matter is studied in connection with the big bang theory in Ch 5-H.

worth noting that our International Space Station (ISS) is located at about 400 km above ground; our GPS satellites at 20,000 km; and our moon at about 400,000 km! We will first study the matter distributed in the outer space and then study the life cycle of a star.

The Outer Space: Once we leave the habitable zones of the earth and its atmosphere, we would be plunging into the outer space, which is not only vast and limitless but is generally empty, extremely cold, and dangerously radiation ridden for human existence. The space we encounter first, as we skip out into the space beyond the earth, is the *interplanetary space* which, of course, spans between the planets in the solar system. This space is flooded with fast-moving particles especially protons (about 5–10 particles per cm³) that take their origin from the 'solar wind' (Sec-D), which, along with cosmic radiation,* make this space potentially dangerous for our survival. It is also laden with particles of cosmic dust consisting of some carbonaceous organic molecules, silicon, and other small atoms (the origin of these particles is from the exploding stars, as we will see below). Apart from this, the interplanetary space is also studded with boulders of matter—large and small—which take the form of numerous *asteroids* (which are larger boulders of metals and rock that make regular orbits within the inner solar system but occasionally trespass into the other regions), a few *comets* (which are relatively small pieces of rock and ice that move in the outer solar system but occasionally pass close to the sun, at which time they release gaseous tails), and other smaller stellar debris. This gritty interplanetary medium gets thinned out as we move away from the sun to the outer reaches of the solar system.

If we go beyond the solar system, we will be landing in the *interstellar space*—the vast empty space that spans between the stars. This space is also really not empty but is composed of *interstellar gas* comprising mainly of hydrogen (75%), helium, and other trace elements. However, in certain regions of space, the interstellar

---

* The 'cosmic rays' which pervade the space outside the earth are efficiently absorbed by the outer layers of the earth's atmosphere and are thus prevented from affecting the life on the earth. If not, this intense radiation would not have allowed complex life to develop on the earth!

gas is dominated by *interstellar dust*, which is mainly composed of tiny particles of carbon, silicon, and oxygen. In some areas, this interstellar dust is concentrated into huge areas of gas clouds called *nebulae*. A few heavier atoms may also be found in these gas clouds which are derived from the ejected debris of exploded stars. The average density of matter in the interstellar medium is estimated to be about 10^6 particles per m^3, but at some places, such as in nebulae, the density may reach far greater concentrations.

As we go beyond the interstellar space, we land up in the *intergalactic space*, which obviously is a vast empty space spreading between the galaxies. Intergalactic spaces are colloquially called *the voids*. This space was considered largely empty in the past, but now it is shown that it does contain some matter (at least a few hundred atoms per cm^3), which exists chiefly in the form of ionic hydrogen in the plasma state—i.e. free-moving protons.

We are no masters of outer space, and we have not explored any of these distant outer spaces in the real sense. Whatever knowledge we have gained so far is mostly from the indirect observations. But then it is indeed an outstanding achievement of the human race that in September 2016, for the first time in the history of space travel, a man-made spacecraft, the *Voyager 1*, has crossed the boundaries of our solar system after journeying for nearly thirty-nine long years to enter into the interstellar space. The fact that the humans could really traverse the breadth of the solar system and could reach such vast distances beyond our planet is indeed an unimpeachable evidence of his versatile genius!

A Note on Temperature of the Universe: The temperature of the universe in general is highly variable. It is super hot in the cores of stars, and it is coolest in the deep space. The baseline temperature of deep space is about 2.7 Kelvin (–270.45 °C) as calculated from the study of CMB radiation (Ch 5-F). The temperature within the solar system is also variable. For example, the exposed surface of the planet Mercury is about 427 °C, whereas its nights are too cold (–173 °C). Pluto has a very low surface temperature of –240 °C. The temperatures that prevailed in the universe at the time of its evolution are described in connection with the big bang (Ch 5-G).

The ambient temperatures on the earth are tolerably comfortable for our existence owing chiefly to the existence of its balanced atmosphere. But sardonically, this delicate balance maintained by the ecosystem on the earth is now being compromised mostly because of our human misadventures!

Having discussed the general features of the universe, we will now see how stars take birth from the gaseous matter in space, grow in size, and then become senescent and eventually die and recycle. We will also see how, sometimes, stars take the form of other interesting celestial objects in the cosmos.

The Birth of a Star: Scientists have understood, after decades of painstaking research, how matter in the interstellar space is transformed into stars and planets and how these celestial bodies are later recycled back into the empty space of the universe. A star is born when a massive cloud of the interstellar gas (nebula) collapses on to itself because of gravity. The event is manifested initially by the clumping together of the interstellar dust—a process called *accretion*. As the clumping proceeds, the constituent particle increases in size, resulting in more gravity, and consequently, more and more matter is pulled towards the centre of the nebula which now becomes the birthplace of a new star. Gradually, the density of matter at the centre of the newborn star increases, thereby increasing the pressure at its *core* which, in turn, increases the temperature because of increased intermolecular collisions. As time progresses, the protostar gets crushed even more and more upon its own gravity, thus building up tremendous amounts of pressures and temperatures within it which again compresses the matter further and further towards the core. This makes the core immensely dense, resulting in very rapid collisions of atoms within it, causing a steep raise in the core temperature. These tremendous gravitational pressures inside the core should theoretically continue to crush the star to become interminably smaller and smaller—if not for an expedient mechanism which counterbalances this gravitational force as described hereunder.

It is the counterbalancing energy of *nuclear reactions* that take place inside the star which checks further gravitational collapse of the star. This crucial process in a star deserves a brief description.

The interstellar gas is chiefly composed of hydrogen, as seen above, and hence the protostar is rich in hydrogen atoms, and so they become the chief source of fuel for the nuclear reactions. At some critical temperature and pressure, the hydrogen atoms inside the core get bereft of their electrons and become free protons. The gravitational force in the star is now so powerful that this force overpowers the positive electrostatic charge on the protons, which then start fusing with each other, forming helium nuclei. This *nuclear fusion* reaction releases monstrous amounts of energy owing to the fact that this process is invariably accompanied by the conversion of some amount of matter into energy (Sec-D). In this sense, a star is really an exploding hydrogen bomb at work! The reader may now see that the tremendous amounts of energy thus released from these nuclear reactions effectively counter the pressure of the star's gravity, and this pushes the core outwards (whereas the gravity pulls the core inwards) (Fig 4.1). Thus, it may be said that the star is kept alive on equilibrium brought about by the inwards pull of the gravitational force that keeps the star contracting and the outwards thrust of the nuclear energy that keeps the star expanding. A young star in this prime period of its life is known as a *main sequence star* (Fig 4.1),* and a typical star would continue to be in this state for a long time.

A star is kept alive on a balance between gravity and nuclear energy.

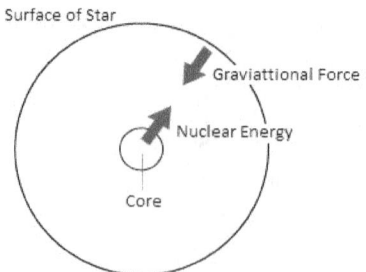

Fig 4.1: Main sequence star.

---

*   Sometimes, a collapsing nebula never reaches the critical temperature and pressure needed to start off a nuclear reaction because of its insufficient mass, and such a star remains quiescent as a *brown dwarf* and will never really reach its 'stardom'. Such stars would eventually fade away into the background of the universe.

But then eventually, all the hydrogen inside the star gets used up, and helium dominates its core. The star is now exhausted, and thus, the main sequence of the star ends. Now because of lack of opposing energy from nuclear fusion, the star continues to collapse under its own gravitational force. However, as this compression progresses, the core temperature once again rises tremendously, and this increased heat would now facilitate a different set of nuclear reactions. The helium now starts fusing into heavier elements because of the existing high temperatures and pressures, and thus, as time progresses, heavier elements get accumulated inside such a star. Eventually, the star consists of a dense core of heavier and heavier elements with immense temperatures which is covered by several layers of matter composed of lighter elements with decreasing temperatures. Because of this differential gradient of temperatures and pressures between the layers of the star, its outer layers expand exponentially rapidly, leaving the core of the star to keep contracting. Thus, these dying stars expand at a tremendous rate (because of expanding outer layers) and become quite large with lesser surface temperatures (so appearing bigger and redder), and these stars are now called the *red giants* (Fig 4.2).

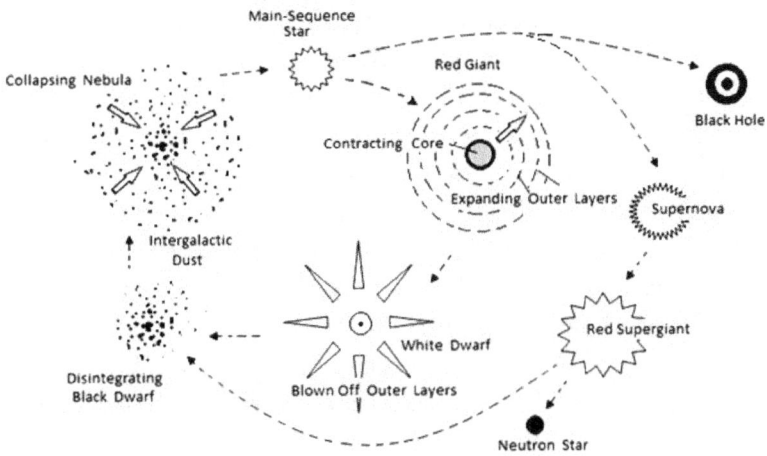

Fig 4.2: Life cycle of a star.

A typical star has sufficient fuel in it initially in the form of hydrogen and helium, and so its inner core repeatedly undergoes the above process of exhaustion and reignition several times in its lifetime. And with each death spasm, the star initiates fresh nuclear reactions which revive the star to life. Finally, this process culminates in a totally exhausted red giant with grossly expanded outer layers, and eventually, its outer layers blow off into outer space completely, leaving only a small dense core. This remnant of the star is nothing but a small cinder of matter comprised chiefly of heavier elements such as carbon and oxygen. Such a dead star is called a *white dwarf*. It has been hypothesized that white dwarfs eventually become very quiescent (which are then called *black dwarfs**), and they subsequently fade away into the background of the universe. This is perhaps the chief source of carbonaceous matter we find in the interstellar space (Sec-A). This is the fate of about 97% of the stars in our universe.

The Supernovae (or Supernovas): However, the story of a minority of stars in the universe (about 3%) is different. In fact, the fate of a star ultimately depends on its size. Most of the stars that end up as black dwarfs (as shown above) are medium-sized stars which dominate the universe. However, a massive star has a different fate. The reader has to note here that contrary to his/her expectation, large stars exhaust more rapidly and die more quickly simply because larger stars have more gravitational pull and consequently need more intense nuclear reactions to counter their massive gravity—more intense nuclear reactions mean more rapid consumption of nuclear fuel, which makes the life of a massive star shorter. Hence, at the end of their lives, large stars are tremendously hotter in their cores with great pressures—a situation more conducive for helium to fuse into heavier and heavier elements. Thus, these massive stars consist of more heavy elements in their cores, which make them remarkably denser, while the outermost regions of the star consist of lighter helium and hydrogen, which are relatively less dense. At

---

*   These derelict stars have much less temperatures, and so they contain elements in the form of atoms, i.e. nuclei surrounded by electrons. These atoms can no longer be compressed further by the existing insufficient gravity because they are supported by energy of the electrons orbiting the elements. Moreover, the atoms cannot be squeezed further into smaller and smaller spaces because of the Pauli exclusion principle, which prohibits such compression (Ch 3-C).

last, the outer regions of these massive stars start expanding (as we have seen above), thus forming a *red supergiant*. We have a record of such red giants in the universe. One such red supergiant, called *Betelgeuse* (of *The Hitchhiker's Guide to the Galaxy* fame), is located in the constellation of Orion, and it is about 500 times bigger than the sun.

However, in the case of massive stars, the core continues to have violent nuclear reactions involving heavier and heavier elements, resulting in the formation of elements such as neon, magnesium, silicon, and finally iron. Generally, the formation of elements stops at the stage of iron because iron is a stable metal (as we can see that iron occupies the middle of the periodic table, being neither too light nor too heavy). However, as pressure mounts up even more, the core of the star gets hotter and hotter, making the situation conducive to create higher (i.e. heavier than iron) elements. This is the reason the supermassive stars of the universe are called the crucibles of heavy metals. This process of synthesis of elements from helium in stars is called *stellar nucleosynthesis* (nucleosynthesis = synthesis of elements). In contrast to this, there is another primitive way of nucleosynthesis called *primordial nucleosynthesis* of which we will be discussing in connection with the big bang (Ch 5-E).

Coming back to the fate of these massive stars, the cores of these stars are eventually subjected to supertumultuous gravitational compression and superturbulent nuclear reactions. And finally, the balance between these two opposing superforces is lost, and the star collapses quite suddenly—and quite violently—under its own crushing gravity. This cataclysmic event happens in a very short period of time. It is calculated that many thousands of kilometres of the core's diameter collapses into a few kilometres in just a thousandth of a second! At this stage, the ultracondensed matter cannot be compressed any further, and the gravitational force is suddenly redistributed inside the star in the form of massive and turbulent shock waves of energy. These violent shock waves result in a cataclysmic explosion of the star with the blocks of stellar matter thrown rapidly into the space not only into the immediate vicinity of the explosion but even to the far-off places in the cosmos. Such an exploding star is called *supernova*. Stars which are at least eight times the mass of the sun are thought to end up their lives in this fashion.

Supernovae present themselves in the cosmos as magnificent flashes of light. A supernova explosion releases diabolic amounts of energies on such a scale that the explosion can be visualized continuously for a few weeks, and this superflash of light would outshine the other stars in the galaxy 'like a billion suns'. This cosmic macabre is a rare but spectacular event of the night sky which may be clearly visible to the naked eye. In fact, the first recorded cosmic incident of supernova was that of the medieval Chinese who had noticed a 'bursting star' in the year 1054 AD that was visible even in daylight for several months! Since then, many supernova explosions were witnessed across the centuries not only in the Milky Way but also in other distant galaxies.

Supernovae explosions release tremendous amounts of high-energy gamma rays, UV rays, X-rays, and neutrinos into the space. These sudden outbursts also spew away huge crusts of matter into the space, and as expected, these chunks of cosmic debris consist of heavier elements. Some of these chunks are so massive that they would become stars by themselves. These are called *second-generation stars* (the *first-generation stars* are thought to have taken birth in the early universe by the collapse of the gaseous nebulae containing mostly hydrogen and some helium). These second-generation stars already contain heavier elements in their cores apart from hydrogen and helium so that these second-generation stars have a head start in nucleosynthesis, and they can now form much heavier nuclei in their cores. Eventually, some of these second-generation stars may also explode to give rise to *third-generation stars*. Now we can see why some very heavy elements, such as radium and uranium, occur in nature. These heavy elements are formed by the compression of elements in the second-, third-, or even fourth-generation stars where they were subjected to massive gravitational pressures.

Neutron Stars and Pulsars: The story of a supernova does not end here; after all, stars are incorrigible beasts! With the exploded portion of the supernova gone, its core remains as a recalcitrant dense object in space. This small but super-compact body is called a *neutron star*. This miniature star continues to become smaller and smaller, which forces the subatomic matter particles to get nearer and nearer to each other. However, we have seen in Ch 2-D that when the distance between the matter particles decreases, they tend to move faster and faster with increasing velocities, which gives the star an outwards thrust. In other words, the balance of forces inside the star is now maintained between the star's gravity and the *repulsive nuclear forces* of the particles inside its core. Because of the high velocities of the particles, the temperature and pressure inside the miniature star ultimately become so extreme that the protons fuse with electrons, forming neutrons (hence the name 'neutron star'). Neutron stars build up immense gravitational pressures within them—a million-million times greater than that on the earth—and are consequently very hot and very dense. Some of these stars have a radius of just a few kilometres with a density of about hundreds of millions of tons per cm^3—just a spoonful of its matter would weigh about 10 million tons (much more than the weight of the Pyramid of Giza!). No wonder a neutron star is considered the densest known body in the universe!

Neutron stars give rise to strange celestial objects called pulsars, which historically have a curious discovery. In 1967, Jocelyn Bell noticed small bursts of light coming from the deep space at regular intervals, and at that time, she claimed excitedly that they were messages from some distant aliens. In fact, she had named these signals as LGM (for 'little green men')! But soon it was found out that they were coming from very small stars at the end of their lives. Further investigation showed that they were in fact neutron stars spinning fast in space, causing them to appear as pulsations of light. Hence they were called *pulsating stars* or *pulsars*. Pulsars occur because some neutron stars keep spinning (because they retain the original spin of their parent stars), and the angular momentum thus generated by the spin creates intense magnetic fields which sweep across the space, causing their light to be emitted as pulsations of energy (obviously, neutron stars which do not spin remain as neutron stars). Scientists have discovered about 2,000 neutron stars in the Milky Way alone, and a majority of them are detected as pulsars.

Formation of Planets: It is now generally believed that planets arise in the same way as stars. Thus, small nebulae collapse under gravity, but they will not reach 'stardom' because of their smaller mass, and eventually they become *protoplanets* which eventually solidify into various types of planets—some become rocky (e.g. Mercury, Venus, Earth, and Mars), some remain gaseous (e.g. Jupiter and Saturn), and some become icy (e.g. Uranus and Neptune). However, it has also been speculated that some of the planets might have taken origin in the supernova explosions of stars, thus explaining the occurrence of a variety of heavier and heaviest elements. However, there is another explanation for the occurrence of heavy elements in the planets as discussed below.

The nebular gas, as we have seen above, is derived from the disintegration of stars or from exploding stars in the cosmos (Fig 4.2). Thus, stardust is generally rich in a variety of elements ranging from lighter atoms such as hydrogen and helium to heavy elements such as iron and lead and heavier radioactive elements such as radium, uranium, and plutonium that were created in exploded supernovae. And when such nebular gas collapses into a planet, these heavy elements get incorporated in their crusts. Our Mother Earth is perhaps an example of such a planet which harbours an assortment of elements which had indeed made our lives on the planet interesting!

We, the human beings, ourselves are of course made up of a variety of elements, such as carbon, oxygen, calcium, nitrogen, iron, and magnesium. Now it becomes clear to the reader that all these elements were, in fact, synthesized long, long ago in the nuclear furnaces of the dying stars. Thus, human beings may be euphemistically said to be made of *stardust*!

> *To the romantic, we are some 'stardust'; to the*
> *cynic, we are but 'nuclear waste'.*

So be it! Now we will continue with the life cycle of a massive star—yes, we have not yet finished with the story of a dying massive star. Dying stars survive in many ways—like some of our memorable movie star giants!

Section C
Black Holes

The saga of every massive star does not end with a supernova. Some massive stars do not die violently but rather pass off 'peacefully' into the background of the universe by becoming black holes! But then what are black holes? A black hole is a region in space where the gravitational field is so intense that *nothing* can ever escape from its gravitational pull—that is, no matter particle, no object, not even light can get away from the grip of its gravity. It is an abysmal pit in spacetime into which everything nearby would eventually fall but nothing can ever come out. They are the endless holes in the cosmos, the points-of-no-return. Thus, black holes are intriguing structures in the cosmos which are of tremendous interest not only to a scientist but also to a layman, perhaps partly because of their elusive nature in the universe and partly because of the eerie 'doomsday' effect the name casts on the human psyche. We will briefly study the nature of black holes in this section.

Formation of Black Holes: We have seen that when a star burns down all its fuel, it becomes quiescent and consequently loses the outwards push derived from its nuclear reactions to counter gravity. Theoretically, any star may continue to crush its core relentlessly because of the unopposed gravitational force and may collapse eventually on to itself. However, as we will see now, black holes are born only in special circumstances.

Not surprisingly, scientists of the past have predicted the existence of black holes far before their actual discovery. In fact, general relativity has already demonstrated that gigantic gravitational force can bend the course of light rays (Ch 1-E) so much so that they can no longer escape the star, and so the light would bend inwards on to the star itself, and such a region in space would become a black hole. Karl Schwarzschild, in 1916, while studying Einstein's general relativity, revealed a surprising implication that when the mass of a star divided by its radius exceeds a certain critical value (called the *Schwarzschild radius*), the resulting spacetime warp would become so acute that anything, including light, will not be able to escape its gravitational grip. He calculated that a black hole of ten times the mass of the sun would have a radius of just about 30 km. These shrinking stars would remain dark and invisible as they do not transmit any light whatsoever into the surrounding space. These cosmological entities were christened black holes by John Wheeler because they were supposed to appear black owing to the absorption of all light, but black holes remained only as theoretical entities for a long time.

In 1928, Subrahmanyan Chandrasekhar made a landmark observation. He calculated that a star 1.4 times more massive than the sun would have an overwhelming gravity which would ultimately win over the opposing nuclear forces and make the star continue to collapse indefinitely further under its own gravity. This critical mass of 1.4 times the mass of the sun is called the *Chandrasekhar limit* (we have already seen that small stars below this limit would eventually stop contracting and become white dwarfs or neutron stars). Chandrasekhar has suggested that there is an inherent mismatch between the generation of gravitational and nuclear forces in a star because of the following reason: the contracting gravitational force of a massive star continues to increase constantly as the mass of the star increases, whereas the repulsive force of nuclear energy will stop increasing its strength after a certain limit. Pauli's exclusion principle says that as the matter particles come nearer to each other, their velocities increase, giving rise to repulsion, but their velocities cannot exceed the speed of light so that the speed of light itself is the limiting factor of nuclear repulsive force. This hypothetical supposition utterly surprised Sir Eddington at that time because this would theoretically make the star shrink to a zero size, and such an outcome was considered practically impossible at that time! However, in 1939, Robert Oppenheimer, basing on the formulations of general relativity, worked out an explanation of what would happen to massive stars when they die, and he showed that these massive stars did acquire huge gravitational fields in their vicinity so that their escape velocity exceeds that of the speed of light. And finally, such a region in space would become a black hole.

Nothing can escape a black hole—not even light.

But then how are we going to observe and study these elusive structures if even light cannot escape its gravitational pull? Soon it was found out that though the black holes themselves are not detectable by our telescopes, their huge gravitational effects on the surrounding celestial objects mark their presence in the universe. The following example illustrates the point: some of the stars in the universe exist in couples. These twin stars coexist in a close companionship, forming a star system called a *binary star system*. Of course, these binary stars revolve round each other in a regular and predictable fashion because they are gravitationally linked together. However, astronomers have long since observed that some distant stars have an erratic behaviour in their orbits in space which could not be properly explained—they behaved as if they are under the influence of some mysterious force. But later on, astronomers have realized that these erratic stars are not individual stars but were really binary stars in which one is a shining star we observe and the other is a black hole which we don't observe. And by taking into account the invisible effects of gravitation of the black hole on its companion star, the astronomers could now precisely calculate the path of the companion star. We have one example of such a binary system in our vicinity. The *Cygnus X-1* is a binary star which is under the influence of the gravity of a black hole which is about 8.7 times the mass of our sun!

With advances in technology in astronomy, we now have many other astronomical observations that would firmly endorse the presence of black holes in the cosmos, and today, the concept of black holes has become an indisputable part of the Standard Model of the universe. In fact, in the April of 2019, scientists have actually documented a monster black hole found in the galaxy M87, which is about 500 km away from us in the April of 2019, by using a virtual telescope which showed a spectacular dark hole at the centre with a bright halo around it (exactly as envisioned by scientists all these years!).

The Event Horizon: We have seen that a massive star at its end stage cannot sustain itself because of its overpowering gravity, and its core is crushed further and further into an infinitely smaller and smaller region in space. The star would theoretically shrink to a point in spacetime where the gravitational effects would attain *gravitational singularity* (see below). This singularity forms the core of the black hole. Thus, in effect, we can say that surrounding this core of the black hole, there are the zones of transition of decreasing gravitational strengths. The outermost boundary of the black hole where the gravitational force begins to show its effects is called the *event horizon*, and this is where the gravity would actually start bending the light. An object falling into the black hole gains increasing speeds as it passes through these transitions, and eventually, as it approaches the singularity, it attains the speed of light. And finally, the object merges with singularity in the core of the black hole. Thus, a black hole can be envisaged as a huge area of the spacetime whirlpool with circles of increasing gravitational force fields which finally leads to a central core of singularity (Fig 4.3).

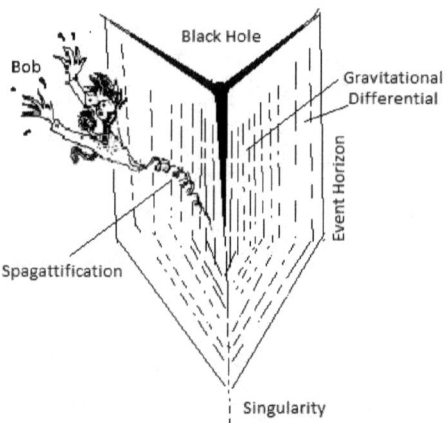

Fig 4.3: Events at a black hole.

But then what is this singularity? The singularity of the black hole is a *one-dimensional point* in spacetime wherein the mass of an object that enters the singularity would become infinite. Its density becomes infinite, and hence, the spacetime curvature at this point becomes infinite. Consequently, at the point of singularity, all the known physical laws of nature would break down completely—a point at which all our physical theories would no longer become applicable and are rendered meaningless.

Singularity is a one-dimensional spacetime entity.

Thus, any intrepid astronaut who dares to enter the influence of a black hole would be dragged inexorably towards its centre with ever-increasing force, and the poor astronaut would ultimately get crushed in its infinitesimally destructive gravitational force. But here is an intriguing affair. Consider this thought experiment: we have a couple of astronauts Bob and Alice trying to study a black hole. Bob decides to enter the event horizon to make a study, and Alice decides to remain at a distance to make a record of the events that take place as Bob enters the black hole. General relativity predicts that to an outside observer (Alice) who is looking at an object (or in our case, Bob) falling into a black hole, the events come to a halt because time stands still at the event horizon. This means Alice would observe that as Bob approaches the black hole, he moves slower and slower, and finally it would appear to Alice that Bob has stopped at the event horizon for an indefinite period of time without ever actually falling into the black hole. In other words, it can be said that for an observer outside the black hole, the ticking of time has stopped completely at the event horizon, and thus, the actual doomsday events of Bob could never be witnessed by Alice.

However, for Bob, the situation is different. For him, time moves on normally until he gets nearer to the event horizon, and then he gets dragged into the black hole to meet his fate. The events of this fatal escapade of Bob can be mercilessly envisioned in theory. Bob, as he travels from the event horizon to the core of the black hole, would become infinitely elongated and stretched out like a string of spaghetti—or in short, he gets *spaghettified*. This needs an explanation. The reader may note that any object lying in a gravitational field exhibits a 'gravitational differential' owing to the object's spatial orientation. For example, as you stand on the surface of the earth (as we do daily!), the head portion of your body experiences a tad weaker pull towards the earth than the leg portion (because gravitational pull of the earth slackens as distance increases). However, this gravitational differential is of no consequence at all on the earth not only because of the weaker gravitational force of the earth but also because of the huge difference in the mass of the two objects (earth and man) involved. But then the gravitational pull in the case of a black hole is exceedingly strong at the event horizon, and it becomes stronger and stronger as it approaches the central singularity (Fig 4.3). Thus, this exponentially strong and increasing gravitational field of a black hole creates a superlative gravitational differential with a tremendous difference in the gravitational effects on the two ends of the object (or in our case, between the head and feet of Bob). This would make our Bob twist and turn and elongate infinitely (or simply, spaghettified) as he travels from the surface to the core! Really, what a damning fate of Bob that could be!

But don't we have a silver lining in this singularly terrifying scenario of a black hole? Yes, of course, we do have! A human being, who chooses to utilize the black hole time magic, may actually become immortal! Consider this: an astronaut spending his/her time in the vicinity of the event horizon of a black hole (taking care not to actually enter it!) for about one earth-year. When he/she returns to the earth, he/she would find that for the earthlings, a time period would have elapsed several thousands of years! But the astronaut would find himself/herself as young as he was when he/she had departed. This is because time for the astronaut would have slowed down inexorably at the black hole, making him/her ageless for that period. Thus, theoretically, black holes can be used as 'time machines' which would enable us to explore the bygone past or the forthcoming future. No wonder, these exotic theoretical possibilities may unleash some fertile ideas in the minds of movie entrepreneurs who could so lavishly portray these unrealistically realistic time machines in their sci-fi movies.

Now coming back to our affairs with black holes, they are shown to emit a large amount of X-rays from their surface which can now be detected by our modern telescopes. The source of this intense radiation is found out to be from the matter which is being rapidly dragged into the black hole. For example, it has been demonstrated that in a 'binary-star-black-hole system' (as described above), the black hole sucks in the matter from the companion star, and the matter gets drawn in a spiral fashion. And as the matter gains increasing speeds, it gets heated up intensely, and this intense heat is responsible for the emission of high-energy X-rays. This spiralling of massive amounts of matter around a black hole would make the black hole whirl rapidly, which, in turn, would create an intense gravitational field along the axis of its rotation—just like the earth's rotation creating magnetic fields along the north and south poles. These rotating magnetic fields would direct the collapsing matter particles to be ejected in jets, and these jets would be seen as spewing jets of matter and radiation along the axis of its rotation. Such jets are indeed discovered near a number of black holes. In fact, the presence of these jets would strongly indicate the presence of black holes in the vicinity. Of course, the black hole would eventually engulf the companion star completely, and then it becomes quiescent.

In 1974, physicist Stephen Hawking proposed that just outside the event horizon, the subatomic particle pairs (photons, neutrinos, and some massive particles, Ch 3-C), as they fall into the event horizon, split into particles with positive and negative energies. The particles with negative energy fall into the singularity and finally disappear, and the other particles escape out and are reflected as a burst of radiation called *Hawking radiation*. It is thought that this radiation is emitted just outside the event horizon because black holes themselves cannot radiate any energy. So far, astronomers could not discover any such radiation, but many researchers are hopeful that they could find some evidence of Hawking radiation in the coming times.

Types of Black Holes: There are two types of black holes—rotating and non-rotating black holes. Rotating black holes (called *Kerr black holes*) are the commonest in the universe and are formed when a spinning star becomes a black hole. These rotating black holes bulge out and take oblong shapes owing to the angular momentum they experience. A non-rotating black hole (called *Schwarzschild black hole*) is a rare occurrence in the cosmos, but such a black hole is mathematically shown to exist in a perfect spherical shape.

Black holes vary in their sizes. Some black holes are massive. For example, Sagittarius A, which is at the centre of our Milky Way, weighs about two and a half million times that of our sun; some are 'supermassive black holes' which weigh billions of times the mass of the sun!

Stars may not be the only source of black holes. It appears that there is a non-stellar origin of a black hole where a massive nebula may collapse on to itself and become a black hole without forming a real star; perhaps Sagittarius A has such an origin. Stephen Hawking proposed the existence of another source of black holes. According to him, some of the black holes, called *primordial black holes*, are generated at the time of the big bang. They are thought to be very tiny in their masses and would have lost their mass sooner by the way of Hawking radiation and would have disintegrated and disappeared into the background of the universe. He further speculated that the number of these tiny black holes may far exceed the number of visible stars in the universe and may even contribute to the bulk of the mysterious 'dark matter' (Ch 5-H) of the universe!

Now consider an eerie question: do we find any black holes on the earth? Speculations are rife that such simulated forms of primordial black holes are generated in our particle accelerators such as our LHC (Ch 3-C). Some scientists have actually opposed to the construction of massive particle accelerators, fearing that these artificial black holes may have some immediate diabolical effects on our existence. These scientists have speculated that these black holes on the earth may trigger monstrous earthquakes or may cause some astronomical disturbances or may even engulf the earth eventually! But then the evidence is to the contrary, and it appears that these tiny black holes, even if they were actually effectively generated in our accelerators, have not caused any discernible effects so far.

Another type of black hole is presumed to exist. We have seen that a black hole has a singularity at its core with layers of differential gravity and with the outermost layer comprising of an event horizon. However, theoretically, a 'naked singularity' is predictable, in which the gravitational singularity has no such event horizon. But so far, no such naked black hole is observed in nature. And it is speculated that the big bang had started off in this sort of naked singularity at the time of the origin of the universe.

So far, we have seen some of the most outstanding structures of the known universe. But then the universe is a mysterious place. The celestial objects described so far are only the known constituents of the cosmos. It is certain that the universe harbours many more interesting structures which are awaiting our discovery! Moreover, we have no proper understanding of what actually exists *outside* our universe. There are many theoretical sketches and speculations in this regard, which we will discuss in Sec E. But before going into those smudgy theories, we will take a look at our sun and see what exactly is happening inside it.

SECTION D
The Sun and Its Structure

Our sun, a prototype of medium-sized, middle-aged star, is a massive gaseous ball of fire which is presently in the main sequence phase. The sun is considered a third-generation (or perhaps a fourth-generation) star. The age of the sun, according to the recent estimates, is about 4.5 to 5 billion years, and it would live for another 4.5 billion years (Ch 5-F). As it turns old and senescent, the sun becomes a red giant (Sec-B), and then it would swell to such an extent that it would possibly engulf the nearby planets, including the earth. But that should happen only in another 4 billion years from now. So that fateful event should not worry us more than the self-destruction mode we are in by the way of ecological destruction and nuclear warfare!

In the early times, the sun was considered a perfect orb with no imperfections, but we now know that the sun has many spots on it along with a significant equatorial bulge. The sun is the most massive body in the solar system with a mean radius of 695,500 km and a mass of about 1.98×10^{30} kg. No wonder, 99.8% of the mass of the solar system is contributed by the sun! It is composed chiefly of hydrogen (74%), helium (24%), and oxygen (1%); and the remaining 1% is contributed by iron, nickel, silicon, sulphur, magnesium, carbon, neon, calcium, chromium, etc.

Structure of the Sun: The sun has a dense *core* at its centre which is exceedingly turbulent with relentless nuclear reactions (the reader is advised to look at the details supplied in Fig 4.4 as he/she keeps reading the following account). These nuclear reactions release colossal amounts of energy (see below), generating tremendous amounts of temperatures and pressures. It is estimated that the sun's core converts about 4 million tons of mass into energy each second,[*]

---

[*]   Our sun is so enormously massive that it has a store of hydrogen to the tune of 2 ×

taking the core temperatures to about 15,700,000 K with pressures reaching up to 150 times the density of water! These astonishing amounts of energies inside the core are responsible for keeping the sun alive without collapsing under its own gravity as noted in Sec-B.

The core is surrounded by several layers of hot and dense gas with temperatures decreasing from inside to outside. The immediate layer covering the core is the *radiative zone* (Fig 4.4), and this is the thickest layer of the sun which spans for about 300,000 km. This is the conducting medium of the sun in the sense that the high-energy gamma rays that are released in the core (see below) are absorbed and re-emitted repeatedly in the radiative zone.

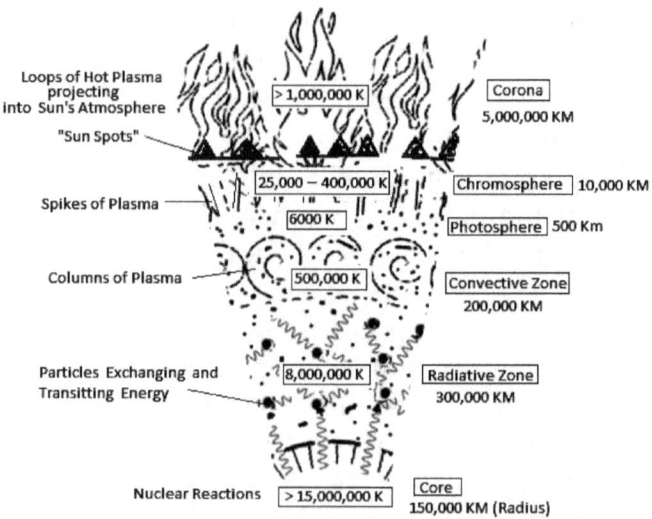

Fig 4.4: The sun, its layers, and other details.

The next layer is the *convective zone*. Here, the process of convection of energy takes place—the heated gas from the radiative zone comes to the surface in great columns of hot plasma, transmitting energy to the outer regions, and as it cools, it dips down again into the depths, only to get reheated and then to come to the surface once again. This process of convection continues in a cyclic manner (Fig 4.4). Each column of convection spreads to about 1000 km, and each cycle lasts

---

10^{27} tons! Thus, even with this gigantic expenditure of fuel, the sun has sufficient hydrogen to sustain its nuclear reactions for another 4.5 to 5 billion years!

for about 8–20 minutes. In fact, these columns make the surface of the sun look granular (when you view the sun using solar filters).

The layer next to the convective zone is the *photosphere*, which is comparatively thin. In this layer, the temperatures become fluctuant, resulting in some cooler areas (6,000 K). These cooler areas appear darker and are viewed by us as *sunspots*. The outermost layer of the sun is the *chromosphere*, which is an immensely dynamic layer with many turbulent spikes of plasma ejecting out into the sun's atmosphere. An interesting feature here is that the temperatures within this layer start ascending again to 25,000 to 400,000 K, whereas the photosphere has much lower temperatures—a really intriguing thermodynamic phenomenon (see discussion below).

Outside the chromosphere is the *corona*, which is the 'atmosphere' of the sun. The corona consists of loops of hot plasma ejected into the sun's surrounding space, which look like gigantic tentacles (Fig 4.4) outside its surface which may spread over for several hundreds of thousands of kilometres (extending sometimes into millions of kilometres above the photosphere). The corona is the portion of the sun that we see during a total solar eclipse. The temperatures here reach to stupendously huge proportions—reaching temperatures greater than 1,000,000 K, which makes the sun a blazing hot star! The actual mechanism of reheating of the chromosphere and corona to reach these superlative temperatures is still an enigma yet to be solved.

The reader may notice a paradoxical phenomenon of temperature distribution in the sun. We have seen that the temperature inside the sun keeps dropping as we go from the core to the surface. But in the atmosphere of the sun (which is its very surface), the reverse happens—whereas the temperature of the photosphere is in thousands of Kelvin, the corona suddenly get heated up to several millions of Kelvin. Now consider this paradox: if the corona is so hot, the heat from here must transfer back to the photosphere because, as per the laws of thermodynamics (Ch 6-B), heat must always flow from higher to lower temperatures. However, in reality, the solar energy flows out into the surrounding outer space, and this process is thought to be thermodynamically intriguing. To explain this phenomenon, some scientists have suggested that the corona's energy is indeed transported back to the interior of the sun, but it is thought that this energy is soon dissipated into the turbulence of the inner layers as it works its way down towards the core. Of course, there are many other explanations to explain this paradoxical phenomenon, but none of them have been met with the total satisfaction of scientists.

Generation of Sunlight: The sun emits all ranges of electromagnetic radiation ranging from gamma rays to infrared rays (see black-body radiation in Ch 2-C). However, it must be noted that the gamma rays liberated by the nuclear reactions at its core are converted into much less powerful rays as they traverse the thickness of its layers so that the sunlight that is finally emitted chiefly consists of visible light, infrared radiation, and some amount of UV rays. Apart from this intense electromagnetic radiation, the sun ejects some matter into the surrounding space in the form of *solar flares*, which are sudden massive ejections of plasma from the photosphere. These solar flares are generally accompanied by the spewing up of subatomic particles, such as small ions, protons, and electrons out into the surrounding space. The solar flares ultimately escape the sun's vicinity in the form of hot gaseous material to become *solar wind* (see 'interplanetary space' in Sec-B). These ejections cause intense temperature fluctuations within the sun which, in turn, create shifts of huge magnetic flux on the solar surface. These magnetic fluctuations are sometimes so intense that they become dangerous to the nearby planets (including the earth) because they may also interfere with the planets' magnetic fields. However, it must be noted that our earth has a mighty shield around it in the form of a thick atmosphere which can, at least partly, mitigate the effects of these solar storms.

Nuclear Fusion in the Sun: Now we will go into the actual details of the mechanism of the generation of nuclear energy inside the sun, which had remained a mystery for many years. In 1929, Fritz Houtermans and Robert Atkinson described their ingenious theory of nuclear fusion as a mechanism of energy production in the sun.

Nuclear fusion is a nuclear reaction involving the fusion of hydrogen nuclei into the helium nucleus. Nuclear fusion is not a thermodynamically spontaneous reaction (Ch 6-B). It needs tremendous gravitational pressures created by the mass of the sun to initiate the required nuclear collisions, which, in turn, create very high temperatures in its core. And again, these high temperatures are responsible for sustaining the nuclear chain reactions that follow. The sun is abundant in hydrogen, and thus, it uses hydrogen as its fuel. Hence, the nuclear reactions in the sun mostly involve hydrogen fusion, and this process is technically called *proton–proton fusion reactions* (because the nucleus of hydrogen is nothing but a proton). Now consider this: the protons bear positive charge on them, so when two protons are brought together, they repel each other. However, they can be brought together when they are subjected to high pressures, and the sun's mass could generate sufficient gravitational pressures in its core to initiate this fusion of protons. Houtermans found out that when this proton–proton proximity reaches a critical distance of 10^{-15} meters, the electrostatic forces undergo 'quantum tunnelling' (Ch 2-E), which assists the nuclei to fuse together, thus initiating the proton–proton fusion reaction. Fusion reaction may be studied in three steps.

Step 1. The first step in nuclear fusion is the fusion of two protons into a nucleus of deuterium. Deuterium has one proton and one neutron. This means that one of the two protons has to undergo *transmutation* into a neutron (Fig 4.5). This initial step deserves some description as it is the most crucial step in the fusion reaction. Here the reader may review the Standard Model presented in Ch 3-C, which come in handy in the following discussion. A proton has three quarks (uud), out of which one of the up-quarks undergoes a weak interaction by emitting a W^+ particle and gets converted into a down-quark. Thus, this proton has now transmuted into a neutron (udd). As a neutron has no electromagnetic charge, it can now readily collapse into a proton, giving us a 'deuteron' (i.e. deuterium nucleus). However, the released W^+ particle is quite unstable and soon disintegrates into a positron (e^+) and an electron neutrino (v_e). This process is called *beta plus decay* (Ch 3-D and Ch 9-C). These released positrons soon come in contact with ambient electrons (e^-) and annihilate themselves into gamma rays (Ch 9-C). The fate of the released neutrinos is described below.

Step 2. The deuterium nucleus fuses with an ambient proton to form helium-3 (^{3}He). During this process, another gamma ray photon is released (Fig 4.5).

Step 3. Two helium-3 nuclei fuse together to form one nucleus of beryllium-6, but this isotope of beryllium is highly unstable and soon disintegrates into one helium-4 (^{4}He) nucleus and two protons. These released protons are recycled into the proton–proton cycle (Fig 4.5). Thus, the end product of nuclear fusion in the sun is the formation of a helium nucleus concomitantly liberating gamma rays and neutrinos.

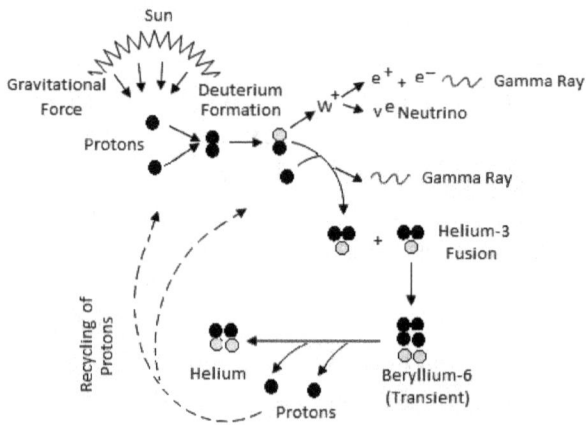

Fig 4.5: Proton–proton fusion nuclear reaction.

The released neutrinos from the core do not interact with anything in ambience and move unhindered (Ch 3-C). They travel at almost the speed of light and reach the surface of the sun in just about 2 seconds and escape into the space from there. However, the story of gamma ray photons is different: the transmission of gamma ray photons from the core of the sun to the surface is a very slow process, and it is estimated that for a gamma ray photon to travel a distance of about 0.7 million km from core to surface, it would take nearly 200,000 years! The reason for this tremendous lag is that the journey of these gamma ray photons to the surface is not a smooth sail as that of neutrinos but rather a zigzag one. This is because the gamma ray photons generated in the core hit the various subatomic particles over and over again as they traverse the whole thickness of the radiative zone (Fig 4.4), which makes their course irregular. But even more importantly, as a consequence of this zigzag travel, the gamma ray photons lose their energy and become less intense with longer wavelengths as they ascend up to the surface. And finally, at the surface, the sun ejects photons of much gentler light (mostly in the visible range) into the photosphere. However, the reader may note that once the photons reach the corona, the photons take only 8.3 minutes to cover a distance of 150 million km to reach the earth! And blissfully, we receive mostly the gentler white light from the sun's surface instead of the highly dangerous gamma rays.

The reader may note that in these proton–proton fusion reactions, some amount of mass invariably disappears and is converted into energy. It is shown that the combined mass of four nuclei of hydrogen (i.e. protons) is more than the total mass of the helium nucleus, and it is estimated that for each kilogram of hydrogen that is processed into helium, about 0.007 kg of matter is converted into energy. If we go by the equation $E = mc^2$, we can say that the energy liberated from this conversion would be enormous: $0.007 \times (3 \times 10^8)^2 = 6.3 \times 10^{14}$ joules of energy!

In summary, the reader may realize that the ideas presented so far in the above sections are more or less well-documented facts. We will now tread into the unknown worlds where we are not so sure-footed in our speculations.

Section E
The Newer Worlds

We will now come back to the questions we have posed at the beginning of the chapter: How big is the universe? Is the universe finite or infinite? And are there any other universes apart from our own? In the sections above, we have examined many of the known facts of the universe and discussed many of the events that occur in the far and farthest reaches of the universe. But then we are no better now than ever in answering these fundamental cosmological questions. However, scientific speculations related to these questions are rife. Though some of the ideas are based on the known facts of the universe, many of the propositions have flimsy scientific footing with no solid factual evidence. Here we will study the following three topics of some importance to us at present: (1) parallel universes, (2) wormholes, and (3) extraterrestrial life. Some of the 'facts' presented below are the products of fertile imagination and the guarded guesswork of many scientists, including some of the leading figures in science. But of course, we must not slight off these 'guesstimates' (= guesswork + estimates) of our scholars because such ideas have been proved, time and again in the history of science, to be the torchbearers of scientific progress!

Parallel Universes and the Multiverse Theory: Are there any other universes other than our own? If so what do they look like? We do not really know! Our answers, however, may range from some amusing tales of folklore to some religious accounts of parallel worlds (almost all religions are replete with stories of the other worlds) but none with any credible scientific facts. However, here we will try to conduct a brief survey on this subject.

The crux of problem with the idea of the existence of parallel universes, as can be guessed, is our lack of knowledge of the extent of our own universe. We do not know if the universe is finite in its extent or infinite. A finite universe, as we can see, could be imagined to be localized, and we may then think that our finite universe is contained at some place in the vast universe, and that is all we have anything out there! On the other hand, the idea of an infinite universe has intrigued the scientists even more with many of its exciting possibilities, and this idea, in general, is in favour of most scientists now. We will discuss an infinite universe here with some of its infinite possibilities.

We have known that our universe has started in a big bang, and it is now expanding. In fact, the universe is now shown to be expanding very rapidly in its outer edges (Ch 5-H). But then we do not know what it would be like outside the edge of the universe. If there is an edge, there must be something beyond, and if something exists beyond, then that would not qualify to be called an edge? What a theoretical impasse! However, it is a known fact that infinity is an issue which cannot be imagined by the human mental faculties

(Ch 11-C). To avoid this perplexity, some scientists have suggested that the ever-expanding universe would break up into an infinite number of finite bubbles at its periphery, each bubble segregating from the parent universe and becoming a universe on its own, perhaps with a different set of physical laws operating in them.

Alan Guth and Andrei Linde proposed a *chaotic inflationary model* in which there was a slow breaking of the symmetry of fundamental forces (Ch 5-G) with the formation of *bubble universes* from the background inflation. Each bubble universe will have its own timeline, its own history, its own laws governing the local events of the universe, and its own big bang and big crunch (Ch 5-I)—and our universe is one such bubble! These bubbles are surrounded by the general universe, which is thought to be full of undischarged energy, and it has the potential to keep expanding and forming new big bangs and new bubbles on a continuous basis. This is the *multiverse theory* (Ch 11-A). It is thought that in a multiverse, the number of possible bubble universes is infinite. Some researchers have calculated as many as 10^{500} possible worlds in existence, each with a different set of laws, perhaps with different physical constants operating in them, consequently with different sizes and shapes of forces, particles, and atoms. In fact, the string theory (Ch 3-I), with all its manifold dimensions, gave a fillip to the possibility of such a multiverse.

All these hypothetical possibilities are fine, but is there any evidence at our disposal to prove the multiverse theory? In 2004, NASA detected a 'cold spot'—or a 'super-void'—in the outermost regions of the universe (which was later confirmed by the European Space Agency's Planck mission in 2013). There has been a flutter of excitement with this discovery among the scientists because this super-void was hard to explain by using the current model of big bang explosion. The only other explanation (or so it was claimed!) that could be offered by the scientists is that of the collision of a parallel universe with our own universe—something like a soap bubble colliding into another bubble! However, some researchers have claimed that the cold spot is an optical illusion created by gravitational lensing (Ch 5-H). But this explanation, by itself, has turned out to be a lame duck! Whatever these claims are, the reader may note that the evidence for (or against) a parallel universe, at present, is flimsy at best.

It is worth noting here that there is a theoretical impasse in our investigation into parallel universes. In Ch 5-H, we will see that the periphery of the universe is moving away from us at a pace far greater than the speed of light, but because our cosmic vision is restricted to the speed of light itself, this renders our observation beyond the edge of universe utterly impossible! Nevertheless, this theoretical hurdle may someday be circumvented by us by using a hypothetical concept called wormholes! But then what are these wormholes?

Wormholes as Time Machines: In the above discussion, we have seen that our cosmic vision is severely restricted by the speed of light, and this has been our chief impediment to explore our universe to the end and beyond. However, the scientists are optimistic in the idea of covering these infinite distances in the shortest periods of time by utilizing our theoretical knowledge of black holes and turning them into time machines! Here we will briefly see how this is possible.

We have seen in general relativity that gravity is a distortion of the spacetime fabric, and we have also seen that light takes a bent course in the vicinity of strong gravitational fields (Ch 1-E and 1-F). While discussing black holes (Sec-C), we have stated that if the spacetime distortion is severe enough, then a light ray, which has trespassed into that region of spacetime, would acutely bend upon itself, and it then fails to escape from the distorted region at all. We also have seen that time would slow down as an object approaches this distorted region and would eventually stop altogether at the black hole.

Now consider this: if the spacetime warpage happens beyond a critical limit, then would it really cause a 'hole' in the fabric of spacetime? Perhaps yes! And if two black holes far apart in the universe come together (Fig 4.6), then that would theoretically create a 'bridge in spacetime' (the so called *Einstein–Rosen bridges*) which may allow us to slip from one region of the universe into another region of the universe within the shortest period of time!

It is suggested that at one end of this spacetime conduit, there is a black hole, and at the other end, there would be a *white hole* (an essential reverse of black hole which would keep nothing but emit everything out)—so that we are spaghettified as we enter the black hole, and as we emerge from a white hole, we will reassemble again! This hypothetical tunnel in spacetime is called a wormhole, and this would prospectively curtail the time taken for space travel. A wormhole may be imagined by folding a paper with two black holes with a connecting tunnel in between (Fig 4.6). Though wormholes were suggested by John Wheeler way back in 1957, these theoretical conduits of spacetime are still in the realm of science fiction.

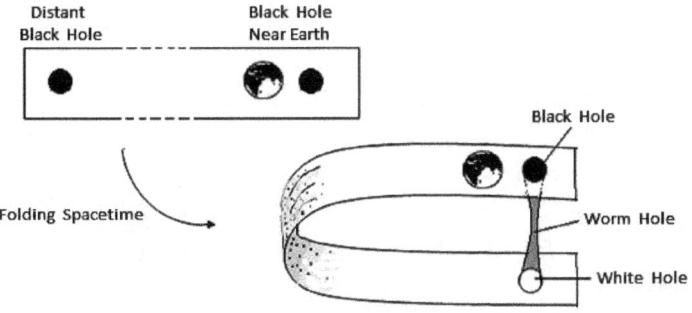

Fig 4.6: A wormhole.

Some scientists have supposed that wormholes open up transiently and close down spontaneously all the time in the universe. Whatever these presuppositions are, the reader may now see that these wormholes may be used in the future as time machines to circumvent the speed limit of light in our intergalactic communication and exploration.

Extraterrestrial Life: No discussion on the universe is complete without touching upon the subject of extraterrestrial life! However, to some of us, the idea of 'aliens' presupposes the fantastic notion that 'out there', there are certain exotic and weird 'beings' who are supposed to be far more intelligent than us and are somehow endowed with some unexplainable supernatural powers (remember Spielberg's *E.T. the Extra-Terrestrial* or *War of the Worlds*!), and we somehow 'cherish' the idea that they always try to invade the earth and harm the human race for some unrealistic gains. And what more, we, in excitement or in some sort of difficult-to-understand feeling, dote on the 'thrill' of this alien invasion (the excitement of UFO sightings is an example)! This has been the recurring theme of many sci-fi movies across the decades of the movie industry. But none of these imaginations have any rigorous scientific foundation, and the subject of extraterrestrial life can be discussed without attaching any fantasy to it at all. But don't worry. To a science enthusiast, such a 'spiritless' presentation would not necessarily make the idea drab. In fact, understanding science in the right perspective actually boosts up a science reader and makes it rather more exciting than fiction! And in any case, here we are doing some real scientific exploration on the subject.

The term *extraterrestrial life* encompasses any form of life that is likely to exist anywhere in the universe but outside our earth (*extra-terrestrial* = outside-earth). And before going further, the reader must realize a singular fact that we, the human beings, have no knowledge of *any* sort of cellular life other than that exists on the earth. In short, we have one, and only one, 'sample' of life form in the universe for us to study and none other to compare with! This feature makes the idea of extraterrestrial life a hypothesis with a severely restricted speculative value. But then some scientific guesstimates have shown us that the possibility of existence of life outside the earth is almost certain. Of course, these life forms that exist outside our earth may or may not resemble our cellular life on the earth. In short, life in any form is likely to occur somewhere in the universe. Or in other words, the grand idea of present-day understanding is that 'life is a cosmic imperative' (as the biologist Christian de Duve had put it once), meaning that life is a universal phenomenon!

There is a general term that is connected to the subject of origin of life in the universe—*panspermia*. Panspermia connotes a different, albeit an extravagant, idea that life has originated somewhere in the early stages of the universe (in the form of certain exotic microorganisms or even perhaps some 'nano-organisms'), and subsequently, these early life forms were 'dispersed' and 'seeded' into many regions wide across the cosmos aided by the way of scattering asteroids, meteorites, or some sort of exotic 'space vehicles'—and one such seeded region is our Mother Earth! By this we mean that all life forms in the universe have a single origin and thus have some gross features similar to what we are made up of. Of course, all these ideas are conjectures without a proper validation as noted above.

There is another important concept in this connection which is, in fact, more scientifically oriented than the above concepts. *Abiogenesis* is a concept which assumes that life has started *de novo* on the earth from inorganic substrates. The reader may see that the only way for our scientists to prove this is to 'synthesize' cellular life on the earth starting from scratch in a lab! Lots of research is going on in this direction but with a limited success. Anyway, this subject is outside the scope of this book; hence, we will not go into these issues here.

Coming back to life in the universe, we may summarize that alien life in any form may occur in any of the various planets of our solar system or in our Milky Way or in the Andromeda Galaxy or way beyond our local galaxies anywhere in the vast expanses of the universe. And the extraterrestrial life could take any life form ranging from a mysterious cellular microorganism to an undefinable nano-organism to a grotesque superintelligent 'humanoid'—or anywhere in between.

The search for life outside the earth has started along with developments in astronomy. The most important lead to look out for life has been the search for water. Water has a unique place in the development of life on the earth, and it is supposed that all life in the universe is water dependent, and hence, the evidence of water on a planet is thought to carry a strong possibility for the existence of life (at least sometime in the planet's past history). But it should be noted that the earth is the only planet we have known so far in the solar system which is loaded with large terrestrial repositories of water in *liquid* state, which is so essential for life's sustenance. Current astronomical observations have identified many planets and large satellites in the solar system with water in some form, and all these bodies *may* lodge life (or might have lodged life in the past) in some form, such prospective candidates being Mars, Venus, Europa, Ganymede (moons of Jupiter), Enceladus, Titan (moons of Saturn), etc. And thus, these planets are the candidates not only for extensive search for life but even for an exuberant science fiction.

Nevertheless, the scientific reader may consider the fact that there is still a possibility that certain totally different forms of life could exist in some place in the universe which are based not on water but on molecules, such as ammonia, methane, or other molecules.

Another important prerequisite for life is the presence of many elements that constitute life (such as carbon, hydrogen, oxygen, nitrogen, and phosphorus) in right proportions and at suitable atmospheric pressures and temperatures. It is proposed that many stars may have zones around them which are likely to possess planets in them which harbour the right conditions for the life to originate. These hypothetical zones are called *life zones* (or also called *habitable zones* or *Goldilocks zones*—'not too hot, not too cold, but just right!' as the Goldilocks girl finds her porridge in the *Goldilocks and the Three Bears*). Of course, our earth, at present, is in the Goldilocks zone!

It can be said that it is entirely rational to believe that there *is* extraterrestrial life in the universe. Consider this: there are billions of trillions of stars in the universe, with each star with at least one planet on average (even by the most conservative estimate) and with at least some of them in their Goldilocks zones with appropriate chemistry to start life in them—with all these facts, it is statistically possible that life does certainly exist at many places in the cosmos. One estimate stretches this figure up to 200 million habitable planets in the Milky Way alone! But then could there be some sort of *intelligent* life existing somewhere else in the cosmos? Yes, this must also be statistically possible! To suppose that there could be no intelligent being in the universe other than the human race is utterly egocentric and severely short-sighted.

Even if there are intelligent beings out there, could they really communicate with us, or as some proclaim, did they already have? Given the great interstellar and intergalactic distances and considering the speed limit of possible communication to the speed of light (the speed of light at 299,792,458 m/sec may be a superlatively fast thing—but only from a human standpoint; this speed would become begrudgingly small and inadequate when it has to travel for billions and trillions of light years to cover distances across the length and breadth of this gigantic universe), the possibility of an alien contact by the human race can be said to be almost negligible or even non-existent. Unless, of course, an advanced alien civilization has circumvented the delay in their travel time by journeying through wormholes!

Anyway, we have one international agency to look out for aliens: SETI (acronym for *Search for Extraterrestrial Intelligence*) is an ambitious organization for the search of advanced civilizations in the universe. It attempts to continuously monitor the skies for some sort of exotic interstellar electromagnetic messages from some highly advanced alien civilizations. Though, thus far, there has been no actual success, the researchers are on a constant watch for some weird signals. Of course, there have been some false alarms over a period of decades. Way back in 1896, Nikola Tesla thought he had received signals from planet Mars, but a detailed analysis of the data later showed its fallacies. As another example of false alarm, we have already seen in Sec-B that in 1967, Jocelyn Bell had named bursts of light from deep space as 'little green men', only to rename them as pulsars after investigation. However, in 1977, Jerry Ehman noted a strong narrowband signal received by the 'Big Ear' telescope at Ohio State University which was interpreted to contain some queer communication signatures and thought that it could have come from the Sagittarius constellation (and he, in excitement, quickly made a red-ink note of 'Wow!' by the side of the signal). Soon this 'Wow! signal' became popular both in media and with researchers. Many explanations were offered ranging from natural to man-made sources—but none satisfactory—and by far, this remains the strongest candidate of a possible ET communiqué, if at all!

And interestingly enough, in the past half-century, several space agencies have also sent messages *from* the earth *into* the deep space (they go by the general term *IRM* for *interstellar radio messages*). They are directed at various constellations in the deep space in the prospect of some alien species coming to know of our existence on the earth (e.g. the 'Arecibo message' sent out in 1974). Many of the leading scientists, who have believed in ET intelligence (based on statistical probabilities), have raised concerns over the wisdom of this unsolicited correspondence, and Prof. Hawking had even dutifully warned us to stay away from the aliens!

And of course, UFOs (the so-called flying saucers) have stayed with us as a part of our popular culture for a long time. Some authors (such as Erich von Däniken and Zecharia Sitchin) have even cited evidence for alien visitations to the earth in the bygone past. But mysteries apart, the subject of extraterrestrial life is a deep enigma and is likely to remain so for much more time!

We will conclude this chapter by asking this question: so far, we have learnt about the vastness of our universe, but from where did all this vastness popped up into existence? To answer this question, we will study the essentials of the big bang in the next chapter.

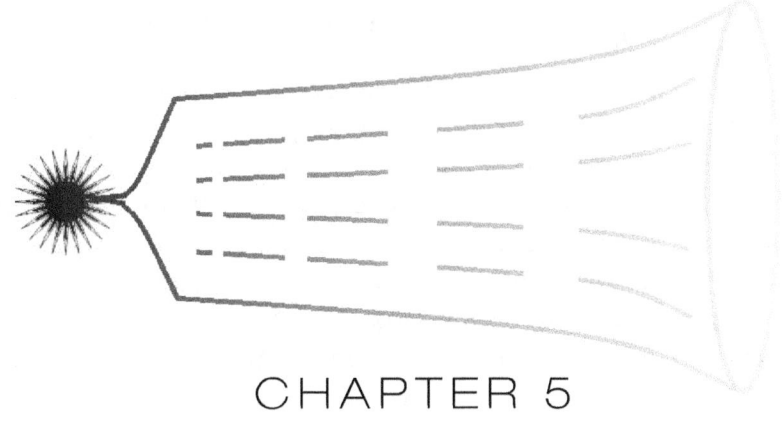

CHAPTER 5

The Big Bang
And the Ever-Expanding Universe

Overview

So far, we have learnt about atoms, we have studied stars and planets, and we have discussed about the laws that operate the world of the small and big. But then how is our universe born in the first place, and what is its date of birth? Of course, this question is not a recent one. It has bothered the ancient religious folk as vehemently as it has troubled the best scientific minds of the twentieth century. The evolution of a scientific answer to the question of birth and development of the cosmos is encompassed in the great saga of 'cosmology', as this branch of science is now called.

To begin with, we will examine the models of the universe in a historical perspective, and then we will see how several scientists, down the decades in the twentieth century, have contributed to the development of the idea of an expanding universe which, in turn, has led to the big bang theory. We will see how the other competing models of the universe were snubbed away in the wake of an avalanche of evidence that has piled up in support of the big bang. In the subsequent sections, the reader will see that the science of cosmology has finally given us a date of creation, and this has allowed us to categorically look at the timeline of events from the big bang to the present day.

In the final sections, we will learn that the big bang theory, though the pinnacle of collective human genius, by itself is not a complete model but it has opened up a Pandora's box of even more fundamental questions. In this connection, we will also learn about some sort of unknown dark matter and dark energy which appear to utterly dominate the universe, and these findings allow us to realize the inadequacy of our present-day knowledge about the universe. At the end, we will speculate over the fate of the universe itself and see the many possible ways by which it may end.

With this brief note, we will jump into the maelstrom of cosmic controversies!

Section A
Models of The Universe

It was traditionally believed by the ancients that the earth is stationary and fixed at the centre of the universe. This ancient *geocentric model* had survived throughout the medieval period of human history because it was very obvious to our observation that the earth we live in is steady, and the moon, the sun, and the other celestial objects move in relation to it. The medieval church also concurred with this geocentric model simply because it was in agreement with the old scriptures which had divinely placed man at the centre of the universe, and the church has promptly decreed this notion into an indomitable religious dogma and indicted whoever said anything against it with its manifold curses.

But then the ancient stargazers did realize that the observed movements of the celestial objects were erratic and haphazard. The paths taken by many celestial objects could not be explained by some simple circular orbits around the earth—the celestial objects had to take some really complicated paths in order to account for their movements. Many thinkers of the past, such as Pythagoras, Aristarchus, Eratosthenes, and Ptolemy, to name a few, have realized this. But the dogma of the earth-centred model of the universe remained so rigid on their minds that they had tried to explain these erratic orbits by some complicated paths (e.g. Ptolemy's epicycles) instead of searching for some simpler explanations.

Nevertheless, despite their rigid dogmatic adherence to old views, these ancient thinkers did contribute significantly to our present-day understanding of the universe, either directly or indirectly, by making some remarkable observations and calculations (e.g. Pythagoras theorem, Eratosthenes's calculation of the earth's circumference, Aristarchus's measurement of the distance to the sun, only to name a few). However, gradually, over centuries, rational thinking of men has prevailed over the dogma, and this has ultimately brought down the mystic 'astrological' speculations and replaced them with 'astronomical' certitudes, and subsequent advancements in science and technology have eventually led to giant leaps in astronomy, as we will see in this chapter. Not only that, but these developments have culminated in the beginning of an exciting and elegant branch of modern science called *cosmology*, which deals with the origin and fate of the universe, as we will see in this chapter.

Astronomy as Science: The history of astronomy as science has really begun with Nicolaus Copernicus (1473–1543) when he had outrightly denounced the imprecise geocentric model with a more rational and simple sun-centred model, called the *heliocentric model* (*Helios* was the Greek sun god). It was perhaps the first evidence-based understanding of the universe. But at that time, scientific progress was stymied by the undue interference of religion with reason, and thus, Copernicus was promptly indicted by the church, and this model was thrown into oblivion.

However, science progressed despite theological hurdles, and the work of Copernicus was rediscovered and openly published by Johannes Kepler in 1609, who realized the importance of this simple sun-centred Copernican model. But at the same time, he showed that this heliocentric model had its own faults—the Copernican model simply assumed that the planets moved in *perfect circles* around the sun with a *constant velocity* and that the sun was *exactly in the centre* of these orbits. But then this model still did not properly account for the movements of many planets. Kepler, based on the immaculate observations of his master Tycho Brahe, made some significant improvements to the Copernican model. He asserted that the planets moved in *elliptical orbits* around the sun (not in perfect circles) with *variable speeds* (not at a constant speed), and the sun is located in a slightly *eccentric* position (not at the exact centre of the orbit). Now that this model is perfect, it has explained nearly all the 'erratic' movements of the celestial bodies in the heavens. However, this model also faced resistance and opposition from the religious establishment of the time and met with the same fate as the Copernican model.

It would greatly astonish the modern science student should he/ she learn that all these conclusions of the celestial bodies hitherto made by these medieval observers were drawn by naked-eye observations of the night sky without any optical aid. But when Galileo Galilei (1564–1642) turned his telescope to the night sky for the first time, he could now look into the heavens far more clearly and deeply, and this was perhaps the first step of the modern man's voyage into space. Galileo's two important discoveries stood as an unimpeachable evidence for the heliocentric model. *One*, hitherto, it was believed that every celestial body in the sky orbited round the earth, but he demonstrated that at least four bright celestial objects did not orbit the earth but orbited around the planet Jupiter (they were really the moons of Jupiter), proving that not all bodies moved around the earth, which, in turn, had proved that the earth is not the centre of the universe. And *two*, he had clearly charted the changing phases of the planet Venus, which could only be explained using the sun-centred model (but not by the earth-centred model). However, even this time, this evidence-based heliocentric model did not catch up with the establishment, and Galileo's works were also promptly banned.

However, despite all this resistance from religious dogma, science progressed. There was a gradual change in the climate of thought across generations with the dwindling influence of religion on the human lives, giving way to a more and more rational approach to problems in general. And eventually, the geocentric model was thrown into disrepute, and the Galilean heliocentric model came up to the forefront of science and started to enjoy wider acceptance. And finally, by the turn of nineteenth century, the idea of the *planetary model* of our solar system has become our established scientific model.

The Big Question Remained: As science progressed, technology improved. This has helped in building bigger telescopes with wider dimensions, and our view of the universe has now extended well beyond the solar system. Now the astronomers could see deeper and deeper into the cosmos, identifying newer and newer stars farther and farther away in the universe arranged in galaxies and galactic clusters (Ch 4-A). This improved view of the cosmos has rightly answered some of the age-old questions regarding many features of the universe, but more importantly, it has brought in more intriguing new questions about the universe. One major question that bothered the scientific world at that time was the origins of the universe: what was the beginning of the universe, and when exactly did it take place?

But then again, this question was already answered irrevocably and inarguably in the Bible! Bishop James Ussher in 1624 conducted a thorough 'historical research' into the Holy Scriptures and pronounced the precise date and time of birth: God has created the universe at 6 p.m. on the Saturday, the 22nd of October, 4004 BC! Now that this verdict was rather an erudite one, how could a modern scientist either consider this view or contest it? Did the scientist possess any gadgets to gauge the age of the universe either to prove this observation or disprove it? Hence, many scientists considered the question of the age of the universe outside the realm of science, and thus it was reckoned a technically irrelevant topic for a rational scientific discussion. And thus, this question was conveniently left out of scientific debate for centuries.

As it turned out, ironically, the question of the age of the universe did become relevant to the scientists in the coming years with new concepts and discoveries. Specifically, three discoveries made scientists ponder over the issue of age of the earth. In 1859, Charles Darwin published *On the Origin of Species* wherein he proposed his theory of biological evolution. It was soon realized that evolution was such a laboriously slow process that the origin of species on the earth should date back to some millions of years (even by the most conservative estimate) but not merely thousands of years as was generally proclaimed. Obviously, if the earth is millions of years old, then the universe must be much older than is generally thought. Another discovery also had a bearing on the age of the earth. In 1897, Lord Kelvin made thermodynamic studies and said that the earth was initially a hot molten blob of matter, and he calculated that for the earth to cool down to solidify and to become rocky, as in its present form, it would take at least 20 million years. Yet another indication of the age of the earth came from the study of fossils which had pushed the age of the earth further back to more than a billion years! Gradually, as science progressed, it also became evident that even our sun is just a medium-sized star (Ch 4-D), and much bigger stars are in existence outside the solar system, which meant that the outside universe is immensely bigger and older than was previously thought. All these new findings indicated that the vast universe of ours must be really very, very old.

Now this has become a vexing problem for the scientists to determine the exact age of the universe, but then there was a way out of this perplexity! Gradually, it dawned upon the scientists that the simplest assumption to counter this problem is to consider that the universe is *eternal*—just that the universe has existed forever! This magnificent assumption has an additional benefit. If there was no particular beginning to the universe, the question of a creator would not arise at all. So there is no need for us to invoke God into the mechanics of creation, and so we can do away with religion once for all. Thus, the scientists could now avoid any theological confrontation from the church, and so they were contented with this supposition of an ageless universe—and such an idea prevailed among the scientists for a long time. However, many scientists realized that there were no theoretical grounds for the scientists to support this view, and they considered that the problem of the age of universe has really reached an impasse, pending further developments.

Section B
Preparations for The Big Bang

The Grand Extent of the Universe: The early astronomers were inquisitive about the extent of the spread of stars they see in the night sky. They knew that the world above was big but did not know how big. Wilhelm Herschel (1738–1822), the discoverer of the planet

Uranus, studied the brightness of the stars and laid out a formula. The brightness of a star fades away with the square of its distance from the earth. With this formula, he tried to measure the distance of Sirius, the brightest star in the night sky. Unsuccessful though this attempt was, his formula of the star's brightness nevertheless became useful to the later astronomers in their calculations, as we will see a little later.

With the improvement of the telescope, our view of the cosmos improved, and the scientists could now see stars more clearly. The astronomers could now observe some interesting smudges of light in the night sky which contrasted well with the pointed specks of the light of stars. In 1764, Charles Messier identified hundreds of such smudges using his telescope and painstakingly numbered and catalogued them for future reference. A little while later, Wilhelm Herschel identified thousands of such smudges, and further observations revealed that these smudges are nothing but huge collections of stars! But the scientists at that time were unsure of the exact location of these smudges. Here's an important note: the prevailing idea at that time was that the whole of the universe was contained in the Milky Way, and nothing existed beyond it. Thus, it was considered that the extent of the Milky Way is the extent of the entire universe. So naturally, Messier presumed that all these smudges must lie somewhere within the Milky Way.

However, there was another camp of astronomers who believed that the Milky Way was only a local group of stars, but there were many stars located far beyond the Milky Way. Thus, according to these scientists, the cosmos was a much bigger place than it was supposed. But then scientists at that time had no way to decide which view was correct. In any case, this conflict has set a great debate among scientists which persisted for many decades, only to be resolved by sophisticated astronomical observations, as we will see below.

In 1838, Friedrich Bessel measured the *stellar parallax* (the infinitely minuscule change in the position of a star with the earth's revolution) and estimated the distance of the star *61 Cygni* from the earth. This distance was found out to be astoundingly enormous, and scientists realized that to express such a distance in miles/ kilometres would be too cumbersome, and they soon developed a convenient astronomical scale of distance which was named the *light year*,* and that is the distance covered by light in one year.

The modern science student would again be greatly surprised to learn that hitherto (from the time of Galileo), the telescopic observations were only subjective findings but not objective recordings; there was no way to record what we see through a telescope. For example, to describe a star by its brightness, one has to say it by some elaborate description or draw the sketches, which were also subjective, and thus, these descriptions were so prone for individual inclinations and bias. Hence, the results of these observations could not be compared among astronomers in any accurate way. Furthermore, these valuable astronomical observations could not be stored and passed on to the future generations in any reliable manner.

In 1839, an important development took place. Louis Daguerre discovered photography, which soon replaced the subtle art of portrait painting with impersonal replication of a person just by the click of a button! Anyway, photography could now record our subjective experiences permanently on photographic plates as objective observations, and Sir John Herschel (son of Wilhelm Herschel) used this facility as a tool of science in astronomy and developed the 'long-exposure technique'. Now with this technique, the faintest of the stars in the sky, which had hitherto escaped detection, popped up on the photographic plates—the camera has at last replaced the human eye. And this technique has not only improved the precision of observation but enabled proper documentation as well. We can cite the following excellent example in support of long-exposure photography: the naked eye of the ancients could pick up only 7 stars in the constellation *Pleiades* (the so-called 'Seven Sisters' constellation), whereas Galileo's early telescope could pick up 47 stars; the long-exposure camera, to everyone's surprise, could now reveal 2,326 stars!

---

* A light year is defined as the distance covered by light in one year. One year has 31,557,600 seconds, and light travels at 299,792 km/s, and thus, one light year equals 9,460,700,000,000 km (i.e. 9.46×10^{12} km)! And 61 Cygni, one of our nearest stars, is 11.4 light years away from us—a great distance indeed! But we will soon see that this astronomical distance would appear begrudgingly small when compared to mightier distances of other stars and galaxies!

But there appeared an additional important benefit with photography in astronomy. The photographic method has allowed the astronomers to accurately chart the brightness of a star. The *luminosity* of a star could now be best documented by using the photographic method! And consequently, this objective feature of luminosity of a star has become an important astronomical tool to measure the distance of stars from us, as we will see.

In the late eighteenth century, John Goodricke and Edward Pigott discovered some special stars called *Cepheid variables* which would grow bright for a few days and become dim for the next few days in alternating cycles, but they had not known the significance of this twinkling mechanism (we now know that this happens because of alternate contraction and expansion of a star, Ch 4-B). In 1912, Henrietta Leavitt studied several Cepheid variables by employing the photographic technique, which revealed a very interesting phenomenon—there was a relationship between the brightness of a Cepheid and its period of twinkling variability. This remarkable observation had profound implications—it allowed the scientists to compare the brightness of various Cepheids, which, in turn, allowed them to calculate their relative distances from each other (because Herschel has already shown that brightness of a star fades away with square of its distance). Soon after, the distance between one such Cepheid and the earth was estimated by Harlow Shapley and Ejnar Hertzsprung by using methods such as stellar parallax and other painstaking procedures. Combining these two methods, it now became possible for us to estimate the distance of various stars from the earth. And finally, these calculations have enabled us to achieve the seemingly impossible task of sizing up the vast extent of the universe! But even at this stage of astronomical development, most scientists were inclined to think that the known universe is all within the Milky Way and thought that nothing existed beyond— just because this concept could not be falsified in any way!

However, in 1924, a real breakthrough happened. Edwin Hubble studied a smudge of light in the night sky which was labelled *M-31* (*M* stands for *Messier*, and the number in his catalogue). At that time, as already noted, all these smudges were considered to be part of the Milky Way (which, of course, represented the whole universe). Hubble identified one Cepheid variable located in the M-31 and diligently estimated its distance from the earth using the Henrietta Leavitt method, and to the surprise of everyone, this turned out to be no less than a staggering 900,000 light years away! Astronomers had already estimated the extent of the Milky Way, and it was only about 100,000 light years across and 10,000 light years deep. We can now see the obvious conclusion. Because the Milky Way has a span of just 100,000 light years, it means that the M-31 must be a separate galaxy situated much farther away from the Milky Way. This was a remarkable conclusion at that time, indicating that the universe is far larger in extent than just the Milky Way. From then on, this M-31 gained the status of a separate galaxy, and subsequently, it was christened the *Andromeda Galaxy*—the first named galaxy in modern astronomy.

The cosmic smudge, M-31, became known as Andromeda Galaxy.

Upon further research, Hubble identified many more such galaxies which are located millions and millions of light years away from the Milky Way. Today, we know that these numbers run into *trillions* (not millions). There are about two trillion galaxies in the universe which are vastly larger than our Milky Way, and they are situated billions and billions of light years away from us (Ch 4-A)!

Thus, looking at this protracted history of astronomy, we may say that man has progressed from his self-centred geocentric model (with the implied egocentric primacy of man) to show that our Mother Earth is just the merest of the mere speck in the vast expanse of the universe. And now it became clear that our earth enjoys no special status in the universe—except that 'we' live here (not really knowing how many other exotic civilizations are in existence 'out there' in the vast cosmos, Ch 4-E). Ironically, after all, man has subverted himself of his position in the universe with his own genius! We leave it at that and continue with the story of the big bang.

The Idea of the Big Bang: So far, we have seen that the universe is immensely vast, but then the scientists had no idea as to how this vastness could have come into existence. Gradually, with the development of fresh insights into the problem of the origin of the universe, a new field of science called *cosmology* emerged, which exclusively dealt with theories related to this issue. Perhaps it is no exaggeration to say that no other field in the history of science is chequered with so many debates, personal innuendoes, and awkward dissents, and for such protracted periods of time across decades, than this field of cosmology! We will not go into the details of these historical debates, which have collectively led to the theory of the big bang (which the interested reader may get from an excellent account of Simon Singh in his book *Big Bang*), but we will only see a brief outline of these affairs.

The beginnings of the big bang theory have deep historical roots. Newton had already described gravity as a universal phenomenon, which meant that every object attracts every other object in the universe. However, Newton had realized a confounding, yet inevitable, implication of this universal gravity—if all the objects in the universe attract all other objects, then this would lead to an eventual collapse of the universe, an event which is obviously not happening. To overcome this collapse, Newton proposed an *infinitely symmetric universe* in which all the objects in the cosmos are in a state of fine-tuned balance, making the universe stable—each object attracting each other, resulting in a state of perfect equilibrium. But then again, Newton was stymied by the obvious realization that in such a perfectly balanced system, even a tiniest disturbance in the cosmos would initiate an imbalance which would cascade down into an eventual total collapse of the whole universe. We can say that even minor events such as a stray comet entering the solar system (or even if you toss a pebble high up into the sky!) might trigger a minute imbalance in the universe which would lead to its eventual collapse. Newton was troubled by this theoretical outcome of the universe but could not think of a way out of this problem. This puzzle remained unattended until Einstein proposed his general relativity theory which redefined gravity (Ch 1-D). But here again, Einstein was met with the same inevitable fate of the universe—general relativity also, in its simplest form, predicted that the universe has to either keep expanding or contracting. At this juncture, Einstein (who had staunchly believed in a static universe) developed a mathematical trick to the equations of his general relativity (Ch 1-E) to cure this problem of gravitational collapse, and this was called the *cosmological constant* (not to be confused with 'cosmological principle', Ch 4-A). Cosmological constant (represented by the Greek letter, lambda, Λ) is an 'antigravity force',[*] which was supposed to create an outwards force to counter the collapse, and many cosmologists at that time were happy about this cosmological constant with which the universe could maintain a stable and static state. However, none of the scientists at that time had any idea of what exactly this constant could be.

But we do find occasional heretics among us who defy authority. In 1922, Alexander Friedmann rejected the idea of a cosmological constant. He had greater faith in Einstein's equations of general relativity than Einstein himself and did not allow any tampering with it and accepted the theory as it is. He mathematically proposed a dynamic model of the universe (in contrast to the static model mentioned above). He said that the universe need not be static at all and described three models of the changing universe that are possible depending on the density of matter in the universe: (1) in a very dense universe with many stars, the gravity would overpower and eventually cause a contraction of the universe; (2) in a low-density universe, the gravity would be weak enough to cause its expansion; and (3) a universe with a density between the two extremes which would lead to a steady state (not static but a *dynamic* steady state; in Sec-D, we will discuss about the difference between 'static' and 'steady' states). This outstanding refutation of the cosmological constant was considered too radical and did not go well with the scientific establishment at that time, and his dynamic models were utterly disregarded. The static model, with its cosmological toolkit, was still reckoned to be the Standard Model of the universe at that time.

But a few years later, the idea of a dynamic universe resurfaced into the scientific scenario. In 1927, Georges Lemaître proposed an expanding universe, with great insights into the beginnings of the universe. Lemaître developed a mathematical model of an expanding

---

[*]   In order to counteract the effect of gravity, which leads to the eventual collapse of the universe, Einstein added this antigravity constant—$\Lambda g_{\mu\nu}$—to the general relativity equation (Ch 1-E). He had originally tried to correct the predicted curvature of the universe and so added his cosmological constant to the left side of the equation.

universe based on the equations of general relativity, and according to him, if the universe is expanding and reached the present state, then in the past (i.e. 'yesterday'), the universe must have been of a smaller dimension; and before yesterday, it was even smaller; and the day before yesterday, it was even smaller. If you carry back the universe in time far enough, thus argued Lemaître, then the entirety of universe would contract into a very tiny, very compact region. Thus, the beginning of the universe would be an extremely dense region, which he had called the *primeval atom*. Now if the universe has begun with a primeval atom, then retrospectively thinking, the universe must have 'kick-started' with a huge initial explosion. Lemaître had presented several phenomena in the universe in support of this theory, the most outstanding of which was that of the 'cosmic rays' (which were just then discovered to be coming from outside the solar system, and so they were thought to be the relics of this initial explosion—but now we know that it's not the case!). Ironically, this theory had inadvertently implied the concept of 'a moment of creation' with its inevitable theological connotations, which, of course, was not the intention of Lemaître at all—though a Catholic priest he was!

However, when Lemaître presented his theory at the Solvay Conference held at Brussels in 1927, the scientific community (which was now headed by Einstein) snubbed away the idea—rightfully because there was no evidence for an expanding universe but for Lemaître's 'flimsy' calculations and, wrongfully, because neither Einstein's static model had any shred of evidence for its support.* Anyway, science and technology at that time was not advanced far enough, and so no hard experimental evidence could be forwarded to support either of the two models. Nevertheless, the static model won over the hearts of many scientists, and so it continued to be the accepted model at that time, and the idea of an expanding universe was once again relegated to the back seat by the scientific establishment.

---

* However, Einstein, many years later, regretted his mistake publicly (when Lemaître was proved correct) and credited Lemaître appropriately and called his cosmological constant the 'biggest blunder' of his life. It's indeed commendable that Einstein had conceded his own mistake so gracefully and generously. It's a rare case of supreme genius blending with supreme modesty!

Section C
The Expanding Universe

It is interesting to note that in the history of science, progress in one field has often led to a more radical change in other seemingly unrelated field of science. As in our case here, we can see that progress in chemistry has fortuitously led to many developments in the field of cosmology, especially in the advancement of the concept of the big bang!

Apart from the dogma of a static universe, there was another idea which prevailed strong in the scientific milieu of the time. It was generally thought that the elements of the universe differed from place to place—that each star out there in the sky had different composition of elements and that the elements on the earth were unique. But then this idea had just remained as a notion among the scientists for a long time, and it could not be refuted obviously because there was no way we could sample a star to ascertain its composition, as it were.

But then came an important tool of investigation which would give us clues of the composition of distant stars. The advent of *spectroscopy* as a tool of investigation of atoms has led to some important changes in astronomy and helped in solving some of its deepest problems. Spectroscopy is based on the principle that atoms of each element, when heated, emit a characteristic wavelength of light, forming a specific *spectral pattern* peculiar to that element. In other words, an element's spectral pattern is its fingerprint.

So now we can see that the astronomers could use this knowledge to study the starlight and ascertain the elemental composition of stars. And indeed, the spectroscopic studies of stars have shown us that all stars, including our sun, contained more or less the same elements and that almost all these elements were already discovered to exist on the earth. Interestingly, helium, the second most abundant element of the universe, was first discovered in the sun using spectroscopy in 1868, and it was thought at that time that helium existed exclusively in the sun, hence its name (*Helios* in Greek means sun). Only many years later was it discovered to exist even on the earth. Now we know that the universe is made up of the same stuff throughout its expanse, which indicated that a common chemistry is in operation all through the universe. This finding was highly significant; it had radically changed our idea of the universe.

The same kind of elements exist throughout the universe.

More importantly, the spectral analysis of starlight gave us another phenomenon which has radically changed our view of the universe—this is the phenomenon called the *Doppler effect*. We have already seen that most of the scientists at the beginning of the twentieth century believed in a static universe. The stars were believed to be fixed in their position. Even though a very slight mobility was noticed, it was thought to be due to the earth's motion rather than their own proper motion. This view of a static universe was toppled down once and for all by the phenomenon of Doppler effect – we will now see how.

In 1842, Christian Doppler described an important phenomenon of changing wavelengths associated with motion. The essence of this phenomenon is that when an object emitting waves moves towards the observer, the waves appear to have a shorter wavelength (Fig 5.1a), whereas when the emitting object moves away from the observer, the wavelengths appear longer. This is called the Doppler effect. All waves, such as sound waves, water waves, or light waves, are known to obey this principle. The reader will understand this phenomenon better with the following example: it's our common experience that as an ambulance approaches towards us, the pitch of its siren increases, making the sound sharper (its wavelength decreases/frequency increases, Ch 2); but when the ambulance moves away from us, the siren's pitch decreases, making the sound muffle a little (its wavelength increases/frequency decreases). Now consider this. When starlight is studied spectroscopically from an approaching star (moving towards the earth), it would appear bluer (because blue has a shorter wavelength), and starlight from a receding star (moving away from the earth) would appear redder (because red has a longer wavelength) (Fig 5.1b). In other words, when studied using a spectroscope, the starlight shifts its colour to the blue or red side of the spectrum depending on the moving direction of the star. This phenomenon is termed the *Doppler shift*.

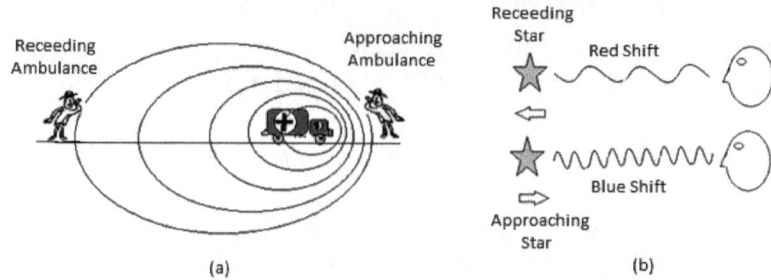

Fig 5.1: The Doppler effect.

A few years later, in 1868, William Huggins and Margaret Huggins studied starlight and found out that many stars in the sky are either blue shifted or red shifted. This finding simply suggested that the stars are not stationary in the universe but are moving. However, in 1912, Vesto Slipher painstakingly studied the light spectra of many stars and galaxies across the universe and categorically documented that *most* of the galaxies are receding away from us. This finding was really amazing—totally unexpected to most of the astronomers at that time, who were steadfast believers of static universe and fixed stars.

Hubble's Law: In 1929, Edwin Hubble (who had already established that the Andromeda Galaxy is located far away from the Milky Way and that a much larger universe exists beyond it) and his assistant, Milton Humason, studied the Doppler shifts of several galaxies far away in the universe and confirmed that almost all galaxies are red shifted, which means that the galaxies are moving away from our home galaxy, the Milky Way. Moreover, their calculations clearly showed that the velocity of the galaxies is proportional to their distance from us—meaning that a galaxy which is twice the distance from us will be moving away two times faster and a galaxy thrice the distance would move away three times faster. This observation is called the *Hubble's law*. The proportionality factor guiding the relationship between velocity and distance of a galaxy is called the *Hubble constant* ($H0$).

The implication of this finding is clear: galaxies are not ambling in the universe randomly but are flying away rapidly from each other—i.e. the universe is expanding. This was an astounding observation, and attached with this is a logical interpretation. If the universe is expanding, it is implied that we can extrapolate and run backwards in time to get a single dense region in space in the distant past. And now thinking in forward direction, we can say that this dense spot in the distant past must have started off with a great explosion at the outset and started off the universe, which is expanding even today. This is the essential concept of the big bang. So Lemaître, after all, was correct all along on his theoretical grounds—there was a beginning to the universe, 'a day without yesterday' to quote Georges Lemaître himself! But when Lemaître had showed his work to Einstein barely six years ago, he was rebuffed with a tart remark: 'Your calculations are correct, but your physics is abominable.' But with this revelation of expanding universe, Einstein corrected himself and lauded Lemaître: 'This is the most beautiful and satisfactory explanation of creation!'

The universe started with a big bang—and is expanding ever since!

Many subsequent observations using sophisticated equipment established the fact that all stars are moving away from us. The currently accepted rate of expansion of the universe is 70.4 km/s/Mpc (as of the 2013 NASA estimate). Mpc stands for *megaparsec* where 1 Mpc is equal to 3,260,000 light years. At this rate, the universe is expanding at an incredibly faster pace indeed!

Now the astronomers can extrapolate the expanding universe backwards and mathematically calculate the time at which the universe was in a single, consolidated state—the age at which the big bang occurred, which, after all, would be the age of the universe. Based on calculations, scientists at that time had estimated that the age of the universe was about 1.8 billion years. However, as science progressed, this figure has undergone great revisions, as we will see.

Thus, the evidence has now shifted gradually from a static universe to an expanding universe, and many scientists were now changing camps to the big bang model of the universe. Furthermore, Einstein, now the messiah of science, openly endorsed the big bang theory, which further gave it an undisputable status in the scientific community.

Finer Notes on the Big Bang: Further theoretical speculation has caused a conceptual change in the perspective of an expanding universe. General relativity had radically changed the very idea of space and time (Ch 1-E), and the big bang theorists now consider that it is not the galaxies that were actually moving through space, but it is the space itself which is expanding. This meant that the galaxies are simply moving along with expanding space. This concept gained support from the newer concepts of quantum theory, which proclaimed that the empty space is no longer empty but is filled with vacuum energy (Ch 2-G). Thus, more precisely, we can say that it is the expansion of the 'spacetime fabric' that is taking away the stars and planets along with it. This subtle nuance was explained with considerable lucidity by Sir Arthur Eddington by comparing the expanding universe with the surface of a balloon wherein the galaxies are represented by dots marked on the surface of a balloon. When balloon expands, it is the surface which expands, thus carrying the dots away from each other, increasing the distance between them. Such an expanding universe would make the expansion a generalized phenomenon, and consequently, there would not be a centre point, and the universe looks the same from everywhere, which aptly explains the present condition of the universe being 'isotropic and homogeneous' (Ch 4-A). Moreover, the red-shifting of the galaxies can also be explained not simply by the flying away of the galaxies but really by the stretching of the intergalactic spacetime, which in turn stretches the waves of light causing the red shift. Thus, in summary, the big bang is now considered *not* something akin to an explosion of a bomb in space but more in terms of spacetime expansion. In a bomb explosion, matter spews into the surrounding space, but in the big bang, the space itself undergoes expansion, and the question of explosion of matter into space does not arise at all!

We will now continue with the saga of the big bang. It is historically interesting to note that though the big bang model was going strong in the scientific circles at that time, there was a small but important clique of scientists who had alternate theories of the universe with some justifiable explanations.

SECTION D
The Steady-State Model

The dissidents of the popular model contested the already much-hyped big bang model with some really pricking questions. But then as we will be discussing below, only time and further research could answer their questions. We will first categorically study the problems encountered by the big bang theory at that time. Firstly, the age of the universe posed a great problem for the big bang proponents, research at that time had showed that the earth appeared to be older than the universe, which obviously could not be the case (see discussion in Sec-F). Secondly, the big bang theorists had the trouble of explaining the question of what has caused the big bang explosion in the first place. Moreover, the big bang theorists have not convincingly explained what came *before* the big bang. And lastly, there was some definite tinge of 'divine creation' in the big bang model, which was a little irksome to the big bang scientists themselves (especially with the church supporting their model!). And in fact, this divine connotation had really shooed away some researchers with a rigorous scientific temper from the big bang theory.

A few other theories had surfaced at that time to cover the pitfalls of the big bang theory, but none sustained for long. One prominent alternate theory was that of the idea of Fritz Zwicky. Zwicky thought that the evidence for the expanding universe is flimsy and sought a totally different reasoning for the red shift of the galaxies. In 1929, he proposed that the light emitted by stars loses its energy gradually as it passes along the vast intergalactic space because of the galactic gravitational forces. The speed of light being held constant and unchangeable, the ambient gravitational force of the galaxies causes the light waves to widen its wavelength, which is manifested as a red shift. This theory became known as the *tired light theory*. In fact, calculations showed that this phenomenon did cause some red shift, but then it was estimated that only a minor portion of the red shift could be attributed for the tired light phenomenon. And so it became certain that the majority of red shift was certainly due to the expanding universe itself. Thanks to Zwicky's irascible temper and improper personal innuendoes, many physicists at that time swayed away from the tired light hypothesis. Whatever the reason, these ideas were soon submerged in the tide of later developments.

However, another valid competing model emerged in the scientific scenario which caused a significant flutter in the scientific community of that time. In 1946, Fred Hoyle, together with Thomas Gold and Hermann Bondi, rejected the big bang model outrightly and came up with an elegantly designed model called the *steady-state model*.

Like in the big bang model, the steady-state model also encompasses a universe which is *dynamic* and *evolving* continuously, but in the steady-state model, the universe is *unchanging*. It needs some explanation to understand this model in the right perspective. It can be imagined that in the big bang model, as the universe expands, it would become less dense with time and leaves voids of space in between. However, in the case of the steady-state model, it was proposed that the resulting voids are filled up by the continuous creation of new matter. Here the reader has to make a clear distinction between the 'steady-state universe' and the 'static universe' (the reader may recollect that static universe was envisioned by Einstein and others in the past). The steady-state model radically differs from the traditional static universe in the sense that the universe in the steady state is dynamically moving over time but remains essentially unchanged by vigorously creating new matter (whereas static universe is a standstill universe). This new matter, as it is generated in the voids, collapses into new stars and galaxies (Ch 4-B), rendering the overall density of matter in the universe to remain the same. The reader can see that the steady-state model also keeps the universe isometric and homogeneous at all times. Though the big bang model was popular at that time, a small group of scientists adhered to the steady-state model ostensibly because of its theoretical elegance.

We will look a little further into the contrasting aspects of these competing models. It was an automatic deduction from the steady-state model that these newborn 'baby galaxies' are interspersed evenly throughout the universe. In contrast, in the expanding big bang model, the baby galaxies were predicted to be distributed exclusively at the periphery of the universe, which means that they are located far away from us. However, at that time, there was no way we could say whether a galaxy was young or old, so this debate has reached a theoretical impasse. Also consider this: whereas the steady-state theorists did not know any mechanism by which matter takes birth in his steady-state model, the big bang theorists also did not know how to account for the age discrepancy of the earth and the universe. Thus, both these models were in theoretical lock horns. With strengths and pitfalls on each side of the theories, these models created a battleground for some decades, which regrettably involved many personal disputes, vituperations, and sarcasms from supporters of each model. Ironically, in one such hostile discussion, Hoyle contested the big explosion heatedly and derisively by calling it a 'big bang!' (mocking at its theatrics). And this snappy name, which was given by the very man who had desisted the explosion concept so fiercely, has elegantly stuck to this model forever!

Anyway, as we proceed further in our discussion, we would know how certain scientific breakthroughs have toppled the steady-state model from the scenario, thereby vindicating the big bang model.

SECTION E
Atomic Connections to The Big Bang

So far, we have seen the evolution of the universe on a cosmic scale. Now we will peep into the microscopic world and see how it supported the big bang theory. Once again, we will see here how research in one seemingly unconnected field has led to advancements in the field of astronomy; the field of research now under consideration is that of the atomic research.

The one question that has pestered the scientists for a long time was that of the asymmetrical distribution of elements in the cosmos. Why is the universe dominated by lightweight atoms—hydrogen and helium—in exceedingly greater proportions than the heavier elements? Research has shown that for every 10,000 atoms of hydrogen in the universe, there are about 1,000 atoms of helium, 6 of oxygen, 1 of carbon, and all other elements put together are even rarer. In other words, hydrogen and helium constitute more than 98% of matter of the universe, and most of this hydrogen and helium is concentrated in the stars and intergalactic space (Ch 4-B).

In 1934, George Gamow, one of the initial proponents of the big bang model, suggested that if there was a primeval atom from which the big bang exploded into smaller fragments (as suggested by Lemaître), then in all its probability, the result would be an abundance of elements in the middle of the periodic table (in the vicinity of iron, cobalt, nickel, etc.). These middle elements are neither too small nor too large, hence more stable. The reasoning for the above conclusion was simple. It needs more energy for the primeval atom to be broken down into lighter elements, such as hydrogen and helium, and this process would not be thermodynamically viable. In contrast, for the heavier elements to form, they are inherently unstable, which also makes them thermodynamically unsuitable, and so they could not dominate the universe (but iron is a stable element, so theoretically, it must be the most plausible candidate). However, it was a stark fact of nature that the lighter elements overwhelmingly dominated the universe, and this was inexplicable. Therefore, Gamow disregarded the idea of a primeval atom and thought that some other explanation was in order. A proper explanation of this cosmic phenomenon incidentally came up owing to the advancements in atomic physics.

The First Moments of the Big Bang: The correct explanation of the preponderance of lighter elements in the universe came out indirectly from the work of Fritz Houtermans, who had just then described the mechanism of nuclear fusion in the sun (Ch 4-D). He showed that the inside of the sun has sufficiently high temperatures and pressures to fuse hydrogen nuclei into helium, and this fact has prompted astrophysicists to explain the origins of the big bang and consequently the abundance of lighter elements in the universe, as we see below.

During 1940s, George Gamow along with Hans Bethe and Ralph Alpher worked together and brought the hitherto vague big bang theory into a mathematically elegant model of the universe by marrying the equations of nuclear physics to the initial conditions of the early universe. Their calculations have estimated that 'primordial nucleosynthesis' (= formation of initial atoms, see below) took place in the first 300 seconds of the big bang, during which time much of hydrogen has formed (Table 5.1). This was a sensational idea at that time—the media extolling the three scientists upon their discovery that 'the world was created in 5 minutes' and deprecatingly calling out that 'the world could also be destroyed in 5 minutes', paraphrasing the power of the immensely destructive atom bomb, which had just then become popular!

The trio envisioned that in an expanding universe, if you extrapolate it backwards in time, the density would keep on increasing as the age of the universe decreases. The universe at these early phases would be hotter and hotter with atoms moving faster and faster. They proposed that the early universe at the beginning had extraordinarily high temperatures, and at this state, it consisted of matter in the plasma state—i.e. matter existed not in the form of atoms but in the form of a boisterous soup of rapidly moving electrons, protons, and neutrons named the *primordial soup*. Now looking in the forward direction, these researchers have envisioned that this early universe has started expanding ever since and reached the present state. As the early universe expanded, the temperatures dropped down, thus lowering the kinetic energy of the matter particles, which enabled the electrons to latch upon to the protons and start orbiting round the protons, forming hydrogen atoms (and some amount of hydrogen itself started fusing into helium). This process of formation of elements is called the *primordial nucleosynthesis* (in contrast to the *stellar nucleosynthesis* we have studied in Ch 4-B). This initial process of formation of elements from the primordial soup is called *recombination*. It was later estimated that it needed about 300,000 years (after the big bang) for the universe to cool down to a temperature of about 3000 °C when this process of primordial nucleosynthesis became possible (Table 5.1). These proposals were presented in 1948 in a paper called *Alpher–Bethe–Gamow paper* (a metaphor for Greek letters a–b–γ). This paper had a great impact in understanding the origins of the universe at that time and created flutters in the scientific milieu for some time.

In 1949, Ralph Alpher and Robert Herman went a little further and showed that high-energy radiation was released into the cosmos after recombination 300,000 years after the big bang and calculated that the wavelength of this initial light would be very short (about one-thousandth of a millimetre, i.e. in the range of gamma rays). With a great theoretical insight, Alpher further proposed that this emitted short-wave electromagnetic emission would have stretched its wavelength as it passed through the expanding universe so that it would now measure a wavelength of about 1 millimetre (this long-wave radiation falls in the range of *microwaves*, Ch 2, Table 2.1). They concluded that this microwave radiation must be the relic of the bygone big bang and hence would be distributed equally in all the regions of today's universe. However, nobody at that time had ever heard of any microwaves anywhere in the cosmos, and thus, this remained just a vague theory and was soon lost into the mists of scientific progress. Ironically, this microwave radiation (later to be called CMB radiation, see below) stood as one of the solid strongholds of the big bang theory in the times to come!

Section F
Confirmation of The Big Bang

However, it must be noted that the steady-state model still had its followers. In fact, for a long time, scientists were divided between the two competing models, both of which seem to have almost equal theoretical strengths and pitfalls. However, gradually, the big bang model gained ground with experimental evidence piling up in its support. The main arguments in favour of the big bang are presented below.

Readjustment of the Age of the Universe: One of the major drawbacks of the big bang theory, as we have seen, was that of the age of the universe. Astronomical calculations based on galactic distances have placed the age of the universe at only 1.8 billion years, whereas the age of the earth, based on the radioactive rock age estimation, was at least 3.5 billion years. This discrepancy obviously did not make any sense. However, in 1940, Walter Baade made some accurate observations using some advanced telescopic techniques and doubled the distance of the Andromeda Galaxy, thereby doubling the age of the universe (because when we extrapolate it back, it now takes double the time to reach the stage of primeval atom, thus increasing the age of the universe). About a decade later, Allan Sandage discovered galaxies called *quasi-stellar radio objects* or *'quasars'*, which are powerful radio galaxies with a central supermassive black hole. Quasars emitted intense radio waves which were intense enough to be seen today by us, and thus, they were thought to be situated extremely far away from the Milky Way—a feature which, by itself, has increased the age of the universe. In addition, scientists have also realized that these quasars require the hottest of the possible temperatures to form, and such extreme temperatures are possible only in the case of an explosion of the big bang but not with the gentle pace of a steady-state model. In any case, all these improvements pushed up the age of universe, which could now be placed anywhere between 10 and 20 billion years, which has now given a really comfortable margin for the big bang theorists to accommodate not only for the age of the earth but also for the other cosmic events!

The Abundance of Helium: The big bang model had to jump a few more hurdles to gain ground over its competing model. We have seen that the universe is dominated by the lighter elements hydrogen and helium (which are known to constitute more than 98% of the elements of the universe). Big bang theorists have claimed that hydrogen is formed from the primordial soup by the latching up of electrons on to protons in an early stage of the universe, whereas steady-state contenders said that hydrogen is formed *de novo* in the interstellar/intergalactic voids as the universe expanded. Whereas both these theories could reasonably explain the preponderance of hydrogen, the problem lies with the abundance of helium. Calculations clearly showed that the rate of formation of helium by fusion of hydrogen atoms in the stars is too slow to account for helium's second-vast abundance in the universe—i.e. the stellar nucleosynthesis of helium (Ch 4-D) alone cannot account for its abundance, but there has to some other source of helium formation in the universe.

The scientists were also puzzled by an apparent paradox. If the temperatures and pressures at the time of the big bang are so intense, then it could be expected that these intense conditions would have transmuted hydrogen into heavier and heavier elements (copper, iron, etc.). Why would the fusion mechanism simply stop at the helium formation? Thus, according to the big bang model, the elements that are in the middle of the periodic table must dominate the universe, which is clearly not the case. The steady-state model fared still worse in its explanation. As per this model, there was no reason, in the first place, why the freshly formed hydrogen has fused into helium (not to speak of other heavy elements). There was indeed no drive for this process at all.

Eventually, big bang theorists came out with an ingenious proposal. They theorized that the big bang event had lasted only for a briefest period, and it was calculated that in that ultrashort duration, only elements such as hydrogen and helium could have formed. Furthermore, their calculations showed that during this short period, a few atoms of lithium and beryllium could also have formed. And further astronomical observations had really shown that almost the same amounts of hydrogen, helium, lithium, and beryllium exist in the present-day universe as were expected from these calculations. These observations have swayed the wind in support of the big bang theory to a very great extent.

The Mystery of Radio Galaxies: The steady-state model faced defeat year after year with each new discovery in the field of astronomy. In 1940, Karl Jansky detected that a small portion of radio waves detected on the earth were actually coming from outside of solar system—a really surprising revelation at that time! In 1948, Stanley Hey discovered that certain galaxies, called *radio galaxies*, emitted strong radio signals instead of visible light, and this discovery soon led to the development of the field of *radio astronomy*—a method of observing the universe through radio waves (instead of visible light).* It was shown that these intensely radiating radio galaxies could only be created by the extremely hot and dense conditions

---

* Scientists use various types of electromagnetic radiation to probe the universe. For example, X-ray telescopes use short-wavelength x-rays to probe the most energetic cosmic events in the universe, whereas infrared telescopes use long-wavelength infrared rays to detect cosmic events at longer distances. Even gamma ray spectrometers are now available to study very-high-energy emissions in the cosmos!

that could have prevailed at the time of the big bang but not in the conditions of a more genteel steady-state model. Whatever the case, scientists have now realized that they can consider that these radio galaxies as newly formed galaxies in the universe—meaning that radio galaxies are the young galaxies!

We can recollect from the steady-state model that the 'baby galaxies' are predicted to be distributed evenly throughout the universe (because of the continuous collapse of the intergalactic gas clouds into new stars and new galaxies), whereas the big bang theory suggests that newly formed galaxies are situated far away from our galaxy at the periphery of the universe. In 1961, Martin Ryle introduced a technique called *interferometry* in radio astronomy and catalogued 5,000 radio galaxies. The analysis of this data irrevocably showed that radio galaxies are not distributed evenly in the universe but rather situated exclusively in the farthest reaches of the universe. This observation, as you can see, is in complete accordance with the big bang model but not the steady-state model. Subsequently, in 1963, Maarten Schmidt estimated the distances of various quasars, which further confirmed that every single radio galaxy is situated in the periphery of the universe.

The Olbers' Paradox: It may greatly surprise the reader that a completely commonplace phenomenon such as that of the occurrence of 'day' and 'night' on the earth stands out as an explanation for the big bang hypothesis. Consider this: the question of why nights on the earth should be dark is an ancient one (Ch 4-A). Thinkers of the past had considered that the universe is infinite and eternal. But then if such is the case, a problem presents itself—if stars are distributed evenly throughout an infinite universe, the sky should be as bright at night as by day. This is because in an infinite universe, the number of stars is infinite, and it would mean that at any given place and time in the night sky, there would be a star shining through the space, and collectively, they would brighten up the night sky. But still we can see that we have been enjoying—night after night—the serene darkness of the night all through the ages! Then what could be the explanation for night being dark?

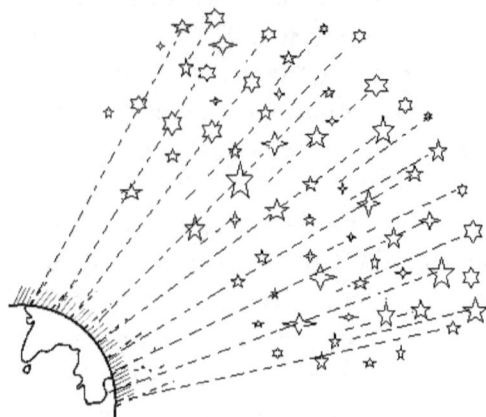

Fig 5.2: Olbers' paradox.

Wilhelm Olbers in 1823 highlighted this paradox, which since then became known as *Olbers' paradox*, but he has left it out without a proper explanation. One prosaic explanation was that the starlight was obstructed and absorbed by the intervening interstellar dust, but it was soon realized that the dust itself would eventually get heated up, which then becomes as intense and shiny as the stars themselves, and this would anyway brighten up the night sky. Another theory was that the light from the distant stars has not arrived yet, which is a brazened sort of explanation because in an infinite universe, the flow of light would be continuous.

By far the best explanation that is offered is by the big bang model—in an expanding universe, the light from distant stars becomes progressively red shifted, which would render light dimmer and dimmer and may eventually disappear. And moreover, the big bang model proclaims that the universe is finite, and thus, we may suppose that the light from many of the stars that are far too away from us has not yet reached us.

Cosmic Microwave Background Radiation (CMB): We have already seen that Alpher had suggested that we must be able to detect evenly distributed microwaves in the universe as a relic of the big bang even today. This relic of a radiation from the afterglow of the big bang came to be called *cosmic microwave background radiation* (CMB radiation). The discovery of this CMB needs a special description not only because it gave a final death blow to the steady-state model but also because it has become an important tool for the scientists to unravel many secrets of the universe, as we shall see.

Though the existence of cosmic microwaves was theoretically proposed in 1948 by Alpher, the actual discovery of these waves had to wait for a couple of decades, and this discovery was actually made more out of luck and serendipity. In 1965, Arno Penzias and Robert Wilson were working on to solve disturbances in a commercial communications satellite which interfered with its performance, and they accidentally discovered that they were caused by faint signals in the range of microwaves. To their surprise, they found out that these disturbing microwaves were not generated on the earth but came from some far-off regions of outer space! However, they could not attribute the waves to any specific cosmic source. At that time, Penzias and Wilson had no knowledge about Alpher's paper on microwaves, but they had anyway published their observations. This paper, however, caught the attention of a few keen scientists who soon realized the importance of their discovery and brought it to the general attention of scientists, which quickly led to further investigation, and finally, this phenomenon of ubiquitous distribution of microwaves was validated and named the CMB radiation.* With the discovery of CMB as a relic and afterglow of the initial explosion in the bygone past, the big bang model attained the status of a complete and consummate theory with full authentication!

Since then, CMB was studied extensively, and it was determined that it is distributed so uniformly in all directions that there are absolutely no fluctuations in its density across the universe. But ironically, this again posed a problem to the big bang theorists. If the CMB represents the fossil of the big bang event, the radiation having travelled all the time to the present, it has to show some telltale evidence of the universe's irregularities (which would be reflected as density variations in the space that occur due to the formation of stars, galaxies, and other tumultuous events of the universe, such

---

* It is interesting to note that about 1% of the 'snow' that appears on your TV's idle screen is estimated to be caused by the CMB coming from space (because of its interference with your TV antenna) – how wonderful it is that we can still witness the feeling of 'creation' at our homes even today!

as supernovae, Ch 4-B). But then how do the scientists explain this uniform distribution of the CMB? Why were the density variations in the universe not imprinted on the CMB radiation? In 1976, George Smoot identified some difference between the CMB wavelengths measured from the two hemispheres, but this finding was soon attributed to observational differences due to the Doppler shift caused by the earth's rotation rather than real variations in CMB. However, it is interesting to note that this observation has achieved, unintentionally but quite sensationally, another great advancement in science: it has calculated the velocity of our galaxy's travel across the space, which was found to exceed million miles per hour!

In 1986, a satellite of NASA named COBE (Cosmic Background Explorer, rhymes with 'Toby') was launched on board the space shuttle *Challenger* to detect the CMB irregularities, but sardonically, the *Challenger* exploded soon after lift-off, killing all its seven crew members. This unfortunate incident, of course, has driven the COBE project into shambles. In the subsequent years, many experiments were conducted with upgraded COBE satellites but proved unsuccessful in detecting any cosmic irregularities. The CMB was shown to be distributed with an 'unbroken blandness' (as was commented once by Marcus Chown). However, finally, the scientists' perseverance paid off. In 1991, COBE did detect a very minor variation in the CMB distribution—1 part in 100,000 (which was likened to 'listening for a whisper during an exceptionally noisy beach'). Nevertheless, the variations were there however minute they were. At that time, George Smoot gave the COBE discovery a journalistic thrust by describing this variation as 'the handwriting of the God', which made an instantaneous hit in the media! However, these CMB variations were later confirmed by an advanced version of COBE called the Wilkinson Microwave Anisotropy Probe (WMAP) satellite in 2001. This ended the long road of debate between the big bang and the steady-state model, and the idea of the big bang was, at last, incorporated into the mainstream science.

The COBE and WMAP satellites gave us, in addition to their main agenda of identifying 'anisotropy' of the universe, new insights into the other problems of the cosmos—it has allowed us to estimate the age of the universe with great precision which, finally, is fine-tuned to 13.77 billion years. Furthermore, the study of CMB radiation has contributed to our knowledge of the total energy of the universe. It is postulated that the CMB contributes significantly to the total energy of the universe. In fact, it is astonishing to note that this CMB radiation energy represents about 99% of photons in the universe sweeping across the voids, and the rest of the minuscule 1% of energy is contributed by the photons of the starlight which we have been observing and studying all the time!

SECTION G
Timeline of The Big Bang and Cosmic Inflation

So now, in summary, what exactly is the big bang? The big bang is a superlative explosion that has taken place at the beginning of the birth of the universe. The term *explosion* must be applied guardedly here in this context. In fact, many steadfast scientists, such as Eddington, desisted the term *explosion* because they considered that the events involved in this initial explosion represent a gradual process more in the form of a 'stately evolution' than a 'violent eruption'. Moreover, the explosion here is *not* an expansion *in* space but, as already noted, an expansion *of* space, driving its constituent objects away from each other. In other words, *both* space and time are considered to be *created* at the moment of the big bang. However, the big bang is thought of as an exquisitely singular event—absolutely uniform in its occurrence and without any irregularities or faults, or whatever, at its outset.

Both space and time are created at the big bang.

Even so, the big bang is a difficult-to-understand concept to the human mind. Imagine an event wherein all the energy and matter of trillions of galaxies of our present-day gargantuan universe concentrated at a single point! This kind of conceptual imagination is only possible when we consider spacetime itself to have evolved with matter at the beginning of our universe. But of course, some very difficult questions remain to be answered even to the best of the imaginative mind: if space and time were created at the time of the big bang, what was there before the big bang? Such fundamental questions may appear to remain with us forever, but the present-day science is not directly concerned with the events *before* the big bang, but rather, it deals with events that have taken place *after the big bang* (usually shortened to ATB).

The very initial events of the big bang, which had occurred within fractions of seconds ATB, are at the smallest of the small scales—i.e. the Planck scales (Ch 2-F). However, it is understandable that the events of the earliest universe at Planck dimensions would be based *not* on the principles of relativity, but are more related to the principles of quantum mechanics. But here is another problem: the rate of expansion of space is chiefly dependent on the slackening of the pull of gravity, and thus, gravity must also exist in its most miniature form at the time of the big bang—i.e. in a state of quantum gravity—and the phenomenon of quantum gravity is an enigma by itself (Ch 3-F)! Understanding these intricate affairs is an intense research topic of which we will not go into the details, but for now, we will concentrate on the current understanding of the events of the universe that took place ATB. The reader may look into Table 5.1 all along the discussion to get an idea of the timeline of the evolution of the universe.

Mathematical calculations showed that at the outset of the big bang, the temperature of the universe is at the highest temperature attainable in the universe—i.e. the *Planck temperature*, T_p (1.416 ´ 10^{32} K). Within a fraction of seconds ATB (Table 5.1), this superhot big bang has started expanding, and this has cooled down the universe dramatically— the temperatures falling from the highest possible temperature to temperatures of the order of trillion degrees. The general idea of the big bang is that at the initial stages of the universe, the matter particles moved with extreme kinetic energy, which was provided by the superhot temperatures prevailing at that time. In this setting, matter did not exist in its atomic structure; rather, it has existed in a plasma state wherein all the subatomic matter particles (quarks, electrons, etc.) are in random motion independent of each other—the so-called *quark–gluon plasma*.

As the temperatures dropped further owing to continued expansion, the kinetic energy of the particles decreased, and they started to come together and started to combine with each other—a process called recombination. This process of recombination of subatomic particles into simple atoms of hydrogen and helium (perhaps some little amounts of lithium), as we have already seen, is called the primordial nucleosynthesis (Sec-E).

Whereas matter particles come closer with decreasing temperatures, the reverse is the case with the fundamental forces. At these extreme temperatures of the big bang, all the energy of the universe was represented by a single 'unified force' (of course, we do not know the properties of such a unified force, Ch 3-F). In other words, all the four fundamental forces of nature (as we know them at present) were unified into this single force, which can be said to be in a state of absolutely perfect *symmetry* of forces. As the temperatures cooled down, the fundamental forces started to dissociate from each other; the four forces eventually acquired separate properties and eventually formed four independent forces as we see them today. This process of dissociation of forces is called *symmetry breaking*, as we have seen in Ch 3-F. The reader may note here that it is the earnest dream of the today's scientists to get a theoretical understanding of the property of this unified force, which, however, appears a long way to go from where we are now! Anyway, let us leave it there and continue with our saga of the big bang.

The Cosmic Inflation: The big bang theory, though being accepted now as a valid theory by scientists in general, presented certain problems. The cosmic microwave background radiation, which is a relic of the big bang, though containing some irregularities (as explored by COBE and WMAP), still has a largely uniform distribution of energy in general across the universe. This gave the scientists one problem: the problem of thermodynamic distribution of energy across the early universe (Ch 6-D). This problem can be understood in simple terms. If we suppose that the big bang is a gradual expansion of the universe from the point of origin, then this implies that the thermodynamic transfer of energy (heat) generated at the start of the big bang would take some time to transfer from one region to the other—energy *cannot* be transferred *instantaneously* to all the regions of the universe (because special relativity has put a limit on the rate of transfer of electromagnetic energy to a maximum limit of the speed of light, Ch 1-B). Thus, in an expanding universe, it would take a lot of time for the heat to transfer from the centre of the early universe to the ever-increasing peripheries. But here is a crucial problem—this delay should have resulted in a differential distribution of energy in various parts of the universe, which means that the temperature of the universe must vary from region to region. Consequently, the very essential feature of our present-day universe being isotropic and homogeneous (looking uniform everywhere in all directions, Ch 4-A) cannot be explained in proper terms. Scientists call this the *horizon problem*.

In order to overcome this hurdle, Alan Guth in 1979 proposed an ingenious addition to the big bang model—the so-called *cosmic inflation*. He proposed that there was a brief but super-rapid phase of expansion soon after the big bang (Fig 5.3). His mathematical calculations showed that in about 10^{-36} seconds ATB, the universe went through a period of a very rapid increase in its rate of expansion. The universe is thought to have expanded by 10^{26} times in a trillionth of a second (that equals roughly to a coin expanding to the size of the Milky Way suddenly!). This exponential expansion is considered to have effectively eliminated the problem of time in heat transfer between different regions in the universe. And Guth suggested that this superlative expansion in a minuscule amount of time would have smoothed out the irregularities in the distribution of energy so that the CMB would now appear almost uniform (as it appears today). Moreover, it is suggested that this gigantic expansion has occurred at speeds far greater than the speed of light (these superluminal speeds, however, would not violate the special relativity because it was concerned with the expansion of the space rather than the movement of the particles themselves!).

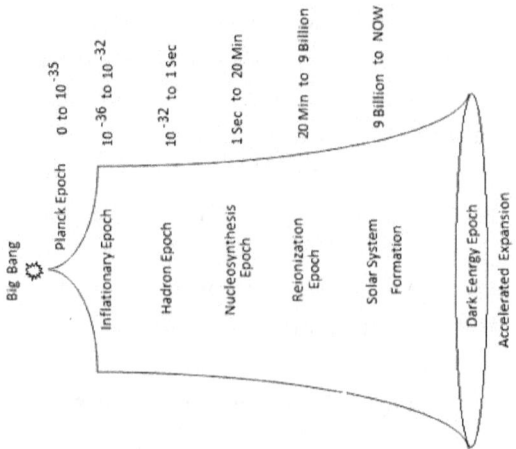

Fig 5.3: Timeline of the big bang.

This briefest moment in the history of the universe perhaps had the greatest impact on its future development. It is now speculated that this sudden inflationary expansion has not only led to the present-day structure of the universe with matter distributed in clumps of stars and galaxies, but this inflation has also accounted for the shape and curvature of the universe at present (thus answering the so-called flatness problem). Furthermore, it is speculated that this inflationary expansion would appear to determine the ultimate fate of the universe as well as we will see later.

But then what has caused this accelerated expansion? More intriguingly, why did this rapid expansion stop after a briefest period? There are many sundry speculations of which only some appear valid enough for discussion, and we will look into them later. But for now we will continue with the timeline of the big bang.

Table 5.1: Timeline of the Universe

Name of Epoch	Time after the Big Bang (ATB)	Characteristic Features
Planck Epoch	0 to 10^{-43} seconds	Absolutely symmetrical, lasted for the briefest period, Planck dimensions, Planck temperature (10^{32} K)
Grand Unification Epoch	10^{-43} to 10^{-36}	Dissociation of fundamental forces, earliest appearance of particles
Inflationary Epoch	10^{-36} to 10^{-32}	Cosmic inflation, strong force separates, universe dominated by quark–gluon plasma
Electroweak Epoch	10^{-32} to 10^{-12}	Appearance of Higgs particles, W/Z bosons
Quark Epoch	10^{-12} to 10^{-6}	Universe cools down to quadrillion degrees, all forces separated, matter–antimatter particles annihilate, a few matter particles remain (that we observe today, see text)
Hadron Epoch	10^{-6} to 1 sec	Temperature below trillion degrees, protons and neutrons appear, neutrinos generated which continue to travel till today
Lepton Epoch	1 sec to 3 min	Proton–electron soup dominates universe, opaque universe

Nucleosynthesis Epoch	3 to 20 min	Temperature drops to billion degrees, universe dominated by nuclei of H, He, and Li
Photon Epoch	20 min to 240,000 years	Cooling of the universe to 3000 °C, radiation dissociation, atom formation, transparent universe, CMB radiation started spreading, 75% H, 25% He, traces of Li
Dark Epoch	300,000 to 150 million years	Atoms of the universe diffusely distributed, no stars, dominated by dark matter
Reionization Epoch	150 million to 9 billion years	Gravitational collapse of atoms, first quasars appear, reionization (splitting of H into proton and electron) takes place, stars form, intense energy released, galaxies form, universe continues to expand, start of all cosmic stellar events such as supernova
Solar System Epoch	9 billion years (i.e. 4.5 billion years *before now*)	Solar system formation
Abiogenesis	9.9 to 11.2 billion years	Life formed on the earth
Humans	13.2 billion years (early humans 5 to 6 million years *before now*)	Humans appeared on the earth
Today	13.77 billion years	Future?

Formation of Atoms and Galaxies: Within the first second after the big bang (0–1 sec ATB), most of the quark–gluon mix has assembled into protons and neutrons, releasing massive showers of neutrinos into the space in the process. These neutrinos have travelled all along for more than 13 billion years to reach us in the present era, and they can now be detected even today as they traverse the earth's atmosphere (Ch 3-C). The reader may take a look at the complicated sequence of events that occurred during this first one second ATB from Table 5.1 and Fig 5.3. Between 1 second and 20 minutes ATB, the temperature of the universe has dropped from trillions of degrees to billions of degrees, and at this stage, the plasma consisting of protons and electrons undergoes a recombination process which has led to the beginning of the formation of atoms of hydrogen, helium, and lithium (i.e. primordial nucleosynthesis), and this process is speculated to continue for a period of 240,000 to 300,000 years ATB.

It is worth mentioning here of an interesting phenomenon that has taken place in the early universe. The universe, prior to the formation of atoms, consisted of matter in plasma state with the freely moving charged particles, and this charged plasma did not allow the surrounding intense energy in the form of light (= EM radiation) to escape far ahead; thus, the light of the universe at this stage was scattered. This scattering of light has made the early universe 'opaque'. In about 300,000 years ATB, the nucleosynthesis is nearly complete with the formation of neutral atoms which now dominated the universe. These neutral atoms would not absorb light any longer (as charged plasma did), but it rather allowed the light rays to sail ahead of them into space. This made the hitherto opaque universe 'transparent' to light. It is something like opaque milk curdling into buttermilk, leaving transparent areas in between the clumps of curd. The intense light is now ready to escape into the expanding universe, which would be later detected by us as CMB.

The atoms of the universe at this period consisted chiefly of hydrogen and helium (with perhaps a little number of lithium atoms). The atoms are free-floating and diffusely distributed in the universe with no evidence of any gravitational force acting upon them. Over a period of 150 million years to about 9 million years ATB, atoms started to clump together due to the effects of mutual gravitational attraction, and this has subsequently resulted in the gravitational collapse of bigger chunks of matter, thus leading to the formation of stars (Ch 4-B). These are the first-generation stars which are composed of only hydrogen and helium as their nuclear fuel (the story of second- and third-generation stars is described in Ch 4-B). And of course, these stars soon started to gather into galaxies and galaxies into galactic clusters.

The solar system with its planets might have formed about 9 billion years ATB (i.e. about 4.5 billion years back *before now*). Life might have started on our Mother Earth about 10 billion years ATB, and our human ancestors might have appeared about 6 million years before now, and modern man might have appeared about 200,000 years before now. And perhaps the human civilization has made its appearance only about 6,000 years before now.

This is the currently accepted picture of the events after the big bang as far as we can deduce from the facts. With this understanding, we will go into some hypothetical speculations of the universe based on the big bang theory.

SECTION H
Implications of An Expanding Universe

So now, at last, the big bang and the ever-expanding universe have come into the mainstream science, and the voice of the dissidents has died down in the avalanche of factual evidence in support of an expanding universe. However, the big bang, though it answered many questions, has opened a Pandora's box of even more intriguing and fundamental questions as we have already seen: What is the cause of this apparently momentous 'event of creation', and what has triggered this 'great explosion'? What existed *before* the big bang? And finally, what could be the ultimate fate of the universe?

The first two questions, as considered by many scientists and theorists, are outside the realm of present-day science. After all, we have only 'theological dogmas' but no theoretical principles to study the 'initial cause' (or so it appears at present), and also, we do not have any theoretical principles to study the events *before* the big bang singularity because the properties of matter and energy (along with all the rules of our physics) break down at the point of singularity, rendering the study of times before the big bang null and void. However, here we may look into the possible fate of the universe and see some of the exciting theories that have chaperoned theoretical science successfully for a few decades.

We will first examine the different factors that could determine the future and fate of the universe. If the universe has started off with a big bang explosion and continues to expand, doesn't it stop expanding at some point of time in the future and then start contracting because of the gravitational pull of its constituent matter? Or is it possible that it would continue to expand forever? Or as another possibility, does the universe reach a stage of equilibrium and remain in a steady state? What parameters would possibly determine one of these fates of the universe? General relativity has showed us that matter induces curvature of spacetime. Then in that case, obviously, the density of the matter in the universe and the curvature of space induced by it would play a significant role in the rate of expansion and the ultimate fate of the universe. But here is an obstinate predicament. The density of the universe is dependent on the total amount of matter in the universe distributed across the universe of a particular size. But both of these parameters of size and amount of matter themselves are unknown to us (as we have seen in Ch 4-A). However, basing on the equations of general relativity, Alexander Friedmann had already calculated the *critical density* of the universe and proposed three possible outcomes of the universe (Sec B).

1. If the density of matter in the universe exceeds the critical density, then the universe would take a *positive curvature*, curl on itself, and assume the shape of a sphere; this represents a closed universe (Fig 5.4a). The sum of the angles of a triangle drawn in such a closed universe would therefore exceed 180°. The fate of this sort of universe, as can be guessed, would be an eventual collapse on its own gravity.

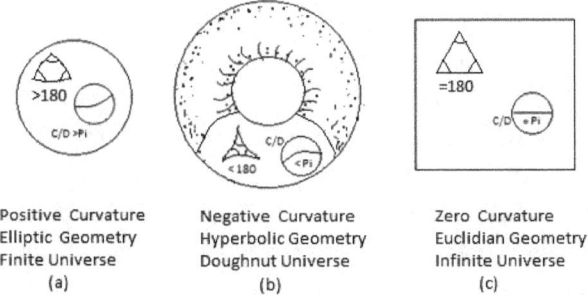

Positive Curvature	Negative Curvature	Zero Curvature
Elliptic Geometry	Hyperbolic Geometry	Euclidian Geometry
Finite Universe	Doughnut Universe	Infinite Universe
(a)	(b)	(c)

Fig 5.4: Possible geometrical structures of the universe.

2. If the matter in the universe is less than the critical density, then the universe would take a *negative curvature*, continue to spread in an outwards direction, and assume a hyperbolic (or saddle) shape. The sum of the angles of a

triangle drawn in such a universe would be less than 180°. This open universe would tend to expand forever. However, it must be remembered that the curved geometries (either positive or negative) have a *constant* curvature—meaning that the curvature *everywhere* within its confines is the *same*. Consequently, it can be argued that the negative curvature ultimately curls upon itself to become a finite space, looking finally, after a complete reconstitution, something like a doughnut or *torus* or bicycle tube (Fig 5.4b). You can really imagine a hyperbola on the surface of the inner portion of a bicycle tube and see that you can extend it to make a full and finite tube as shown in the figure. Thus, we can conclude that *any* curved surface (either positive or negative) is a finite entity, and this represents a closed universe.

3.   If the density is exactly equal to the critical density, then the universe is flat, and the sum of angles is equal to 180° (Fig 5.4c). This is an open universe and would actually represent the dynamic steady-state model of Fred Hoyle as we have studied earlier.

Which one of the above three possibilities is correct? Further theoretical insights into the cosmic inflation predicted that the density of the universe is very close to the critical density and the geometry of the universe is flat. Alan Guth proposed that this is because the actual universe is about 10^{23} times bigger when compared to our *observable* universe, and thus all the curvatures are practically flattened out (the analogy is that of a tennis ball expanding exponentially to the size of the Milky Way when the curvature of the ball would almost disappear, or in more practical terms, the reader may take the surface of the earth as an example, which appears flat to us because of its immense extent). Recent observations of CMB radiation by WMAP also confirm that the universe is flat with only a 0.4% margin of error.

But then the old question remained. A universe containing matter must have a curved spacetime structure—so how is it that the universe is flat? Is it possible that some other force is acting in the universe which flattens out the curvature? We will now see how further progress in cosmology has not only shed some light on these problems but has given some clues into the future of the universe. Of course, all these developments may not be leading us into a firm and definitive theory (at least not yet), but they certainly give us a lead to conduct further research.

Dark Energy and Dark Matter: It was traditionally thought that after a certain extended period after the big bang, the universe may ultimately start decelerating. This means that at some point of time in the distant future, the universe would start collapsing again and would eventually crumble upon its own gravity in an equally dramatic cosmic event called the *big crunch*. This final event was thought to be an essentially reverse process of the big bang. However, as science progressed, it has unravelled new secrets of the universe. In 1973, Jim Peebles and Jerry Ostriker used galactic simulations (using computer numerical simulation techniques) and proposed that some hitherto unidentified matter must lurk in the universe in order to explain the formation of the sort of galaxies we observe today (suggesting that it needed more mass to explain the pattern of matter distribution across the universe). This unidentified matter in the universe was called the *dark matter*. However, the concept of dark matter took deeper roots as science progressed further. In 1975, Vera Rubin determined the rotational speed of our Milky Way, but what really surprised her was the fact that the high rotational speeds of stars in the periphery cannot be completely accounted for by the total amount of mass of the stars in our galaxy, which meant that there must be significantly more hidden mass in our galaxy. We can understand this by the following analogy: in our solar system, the planets that are farther away from the sun move at a slower rate than the nearer planets. For example the farthest planet, Neptune, has an average orbital velocity of 5.4 km/s, whereas the nearest planet, Mercury, has an orbital velocity of 47.4 km/s. Likewise, the stars in the periphery of a galaxy are expected to move more slowly. But on the contrary, they were found to be moving faster than expected. Vera Rubin then extended her study to the rotational speeds of other distant galaxies and made similar observations—the stars at the outer edge of many other galaxies move at a faster pace than expected. The importance of this finding is profound. The fast-moving stars in the periphery of a galaxy need stronger gravitational force to keep them in orbit; otherwise, these fast-moving stars would just fly away from the parent galaxy. And to generate stronger gravity, it needs greater mass, meaning that the galaxy would be expected to contain much more mass than the calculated mass of the galaxy. This simply meant that a significant amount of mass is somehow hidden inside the galaxies. Based on these findings, researchers have calculated that the ratio of total matter of the universe to the visible matter may be in the order of 10:1. This means that a staggering amount of mass remains hidden from our view!

Conceptually, it is thought that the dark matter emits no radiation whatsoever, and so it is utterly undetectable to us. But it is detectable only by its gravitational effects in the galaxy. It is also speculated that dark matter interacts neither with the baryonic matter (i.e. our 'ordinary' matter composed of protons, neutrons, and electrons, Ch 3-C) nor with dark matter of the other regions of the universe (as evidenced by studying certain collapsing galaxies). Thus, dark matter shares none of the properties of our ordinary matter. However, it is now generally presumed that dark matter spreads all through the universe uniformly—in the intergalactic space, inside the galaxies, inside the solar system, in between you and me, and in between the atoms; in short, everywhere!

But then what exactly is this dark matter? A few prospective candidates were proposed in the past, but none of them meet all the requirements of dark matter. We will look at some of them. MACHOs (massive astrophysical compact halo objects) are thought to be composed of ordinary baryonic matter which does not emit any rays, making their detection impossible. Examples are extremely compact neutron stars, black holes, red dwarfs, brown dwarfs, massive rouge planets, etc. But their contribution to dark matter in the galaxies is generally discounted because dark matter is generally thought to be fairly evenly distributed across the galaxies, whereas MACHOs have some lumpy distribution. Another possibility is the existence of WIMPs (weakly interactive massive particles), which are massive particles hitherto not described in the Standard Model which do not interact with ordinary matter particles, but so far, we have no evidence of the existence of such particles. Another queer proposition is that of Mordehai Milgrom. He had approached the problem of fast-moving stars from a different perspective by modifying the classical gravitational dynamics (thus called MOND for 'modified Newtonian dynamics'). The MOND theory states that at very low accelerations, the gravitational force of the galactic bodies varies inversely with the distance but *not* with the *square* of the distance. But then here is a problem: the *equivalence principle* (which is the essence of general relativity) has already shown us that gravity and inertia are one and the same (Ch 1-D). But to accommodate MOND, this fundamental principle of equivalence has to be breached, and this violation would invariably result in the modification of either gravity or inertia, and many scientists would consider such a theoretical tinkering with the equivalence principle an uncalled-for exercise! However, this hypothesis could not stand the test of time because it had failed to agree well with other observations.

Nevertheless, the reader may note that the very concept of dark matter is not universally accepted by all the scientists. Many researchers are still searching for other possible answers for the observational peculiarities of the movements of celestial bodies. One should, however, keep an open mind for newer ideas, reminding oneself of the unfair and unjustified rejections perpetrated in the history of science through the ages!

Dark Energy: While the concept of dark matter is intriguing, the idea of dark energy is even more exotic. Science progressed further with finer astronomical observations which have prompted scientists to shed older theories in favour of newer ideas. In 1998, Saul Perlmutter, Brian Schmidt, and Adam Riess studied the shifting of brightness of *type 1a supernovae* (called the 'standard candles of the universe' for their known constant luminosity) and ascertained that the periphery of the universe is expanding at far too greater speeds than was anticipated by the scientists. This *accelerated expansion* of the outer reaches of the universe surprised the scientists because it could not be explained simply by the initial thrust of the big bang, but it needed some inherent energy to boost up this expansion. Soon it was theorized that this rapid expansion is due to some hitherto unidentified energy called the *dark energy*. Theoretically, it was considered that whatever force this dark energy is, it must be gravitationally *repulsive* so as to give a strong outward thrust to the universe (whereas gravitation itself is *always* an *attractive* force). But the big question remained: in what form does this dark energy exist, and where is it hidden?

We have seen in Sec-B that Einstein had proposed an antigravity force way back in 1917 called the cosmological constant, which he had added to the equation of his general relativity in order to prevent gravitational collapse of the universe (Sec-B). But this cosmological constant was subsequently abandoned because, at that time, evidence piled up in favour of the big bang which caused the expansion of the universe. But now mathematical calculations have shown that the big bang itself could not account for this accelerated expansion; rather, it needs an additional amount of energy to drive stars in the outer reaches of the universe at these tremendous speeds. Now the reader may see that it is a simple matter for the scientists to re-invoke the erstwhile cosmological constant to explain this rapid surge in expansion; and no wonder, some scientists have dutifully reintroduced the cosmological constant back into the cosmological toolkit.

But then how exactly can we incorporate the cosmological constant into the phenomenon of dark energy? Consider this: as already noted in Sec C, it was theorized that in an expanding universe, it is the space itself which expands, carrying the galaxies along with its expansion (rather than galaxies themselves in motion). And we have also seen in Ch 3-I that empty space is no more empty but contains some tremendous amounts of energy within it. These concepts have led to the speculation that this hitherto unidentified energy must be causing the accelerated expansion. Thus, the energy of the empty space, called the *vacuum energy*, has now become the candidate for dark energy. However, there are certain problems. Though the vacuum energy is thought to create a negative pressure in the space, which would result in gravitational repulsion, this negative energy should be very, very small because its cumulative effect in this gigantic universe would soon become large enough to cause an exponential expansion of the universe to the point of ripping away of all its constituent galaxies. Thus, the theoretically calculated energy accommodated by the cosmological constant must be infinitesimally small. But surprisingly, the experimentally estimated value of vacuum energy is quite large, and this theoretical discrepancy is a major unresolved issue in the field of cosmology at present. Could there be any other candidate for dark energy, or could there be a totally different explanation for the accelerated expansion of the universe?

The Preponderance of Dark Matter and Dark Energy: Is there an experimental proof for the presence of dark matter in the universe? If so, how much of this dark matter/energy exists in nature? Further scientific progress has shed light on these matters. A brief account on *gravitational lensing* would throw light in this direction. Eddington's experiment of bending of starlight (Ch 1-F) had already showed us that strong gravity affects the path of light. In 1936, Einstein, based on the work of Rudi Mandl, proposed that space itself could act like a lens, bending light and magnifying it. This happens because of the fact that gravity is all-pervading, and its effects are tangible everywhere in the universe. This idea of Rudi Mandl and Einstein was proved correct because subsequently, it was demonstrated by several observations that as light travels over vast distances in the universe, it undergoes some amount of distortion owing to the bending effects of space. And this distortion would, in turn, generate aberration of the images of the galaxies we observe from the earth (because the light emerging from distant galaxies is refocused somewhere). This effect of magnification and distortion is called the *gravitational lensing*. However, further research has shown that this gravitational lensing could not be entirely accounted for by the distortion generated by the mass of the observed stars. And it was soon proposed that most of this astronomical bending of light was done by the intervening dark matter. Now the reader can see that an estimation of the magnitude of gravitational lensing would indirectly reflect upon the amount of dark matter of the universe!

Basing on such observations, the dark matter and dark energy are now thought to be far more abundant in our universe than ordinary matter (Fig 5.5). It has been estimated that the universe is dominated by dark energy (74%) and dark matter (22%), and only the remaining 4% is comprised of our 'ordinary' (i.e. baryonic) matter (Ch 3-C). Out of this 4% of ordinary matter, only about 0.4% is contained in the stars and planets, and the rest of the 3.6% of matter is constituted by the intergalactic gas (Ch 4-B). In short, all the matter we could directly see and feel and learn and that which comprises of all our stars and planets, including you and me, are made of only 0.4% of the total matter of the universe!

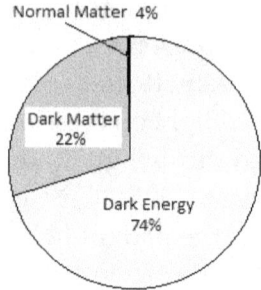

Fig 5.5: Dark energy and dark matter.

However, some deep questions related to dark matter and dark energy remain to be answered: What is the relationship between dark matter and dark energy? Do they have a common origin? Are they interconvertible (with some equation akin to our $E=mc^2$)? We do not know! Many scientists actually believe that both of them have a different origin despite the fact that they both appear to operate together to influence the distribution of matter in the universe. In this connection, it is generally claimed that dark energy was in a very rudimentary form at the early stages of the big bang in order to allow for the formation of clusters of galaxies without being ripped off as they begin to form. However, at some point in the last few billion years, dark energy must have somehow became a dominant factor in the universe, which not only caused accelerated expansion but also prevented the formation of more galaxies and galactic clusters. Another intriguing aspect of dark energy is that it appears to have little or no influence *within* the galaxies where gravitational force is still the dominant force, and we simply do not know why this is so! These profound theoretical lacunae show that we are nowhere near a solid theory, and it appears that our understanding of the expanding universe, dark matter, and dark energy needs some radical conceptual revamping as it were.

SECTION I
Fate of The Universe

If the concepts of dark matter and dark energy did not appear a little speculative to the reader, the concepts on the fate of the universe certainly would. Whereas we have some solid evidence for the beginning of the universe by the way of the big bang, scientific investigation into the future and fate of the universe has not yet provided us with any real evidence-based theory so far. However, scientists generally consider two antithetical outcomes of the universe as pertinent for scientific study: one is the possibility that the universe may end up in a *big crunch* (which may culminate in a 'big fire'), and the other is a *big rip* (which may culminate in a 'big freeze'), as we will discuss below.

The Big Crunch: One of the most prominent determinants of the fate of the universe, as we have seen in Sec-H, is the density of the universe. If it is more than a certain critical density, the expanding universe will come under the gravitation attraction of its matter and will begin to reverse the process of expansion at some point in time in the future and start contracting to collapse eventually on to itself. This is the big crunch (which is nothing but a big bang in reverse). Here the matter gets closer and closer to each other and gets infinitely heated up, and ultimately, the universe is driven into a state of *singularity* (Ch 4-C) wherein the temperature, density, and curvature of the universe would become infinite as predicted by the maths of general relativity. And ultimately, all the physical laws that govern the universe would crumble down into a state of *nothing*. The reader may see that by collapsing into a singularity, the universe will turn into a state of an infinitesimally massive 'unified black hole'. We may presume some really interesting events would happen in the process of the reversal of the present state of expansion because this effectively means a 'reversal of the arrow of time' (Ch 6-E). We may expect that our broken cups would once again fly back to the table and reassemble into a cup, or we may see our dead ancestors come up alive from their graves, or we may start remembering our future, and so on (see Ch 6-E for further understanding).

Of course, the big crunch is speculated to take some trillions of years from now, and it is further speculated that once it has collapsed into the singularity of a big crunch, the universe may once again restart itself in another big bang by some unknown mechanism. And this cycle may be repeated for endless periods!

As can be guessed, other factors may also dictate the future and fate of the universe. The general geometrical curvature of the universe, if positive, may eventually start a phase of contraction (the curvature of the present-day universe seems to be perfectly balanced by the energy of its outwards expansion, thus making it a stable universe). A few researchers have tried to measure the general curvature of the cosmos at astronomical scales, but the results did not draw any big conclusions. Another factor which would possibly dictate the future of the universe is the balance between dark matter and dark energy, but these are enigmatic entities themselves, so we will not speculate any further.

The Big Rip (or the Big Freeze): If the density of the matter in the universe is less than the critical density, then the universe is likely to expand and expand forever, increasing its entropy continuously, moving the stars farther and farther away as time passes, ripping away the galaxies and the star systems, subsequently tearing up and dissolving all its planets, and ultimately annihilating all its atomic matter into subatomic particles and rays of light; this is the big rip. In simpler words, the big rip leaves nothing to exist, eventually making the universe empty of everything. The universe just vaporizes into nothing—what a dishonourable fate for our magnificent universe indeed!

Because of the inexorable increase in entropy (Ch 6), the warmth of this empty universe would conceivably fall to some infinitely abysmal temperatures—temperatures nearest to the absolute zero. Thus, this utterly cold and bleak state of the universe is also called the heat death (discussed in Ch 6-E). Because the universe eventually becomes absolutely cold and frigid, the big rip is also called the *big freeze*. But then what is the time taken for this doomsday to happen? Some really superlative time periods have been proposed, such as 10^{20} years, 10^{100} years, or even $_{10}10^{100}$ years—all beyond human imagination!

Today's scientists are in favour of the big rip over the big crunch because of the fact that the universe is in accelerated expansion. And moreover, recent estimates of the density of the universe, as measured by the WMAP, indicate it to be a tad less than the critical density which also points towards a big freeze!

In the preceding chapters, we have learnt about the theories operative in the macrocosm as well as microcosm; now we will go into our last chapter in Part I, which describes the mechanism of heat transfer in the universe.

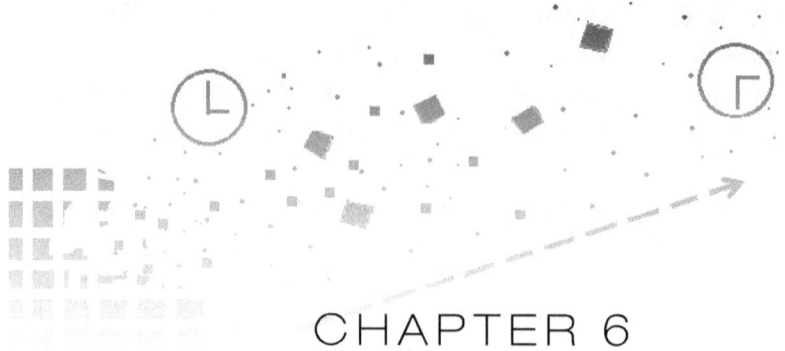

CHAPTER 6

The Arrow of Time
And the Laws of Thermodynamics

Overview

By this time being proficient, amateurishly though, in the general structure and functioning of the universe at a micro and macro level, the reader may now realize that we have left out an important aspect of our understanding of the universe—the exact way by which matter and energy interact in nature. We know that energy is utilized to move matter and to get work done in nature, but what are the guidelines nature follows in these energy transactions? Historically, the intricate issues of energy and work have been encased in a theory called thermodynamics, of which we will study in this chapter. Thermodynamics, putting in a nutshell, is the energy audit of the universe which fundamentally endorses an indomitable principle of nature that 'energy can neither be created nor destroyed' in the universe.

At first, we will try to assimilate the diehard laws of thermodynamics and see how the second law of thermodynamics especially stands out to set in a direction to the flow of energy. In the subsequent sections, we will tirelessly go through various theoretical and practical examples to highlight the unflinching supremacy of the second law in all the physical and thermodynamic events in nature. It is interesting to learn that the second law also plays an inevitable role in the various biological processes that occur inside of all living cells – in fact, we will see how the second law dictates the very rules of life and death on the earth!

The formidable second law is finally shown to have a profound cosmic implication in that it sets in the 'arrow of time', which incessantly takes the universe in a forward direction. In fact, this delineation of the arrow of time is the essential purpose of this chapter.

We will now start our journey from the mundane heat transfer to the cosmic arrow of time!

Section A
What Is Energy, and What is Work?

Energy exists in various forms in nature. Think of an excited electron whirling rapidly in space around an atom, the flapping flagellum of a bacterium, the constant flux of ions across the membrane in a living cell, a boulder of rock falling from a cliff, the earth orbiting the sun, the stars moving across galaxies, and so on. All these are actions in nature, and they represent several changes in the disposition of matter occurring at various levels—at subatomic or microscopic or macroscopic or even astronomical levels. And all actions in nature require energy in some form for their accomplishment. We have seen that energy is represented by the four fundamental forces (electromagnetic force, gravitational force, strong nuclear force, and weak nuclear force, Ch 2-A), and we may say that one of these forces (or more commonly, any of these forces in combination) is responsible for all events in the universe. In this chapter, we will learn about the various ways by which energy is exchanged in nature in order to execute work in the universe.

First we will try to define work and energy in clear terms. Scientifically speaking, any movement (i.e. displacement of objects in space) in nature can be designated as work done in the universe, and to do work, it needs expenditure of energy. In simple words, all movement and all work in nature needs expenditure of energy. We have already seen in Ch 1 that all motion in nature is relative, and all objects in the universe are in a continuous state of relative displacement. Since all objects are in motion eternally, we may state that energy expenditure is a continuous process in the universe.

But then what exactly is energy? Energy is usually defined as the capacity to do work, but the reader may notice that this definition is rather circular because work itself is generally defined in terms of energy changes. Thus, a more useful definition would be that 'energy is the ability to cause *change* in nature'. This change in the disposition of matter is measured as the work done during that interaction. To put it simply, energy is the ability to cause change in nature, and work is the measure of this change.

Energy is the ability to cause change; work is
measured as a unit of this change.

We will look a little further into the fundamental concept of change. More than 2,000 years ago, Heraclitus had intuitively said that *'nothing is permanent except change'*. As already seen in the example of the cosmic journey of a man on cruise in the Pacific Ocean in Ch 1-A, absolute rest is impossible in the universe, which prompts us to empirically assert that Heraclitus's *doctrine of change* is true to its meaning. Because change and movement are universal phenomena, it may also be concluded that work is also a universal phenomenon. This phenomenon of universal work underscores the total dependence of nature on the continuous supply of energy to do work (or to cause movement or to cause change) at all places and times in the universe. Consequently, it may be said that energy transaction is an unstoppable and relentless process in the universe. In the absence of energy, no change is possible, and without energy, all the activities in the universe would come to a standstill—and in that event, the universe would cease to exist.

Energy expenditure and work are universal phenomena.

We will now look into a few examples to illustrate the inevitable and ubiquitous nature of expenditure of energy and work in nature. If we take the case of a stationary body, such as a boulder of rock, we may show that this state of rest is only an illusion on two accounts: one, that the boulder itself is in a constant relative motion in the universe along with the earth in its cosmic journey, and two, that the atoms inside the boulder are in a constant state of flux (which includes the constant jiggle of electrons and quarks inside these atoms!). This makes the rock-steady boulder no more at rest but in a constant maelstrom of change, movement, and work. Here's another example. A man simply standing erect on his feet may also be said to be doing work in the universe because he is doing work continuously not to fall to the ground by way of contracting a group of his muscles (involuntarily, of course), and these muscles, in turn, do work by sliding and displacing the various strings of proteins inside the myocytes! It may be said that work is done even in the case of a chemical reaction by the way of movement of the electrons at an atomic scale. After all, all chemical reactions are based on the exchange of electrons between atoms. It is interesting to note that even an electron cruising in the empty space of a cathode-ray tube needs some energy in the form of electromagnetic force for its propulsion. Not only energy in the form of electromagnetic energy but energy in any of the four forms of fundamental forces may also become operative in causing movement in nature. For example, take the case of a planet hurtling round the sun; it needs energy in the form of gravitational force for its navigation. The tiny movements of quarks inside the protons and neutrons also need a strong force for their activity (Ch 3-C). We will get a clear idea of the work done during these gravitational and nuclear phenomena in Part II, but for now, the reader may simply realize that all activities in the universe (i.e. all the movements in nature) require energy expenditure— without which the universe comes to a standstill!

However, as can be guessed, the interaction of energy with matter in order to get work done is governed by certain fundamental laws of nature, and the branch of science which deals with these principles is called *thermodynamics.* The first theoretical foundations of this science were laid by Sadi Carnot in 1824, and many later scientists, such as Rudolf Clausius, Ludwig Boltzmann, and Willard

---

* The term *thermo* in thermodynamics has only a historical importance because this branch of science was originally concerned with a study of heat exchange in steam engines, but the principles of thermodynamics have ramified since then, and they can be now extended to any form of energy transactions among objects in the universe.

Gibbs, to name just a few, have developed this branch of science to its present-day understanding. Now we will study the laws that govern these matter–energy transactions.

Section B
The Laws of Thermodynamics

There are three fundamental laws associated with work done in the universe called the *laws of thermodynamics*. The *first law* states that energy can neither be created nor destroyed, and the total amount of energy in the universe remains constant, but energy may change its form from one form to the other. This law is straight enough. Simply put:

The total amount of energy in the universe is constant.

The *second law* states that for every physical or chemical process that occurs in the universe, the change that takes place during the process has always a tendency towards causing a greater disorder (or randomness). The reader may find this concept a trifle disconcerting at first, but it will be easily understood with a proper explanation (see below). The second law is by far the most important rule of nature which encompasses the essence of thermodynamics, and the purpose of this chapter is largely to explain this fundamental principle of energy exchange, as we will see in the subsequent sections.

The *third law* states that temperature of absolute zero is unattainable in nature, which would also mean that 'entropy' (we will soon see what entropy is!) cannot be equal to zero (this law can be taken as read at present as the reader will understand this as he/she sails forth in discussion).

There is also an offbeat *zeroth law* which supersedes all these laws and defines the fundamental meaning of temperature and heat flow between bodies, and this law also becomes self-evident as we proceed with the discussion. We will not deal with the third and zeroth laws specifically any more; rather, we concentrate on the second law exclusively in this chapter.

Temperature and Heat: Before moving further, we should make an important differentiation between temperature and heat. Temperature is a measure of the average molecular motion in an object. When energy is supplied to an object, the kinetic energy of its constituent molecules increases, and this increases the object's temperature. For example, water, in liquid form, when supplied energy, increases its molecular kinetic energy and becomes gas (steam), which has a higher degree of temperature. Temperature (symbol T) is measured in *Kelvin, centigrade,* or *Fahrenheit.*

Heat is the total amount of *internal energy** of an object (or a system). The molecules in an object have some internal energy in the form of both kinetic energy and potential energy (Ch 7-C), and thus, the internal energy (i.e. heat) of an object denotes the sum of kinetic and potential energy of its constituent molecules. For example, a block of metal of 1 kg at a temperature of 50 °C has more heat in it than a piece of metal of 1 mg at 100 °C. Heat (symbol Q) is measured in units of *joules* or *calories.* It is important to note here that when work is done in a system (a 'thermodynamic system' is a portion of the universe under our study of heat exchange, and by the term *surroundings,* we mean anything outside this system—a system could be a small cell in a pool of plasma, a human body in the ambient atmosphere, a small isolated oven in a room, the earth in the universe, or the entire universe itself!), it uses up its energy, thereby decreasing its internal energy. Thermodynamics, in general, takes the heat content of a system in its mathematical calculations, as we will see.

Now we will move on to discuss the second law exclusively in this chapter.

---

* The internal energy (U) of a system is the measure of the total kinetic energy and potential energy of that system. Kinetic energy, in turn, may be translational, rotational, or vibrational. Potential energy in a system is by the virtue of the forces acting within the molecules (intramolecular forces, such as covalent bonds and ionic bonds) and forces acting in between the molecules (intermolecular forces, such as van der Waals forces, Ch 10-B). However, internal energy of a system is different from energy harboured within the matter itself, which is the energy derived from disintegrating matter (this is the rest mass which follows the equation $E = mc^2$). Thermodynamics, in general, is concerned only with changes in internal energy.

The Second Law of Thermodynamics: This is the most important law for our purpose here. We will divide the second law arbitrarily into three divisions for our easy understanding. These three divisions are narrated at first without any explanation (knowing well that this may confuse the reader!), and later on, we will proceed with a discussion which would make sense of the second law. The reader is advised to just read these divisions as they are presented below and then follow up with the argument in the examples given below. At the end, all our mists of doubts and scepticism will clear away, giving us some unimpeachable rules of the universe. *First division*. The first and the foremost principle is that each change that occurs in *any process in nature must cause more disorder in the universe*. Nature abhors order in any form and promotes randomness. It follows that in nature, there is a natural tendency for any process which causes more disorder to be preferentially favoured. A measure of this disorder (or randomness) is called *entropy*. And such a process which is favoured in nature is called a *spontaneous process*. What is this order–disorder imbroglio, and why is it spontaneous? The examples supplied below would clarify the point.

Second division. The second law of thermodynamics dictates that when energy changes its form from one form to another while doing work, some amount of energy involved in the process is invariably released into the surroundings in the form of heat. The second law further states that *energy always flows from a high-energy state to a low-energy state*, and *this transfer of energy is accompanied by a release of some amount of energy into the surroundings*. In other words, while doing work, the total amount of energy involved in a system cannot be utilized to do work, but some energy is invariably 'wasted' as heat. This phenomenon of heat loss is *also* called *entropy*. Thus, entropy means not only an increased disorder in any spontaneous process in the universe but it is also a measure of the heat loss, which is invariable during such a process. All too much for the day! But again, anyway, the examples below would be of our assistance in our understanding of this law.

Third division. If the above two divisions were not intimidating enough, this last division certainly is! In the first division, we have confidently asserted that nature will always promote disorder, and this would mean that nature will *not* allow order to take place spontaneously. However, here is a caveat. Nature appears to allow a *local increase in order* in certain situations. While we have termed the increased disorder as entropy, we will name this *increased order* (or decreased disorder) as negative entropy. Now we can see that negative entropy is also possible in nature as a *non-spontaneous* process wherein *order is increased*, and in this case, the *energy is not released but is absorbed*. But here again is an important catch. This local increase in order (or decrease in disorder, i.e. negative entropy) is *invariably* accompanied by a simultaneous *increase* in disorder *elsewhere* in the universe. In other words, this means that locally, the disorder may decrease, but there is a general (or an overall) increase in disorder in the universe. Thus, it is essential to note that no thermodynamic event in the universe shall disobey this increase in the *total* disorder in the universe. This phenomenon has some astounding implications, as we are going to see. But first of all, what do we mean by such terms as order/disorder, local/general, and spontaneous/non-spontaneous? Once again, the following examples would clarify the situation.

The reader is advised to understand the laws of nature not merely as three separate divisions but as a fundamental whole which signifies the basic property of nature which happens inevitably while work is done in the universe. The following discussion would, hopefully, assist the reader to understand the second law as a fundamental whole. We will study the second law under four headings:

- Entropy as disorder
- Entropy as heat loss
- Negentropy and order
- Gibbs free energy

Entropy as Disorder: This is the first division of the second law. We have seen in Sec-A that all physical or chemical processes (or events) that occur in nature involve a change in the disposition of matter, and this change can be measured as the work done. The second law states that for any work done in nature, there is an increased disorder. In other words, nature has a tendency to increase the randomness* while doing work. This is entropy. We will now start with our discussion on entropy by using some common examples.

Let us first take a chemical change. For example, the process of burning a piece of paper involves, at the molecular level, the conversion of large molecules of cellulose into smaller molecules of carbon dioxide and water accompanied by the liberation of energy (in the form of heat and flame). Now if we consider the molecular arrangement of paper, it is composed of a compact and a more or less orderly pattern of strands of cellulose, and upon combustion with oxygen, it is converted into more disorderly (or less orderly) gaseous molecules of CO_2 and H_2O. Suppose we want to somehow convert these gaseous products of the reaction back into a piece of paper? We know that this is nearly impossible—even theoretically. In other words, we can easily run the reaction forward to burn the paper, but we find it difficult to run it back. Now we can call this forward reaction a spontaneous reaction, whereas the other is not. In other words, we can say that nature tends to favour the spontaneous reaction, which increases randomness. This means that the nature tends to convert the orderly cellulose molecules into disorderly gas molecules more readily by liberating energy than to rebuild the cellulose, which is a tedious non-spontaneous process and which is also an energy-consuming process (see below). But here, the reader is advised not to jump into any rapid conclusions right away at this stage; rather, he/she should patiently tread along with the discussion for some more time.

Now let us see some physical changes. A porcelain cup falls from the top of a table and breaks down into pieces. To break a cup this way is but natural, but to reconstruct a cup back to its original shape is practically impossible (if not, the crockery industry would soon go bankrupt!). The cup is in a state of great order with all its perfect shapes and curves, and the broken pieces are in a state of relative disorder. Here again, a process which brings about disorder

---

*   While it is a standard practice to use the term *disorder* or *randomness* here, some authors use the term increased/decreased *rarefaction* in their place, which really would be a better term.

is favoured by nature. Here's another example. A quarry worker strikes a stone boulder into pieces. To break the stone this way is easy and 'natural', but to reconstruct the stone back to its original shape is practically impossible—i.e. nature favoured a change which causes more disorder. In the same way, constructing a building is difficult and time-consuming; bringing it down to rubble is easier and quicker. But still, the reader is advised not to arrive at any judgement right away.

Similarly, think of gasoline which powers the engine of your car to move ahead to take you to your office. In order to do work (i.e. to move the car), the more orderly hydrocarbons in the gasoline have to be converted into more disorderly carbon dioxide and water (Fig 6.1). This process liberates huge amounts of energy, which is utilized to move the pistons and to propel your car. But reassembling CO_2 and H_2O into gasoline molecules is a ridiculously impossible affair. Order in nature is readily disrupted to give more disorder, and these energy-liberating processes are favoured.

We will take in a few more examples. Each morning, your orderly egg in the basket breaks up into delicious pieces of scrambled egg on your breakfast plate quite easily, and it is just impossible for you to organize an egg back into its original order. Here's another example: inside our bodies, the orderly glucose molecules are converted into more disorderly CO_2 and H_2O molecules by cellular respiration (Sec-D), which liberates energy, and this energy is used to do the cellular work. Disorder is easy, quick, and generally acceptable in nature; order is difficult, tedious, and generally abhorred by nature. A measure of this disorder which accompanies various physical and chemical processes in the universe is called entropy.

Entropy is the measure of disorder.

We can notice that all these spontaneous processes result in the release of energy into the surroundings. For example, burning of paper liberates energy (in the form of heat and light), and breaking stone releases energy (in the form of a spark, vibrations, and sound). Similarly, the burning of fuel in your car liberates energy to move the piston up and down, which in turn runs the engine. The nutrients in your breakfast egg break up inside your body cells (by cellular metabolism) to give you the energy to power your muscles to get work done (i.e. your daily chores). You can notice that all these processes increase entropy and liberate energy and are spontaneous. The paper, the gasoline, the carbohydrates, and the proteins all have more energy in them than their final by-products. It means that the *heat content* (called the *enthalpy* in thermodynamics) of the reactants is always higher than the heat content of the resultant products in a spontaneous reaction, and the difference in the amount of energy is released into the surroundings. Moreover, it should be noted that all the spontaneous reactions are *thermodynamically irreversible*. In the parlance of chemistry, these processes are called *exothermic reactions*—the reactions which liberate energy into the surroundings. Nature tends to favour exothermic reactions.

We will now generalize the phenomena. It becomes obvious to the reader that in nature, entropy (i.e. increased disorder) is a favoured change and is a spontaneous process. All spontaneous reactions result in liberation of energy into the surroundings.

The nature of the second law is rather intuitive enough. For example, when you put a hot cup of coffee in a colder room, a gradient of potential is created between the hot cup and the room, and heat flows spontaneously from the cup to the room until the potential difference between them becomes gradually diminished and their temperatures become equal, at which point the to-and-fro heat exchange between the two bodies becomes constant and equal (i.e. the cup and the room can now be said to be in a state of thermodynamic equilibrium). In practical terms, we can say that no further exchange takes place between them.

Energy flows from higher level to lower level.

That is to say that heat flows from higher energy to lower energy until the thermodynamic system reaches a state of equilibrium. Nature tends to minimize the potential gradient between the local system (cup of coffee) and the surroundings in general (the surrounding air). In other words, nature tends to increase entropy of the universe in general; the total amount of disorder in the universe increases as time proceeds. Thus, the essential principle of the first division of the second law is:

Nature tends to maximize the total disorder in the universe.

Entropy as Heat Loss: Now let us turn to the second division of second law—the phenomenon of loss of energy as heat during the process of doing work. We have seen that energy is spent to do work in nature—i.e. while work is being done, energy is transferred from one place to the other, and generally, while doing work, one form of energy is converted into the other form. For example, when we supply electricity to a rotor, electricity is converted into mechanical energy to drive the wheels. The second law states that when energy is converted from one form to another while doing work, it is invariably accompanied by certain amount of heat loss into the surroundings. This means that while work is being done, the supplied energy to do work is not totally converted into work— i.e. 100% conversion of energy in a system is not possible; some amount of energy is inevitably lost (or 'wasted') as heat into the surroundings. A measure of this heat lost into the surroundings is called entropy.

Entropy is a measure of heat loss while doing work.

The following examples would clarify the point better. Take the case of fuel (gasoline) in your car. The energy stored in the fuel is liberated in the form of heat which moves the pistons, and the pistons in turn move the wheels. The second law states that the total energy stored in the fuel cannot be converted into mechanical energy, but some significant portion of it is lost into the environment and is thus wasted, and this energy cannot be utilized to do any useful work (Fig 6.1). This heat loss is inevitable in all engines, however efficient they are—i.e. no engine is 100% efficient. It is estimated that even our most efficient automobiles in general can use less than 50% of energy stored in their fuel to do the desired mechanical work, the remaining energy being lost into the surroundings. In fact, this wasted heat is the reason why our car engine gets heated up gradually as we drive!

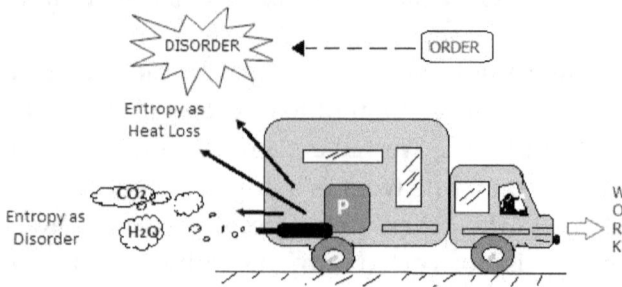

Fig 6.1: Entropy as disorder and entropy as heat loss.

We will now take a biological example. Cells are our machines to convert food (in the form of glucose) into energy to drive our activities. Our cell's molecular machines (which run the metabolism) break down the glucose in order to liberate the stored chemical energy, and this energy is then converted into various mechanical and electrical activities of the cell (i.e. the work done in the form of movement of cilia, flagella, muscle contractions, ion transfer across cell membranes, and various other metabolic activities). However, it should be noted that our cellular machinery also cannot convert 100% of the stored energy of glucose into work, but some significant amount of it is lost into the surroundings. But quite surprisingly, our cells are far more efficient than any man-made machine, and they can yield about 55% of the energy of glucose.* And interestingly, human beings and all warm-blooded animals can exploit this lost heat to keep their bodies warm at an optimum temperature, which, in turn, helps in running their metabolism most effectively. It's a nature's wonder that some plants also use this lost heat to their advantage—for example, the skunk cabbage (*Symplocarpus foetidus*) uses the generated heat to sprout early during the spring season by melting away the overlying winter snow!

---

* It would interest the reader to note that complete oxidation of a molecule of glucose to CO_2 and H_2O outside the cells (e.g. by burning it in a lab) yields about 686 kcal/mol. In a living cell, under optimal conditions, a single molecule of glucose can generate about 36–38 molecules of ATP, which yields a total of 360–380 kcal/mol of energy, which is an energy efficiency of 52–55%, and this is very high when compared to the efficiency of many man-made automobiles.

We will now look into some more examples so that the reader would realize the universality of the phenomenon of entropy. Consider a worker pounding rock in a quarry. The worker delivers his power on to the rock through his hammer strikes. While the energy of the hammer is utilized to break the rock, all energy cannot be converted into mechanical energy, but some of it is lost into the surroundings in the form of heat and light (resulting in heat generation and perhaps a few sparks of light). Here's another example. The electricity we supply to an incandescent bulb converts electrical energy into light, but during this process of conversion, some electricity is invariably lost in the surroundings in the form of heat (which heats up the metal filament and the bulb). Thus, the essential principle of second division of the second law is:

In any energy transfer during work, some energy is lost as heat.

The Order of Negentropy: Now we will proceed further to appreciate the third division of the second law—how the nature allows order to happen. The first division of the second law irrevocably asserts that nature always tends to favour disorder. But then the reader may disagree with this statement by making certain observations. In nature, we keep seeing *order* to happen at all times and at all places, and it is also commonplace for us to see events which take in energy but not give out (see examples below). Does this mean that 'entropy as disorder and heat loss' is not a steadfast rule of nature? Could the nature have defied its own law? Not at all. Entropy in the universe is a staunch principle, all right! But then how do we substantiate these seemingly contradictory events of disorder and order? How could nature allow order formation in the universe in the first place when it so steadfastly prefers disorder?

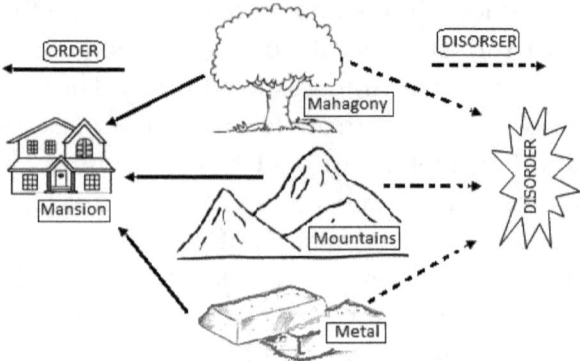

Fig 6.2: There is disorder in order!

The reader may note here that while entropy denotes increased disorder, the phenomenon of 'decreased disorder' (or 'increased order') can be termed *negative entropy*. This term was introduced by Erwin Schrödinger in 1944, and negative entropy can further be conveniently shortened to *negentropy*—a term we are going to use frequently from now on.

We will now see how the nature plays a trick to achieve order. Nature allows order formation (i.e. negentropy) *only* when this order formation at a local region is accompanied by a *simultaneous* disorder formation elsewhere in the universe so that the *total* disorder in the universe in general invariably goes up anyway. In other words, nature allows negentropy at a local region in the universe only when it is accompanied by equal or more entropy in the universe at large. This is the nature's way of bargaining for the best!

Local negentropy is always accompanied by
increased entropy in the universe.

We can understand this uncanny principle of nature using some examples. We will take a simple example at first. To construct a farmhouse, it needs to build up a lot of order—the disorderly stones have to be piled up neat, the irregular woodblocks are cut to fashion, and clods of metal are designed into bars and plates (Fig 6.2). All these efforts would result in an increase in order in the locality of a farmhouse. But consider this—to achieve this order, it always needs the workmanship of several craftsmen and machines. And thus we can see that a lot of energy input is necessary to accomplish the work. And also, to cause this order locally (at a particular place in the farm), we have to create a lot of disorder in the vast surroundings (perhaps in the nearby jungle) by the way of pulverizing the rocky mountain to get stone, chopping many a green tree to get wood, flexing many a human muscle to mould the metal, and pounding many a grain to provide food to all the construction workers (who would burn orderly carbohydrates into disorderly CO_2 and H_2O). Each one of these activities invariably gives out lots and lots of energy and causes more and more disorder in the surroundings (than the order that would be achieved by building the farmhouse). The reader may argue that if we can use stone crushers and electric saws, instead of human labour, we may save upon our energy and cause less disorder. But then the reader may remind himself/herself that the electricity we obtain from various sources doesn't come cheap. We need a thermal station to burn the coal to run the huge turbines which anyway would liberate heat, steam, and smoke into the surroundings (and this would anyway increase the entropy). And more or less the same amount of disorder is generated even if we happen to use hydroelectricity or nuclear energy* because ultimately, the turbines, while generating electricity, cause disorder. After all, there is no free meal in the universe!

Now we will look into the business of paper manufacturing, which involves a highly orderly process, and we will see how energy gets trapped into a piece of paper during its making. And we will also see how this local building up of order is countered by nature to increase the overall disorder in the universe! Paper needs cellulose, and to get cellulose, we need plants, and plants synthesize cellulose little by little in their leaves by the way of photosynthesis (Sec-D). Photosynthesis basically involves conversion of *disorderly* molecules of carbon dioxide (OCO) and water (HOH) into *orderly* chains of carbohydrate (. . . CHO–CHO–CHO . . .). This means that the disorderly molecules of CO_2 and H_2O in gaseous form are bound down into orderly stacks of CHO inside the cells of plant—i.e. the order of molecules has increased. We must also notice here that

---

*   Whereas hydroelectricity uses potential energy of water falling from heights, nuclear energy uses potential energy stored in the nuclei of heavy atoms; these issues are discussed in Ch 7.

during photosynthesis, energy (in the form of sunlight) is trapped inside the carbohydrates—i.e. energy is absorbed into the local system (here the local system is represented by a leaf). The reader may now realize that it is this trapped solar energy which is liberated into the surroundings when the paper is burnt. Thus, we can say that the flame we see when we burn paper is nothing but sunlight!

Those chemical reactions which require energy and cause increase in order are called *endothermic reactions*. Thus, photosynthesis is an endothermic process. However, here in this chapter, we keep using the broader term *negentropy* instead of endothermic reaction. This is because negentropy is a generalized term that can be used to signify any process—either physical or chemical or any other process (as we will be seeing) which results in an increase in order of nature and which requires an input of energy into the local system.

Negentropy brings in order in the universe and stores energy.

Thus, this first step of the synthesis of cellulose has caused an increase of order in the universe and involves absorption of energy from the surroundings into the system (i.e. leaf). Now to get paper, we must press cellulose into sheets. In a paper factory, the tree bark (containing cellulose) is extracted and pulped by using tremendous amounts of mechanical energy, and this pulp is again pressed, dried, and cut by using still more mechanical energy. All these processes invariably result in a further increase of the order of the cellulose in the form of sheets and stacks. The reader may realize now that to manufacture a roll of paper, an enormous increase of order in the universe has to take place and at the cost of input of a lot of energy.

But then the big question is how nature allows order to build up in the form of paper when it is not thermodynamically favoured. As we have seen in the example of farmhouse construction, this local increase in order (i.e. negentropy) is also accompanied by a simultaneous *increase* in disorder *elsewhere* in the universe (i.e. entropy). We will now see how exactly this is achieved here in papermaking.

First we will see the happenings in a leaf. The increase in order caused by cellulose formation in the case of photosynthesis represents negentropy, but this process has an important entropic component, and that is the liberation of oxygen into the surroundings. Thus, it is the release of O_2 gas which counters the negentropy of photosynthesis. Now consider this: the increase in order caused by papermaking in a factory is also invariably countered by an increase in disorder elsewhere in the surroundings. The machinery used to do the pressing and cutting has burnt lots of fuel, resulting in the generation of lots of heat and smoke which, of course, has resulted in increased entropy. This means that the order that was built up 'locally' (in the form of paper) is accompanied by an equal (or more) disorder in the surrounding universe by the way of release of O_2 gas from leaf and smoke and heat from the paper factory. In other words, the overall disorder in the universe has gone up even if some order has taken place locally. Nature made a good bargain—it allowed order at one place just by allowing more disorder to take place elsewhere in the universe. Thus, the essential principle of the third division of the second law is:

Negentropy is invariably accompanied by equal or more entropy.

We may take more examples and see if this invincible principle holds good at other places in the universe too. To make a porcelain cup, it needs a lot of input of energy—sand is first pulverized into silica, moulded into shapes, and then dried, glazed, and designed. All these activities need machinery and expenditure of energy, which in turn cause a lot of entropy. Here's another example. When we cool our food inside our refrigerators, we might be increasing the order by lowering temperature, thereby decreasing the kinetic energy of the molecules (which would allow molecules to pack closely and more orderly). But then our surroundings would simultaneously get heated up in the process (touch the back of your fridge, and you would know the truth in this statement!), and this would create more disorder in the surroundings. So next time when you switch on your air conditioning, be mindful of the global warming you would be causing quite inadvertently!

Now consider another interesting example: a minuscule ovum in the mother's womb develops gradually into a fully grown baby with many organs and structures. This is an example of a tremendous build-up of order of nature indeed! But then how does this sort of biological order create a corresponding disorder in nature? The explanation is that the mother metabolizes lots of extra nutrients in her body during childbearing, liberating extra energy which drives a variety of nutrients across the placenta to build up more order step by step in the form of complex organs of a developing foetus. However, this increased metabolism is invariable, accompanied by the liberation of more CO_2 into the surroundings, which effectively compensates for the building up of the order of the foetus!

We will now attempt to understand negentropy in a broader perspective. If you look at our solar system, it is an orderly system with all the planets circling round the sun in more or less well-defined orbits, and gravitational force is needed to bind them together. On the other hand, if you consider the structure of an atom, it is also an orderly system with all the protons and neutrons stacked together in a nucleus with all electrons moving around in orderly orbits. What is this order, and how do our thermodynamic laws apply here in these vastly different contexts? It's a perplexing question, and we will take up these issues in Part II. For now we will proceed with our examination of the essential thermodynamic principles.

In summary, we may conclude that according the second law of thermodynamics, disorder in the universe is a preferred state. And we may say that in any event of work done in the universe, the phenomenon of disorder (entropy as disorder) increases along with the release of energy into the surroundings (entropy as heat loss). Thus, it appears that the essential purpose of nature is to strive to achieve this goal of entropy, but then we will have a broader understanding of entropy in Ch 7-E.

In the following section, we will briefly study the second law in some mathematical terms, which has a practical bearing in understanding thermodynamics.

Section C
Gibbs Free Energy

We have seen that when energy is used to do certain work in a system, some amount of energy is utilized to do the prescribed work of that system, and the rest of energy becomes 'useless' and is dissipated as heat into the surroundings. But then how do we measure this lost heat? The reader must understand that there is no means by which we can directly measure the entropy of a system itself. Nobody has ever 'seen' any entropy happening in a system. After all, entropy has been only a mathematical concept.

But then there is an indirect way by which we can measure the entropy of a system, and that is by mathematically calculating the amount of work done by that system. And herein comes the concept of *Gibbs free energy*. Gibbs free energy can be understood in a better way if we think of it as 'the amount of "available energy" in a system to do the useful work', and the remaining (i.e. 'lost' or 'useless') energy, as we have seen, represents entropy. Thus, Gibbs free energy can be really thought of as a measure of the amount of work done by a system.

Gibbs free energy is the available energy in a system to do useful work.

Thus, in practice, we can calculate the heat lost by deducting the utilized energy from the total energy of that system, and we can mathematically depict this concept in the following equation:

$$\Delta G = \Delta H - T\Delta S$$

where ΔG is the change in Gibbs free energy of the system, ΔH is the change in enthalpy (i.e. the 'heat content' of the system which represents the total energy of the system at a given time), T is the temperature in Kelvin, and ΔS is the change in entropy. But then we have seen that for any spontaneous process, the entropy of the system (ΔS) should increase. And going by this equation, this should also mean that for any spontaneous process, the Gibbs free energy (ΔG) must decrease. In other words, for all spontaneous processes, the value of ΔG must bear a negative sign. Now the reader may see that this equation can determine the direction in which a reaction can proceed (Fig 6.3). In any reaction, if the ΔG is decreased ($\Delta G < 0$), it is a spontaneous process (also called an *exergonic* reaction); and if the ΔG is increased ($\Delta G > 0$), it is a non-spontaneous process (also called an *endergonic* process). However, by looking at this algebraic equation, we can say that there are three ways of decreasing ΔG—if we decrease ΔH, then the ΔG decreases; or if we increase the ΔS, it will decrease ΔG (because of the equation's negative sign); or if we increase T, then also the ΔG is decreased.

Thus, there are three chief players which determine the spontaneity of a chemical reaction in nature (as indicated by the value of G)—the enthalpy of the reactants/products constituting a system, the entropy of the reactants/products the system, and the temperature of the system. If the result of the interaction among these three players gives a *negative* G value ($\Delta G < 0$), then the system would proceed spontaneously—i.e. no energy input is needed to do the work. If the result is *positive* ($\Delta G > 0$), the system would be a non-spontaneous process—i.e. some external input of energy is needed into the system to make it proceed in this direction. Once again, these issues are understood better by the way of some examples. The following four situations show us how these players shift the value of ΔG, thus affecting the directionality of a reaction.

Situation 1. In a thermodynamic system, if the heat content of the products is lesser than the reactants and if the entropy of the products is more than the reactants, then the process would run forward spontaneously. We have seen several examples of such processes in the above sections. For example, in the case of starch burning into carbon dioxide and water, the heat content of the products (CO_2 and H_2O) is less than starch (the remaining energy is liberated as flame), and also, the disorder of molecules has increased (so entropy has gone up). These two changes in the system are thermodynamically favourable, and thus, it is a spontaneous reaction. If we apply the Gibbs equation: ΔH is decreased, and ΔS is increased, both of which drive the ΔG negative (Fig 6.3a). Thus, it is an exergonic reaction, and it will run spontaneously without any external energy input into the system. There are several other examples in nature. The process of rusting of iron is an exergonic process. When iron oxidizes to iron oxide, it liberates energy and increases disorder (though these changes go unobserved as it is a slow process). Another example is water oxidizing elemental sodium into sodium hydroxide, liberating enormous amounts of heat. Of course, cellular respiration, the basis of life of all organisms, is another leading example of an exergonic reaction and will be discussed in Sec-D.

As an aside, it is interesting to note here that the elements hydrogen and oxygen may be used as a source of carbon-free fuel in future because they combine spontaneously and exergonically into water, liberating significant amounts of energy which can be utilized to power our jets and motors.

Situation 2. We will see what happens if a reverse of the above happens in a system. If the heat content of a system increases (which is thermodynamically unfavourable) and the entropy decreases (which is also unfavourable), then the ΔG would become positive, and the reaction will not run spontaneously (Fig 6.3b). This is because both of these changes are not thermodynamically favoured by nature. However, if we have to drive the reaction forward, we have to supply external energy into the system. This is an endergonic reaction. The leading example of an endergonic reaction is photosynthesis (Sec-D).

Situation 3. There are a few situations in nature where even though the heat content of products increases (which is thermodynamically unfavourable), the reaction proceeds forward because here, the increase in entropy is so much that it overrides the increase in heat content. If we look at the Gibbs equation, we can see that ΔS can be increased exponentially by increasing T, which in turn makes the ΔG negative (Fig 6.3c). In other words, this means that if we increase the temperature of the system, the reaction can proceed forward. Take the conversion of ice to water as an example. When ice melts, the heat is absorbed into the system, which is thermodynamically unfavourable, but the entropy increases because the molecules of water are more disorderly when compared to molecules in ice (which is favourable). Thus, the change in this situation is favoured only at higher temperatures. Therefore, we can observe that if we put ice at low temperatures, it will not melt, but it melts at higher temperatures. One would never have thought there could be so much of cool science in an ice cube in a glass of whisky!

Situation 4. We will now reverse the situation wherein the heat content of products of a reaction decreases (which is thermodynamically favourable) but the entropy also decreases (which is unfavourable). The reaction in this situation can proceed in this direction only if the temperature is lowered. An example is the conversion of water to ice. In order to make ice, we have to take away heat from water ($\Delta H < 0$), but then the formation of ice decreases disorder ($\Delta S < 0$). This reaction is possible only when T is lowered, which, in turn, lowers the value of ΔS so that a small increase in entropy will not affect the ΔG (Fig 6.3d). Thus, the formation of ice is favoured at low temperatures.

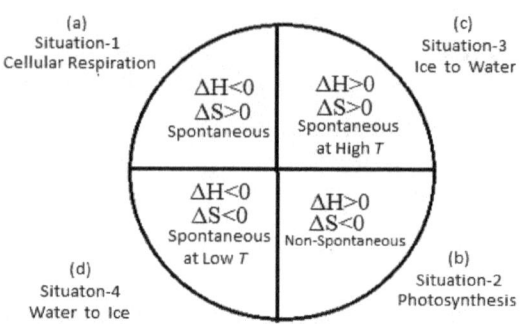

Fig 6.3: Gibbs free energy change in various examples.

Having studied the theoretical background of the principles of thermodynamics, we will now look into some of the implications of thermodynamics in nature.

Section D
Thermodynamics and Life

How do these theoretical laws of thermodynamics affect our lives? The implications of thermodynamics may be studied in three ways— technological (or industrial), biological, and cosmological.

Thermodynamics has, in fact, started originally as a practical science. Our lives in the present generation are so comfortable and efficient owing largely to the industrial and technological developments which, in turn, are directly related to advances in thermodynamic applications used to develop engines and machines. We may even say that the heart of the industrial revolution lies in the laws of thermodynamics. However, the ecological side effects of the industrial revolution in the causation of pollution and global warming have been due to the machinations of human misadventure rather than the advancement itself! We will not go into the details of all these affairs here in this book, but we will discuss the other two important implications of thermodynamics which are of direct relevance to our purpose of study here. One is the theoretical understanding of energy transactions in life processes on the earth, and the other is the energy transactions in the universe at large. We will briefly study the basic principles of energy transfer in living cells (*bioenergetics* is the name of the subject) here in this section and study about the thermodynamic arrow of time in Sec-E.

The Thermodynamic Cycle of Life: We have already seen that movement and change in the universe are continuous and unstoppable processes, and energy is needed to cause any change (or movement) in the universe. Life is no exception. In fact, all life processes are fundamentally based on a continuous expenditure of energy in the cells. When movement and expenditure of energy cease in a cell, then it can be considered that life processes have stopped, and it is a dead cell. What is this movement in a cell, and where is the energy spent? What sort of changes life brings about inside the cell? Hereunder, we will have a brief discussion on life processes in a cell from the thermodynamic point of view.

First we will see how energy is spent inside a cell to bring about changes that are vital to its survival. Living cells, in order to survive, have to spend energy to maintain a steady *internal milieu* (internal environment) by an important mechanism called *homeostasis*. To understand homeostasis, we must know about the internal milieu. All living cells contain water inside them in great proportions (about 70%), and all cells also are continuously bathed in a medium of water and electrolytes outside them. However, there is a difference in the electrolyte concentration on both sides of the cell. Grossly speaking, a living cell contains more of potassium and less of sodium inside them, and the outside medium contains more of sodium less of potassium. All cells, without exception, follow this rule as long as they are alive and will have to maintain this status.* Now consider this: the cell membrane which covers the living cells is a *semipermeable structure*, and thus, as the cells keep operating in the environment, potassium escapes out of the cells, and sodium drives in. This is a continuous and unstoppable process in all living cells, and it is the inevitable consequence of the membrane's permeability. This ionic transfer in this direction does not require energy; this is just a simple diffusion process towards the concentration gradient. Nevertheless, to restore homeostasis, the cells have to pump Na^+ ions back to the outside and bring K^+ ions back into the cells by using the *sodium–potassium pumps* situated in the membranes of all cells (whether it be a bacterial cell or a plant cell or an animal cell). This is the most vital step in a cell. But then this ionic shift to *status quo* can only be performed by the expenditure of energy because these ions have to be shifted thermodynamically across the membrane *against* their concentration gradients.

Now consider this: since this disturbance in the ionic distribution in a cell is a constant and continuous process, the energy needs of the cell to restore homeostasis must also be constant and continuous (i.e. as long as the cells are alive). The only way cells could get a continuous supply of energy is to store energy in them and use it bit by bit to maintain homeostasis.† The process of storage of energy is important not only because the availability of energy in nature is at a premium but also because energy is not in continuous supply in nature (even solar energy is not continuously available to the plants, e.g. at nights).

Thus, we can say, thermodynamically speaking, that the essential purpose of life is to procure energy from the surroundings, to store it, to spend it on a continuous basis, and to return the energy back

---

* This electrolyte distribution is a gross simplification, but in reality, there are several other players such as calcium (Ca^{++}), chloride (Cl^-), hydrogen ($H+$), sugars, amino acids, and nucleotides which influence the transport process across membranes. However, the most fundamental elements of all life processes are sodium (Na^+) and potassium (K^+).

† However, it is important to note here that energy is also needed in a cell for other purposes apart from doing this vital work of homeostasis. For example, a cell also spends energy to perform activities, such as locomotion, growth, heat maintenance, reproduction, and defence.

to the surroundings. After all, life too has to follow the principle of conservation of energy. Thus, life can be defined thermodynamically as an effort of cells to balance these events.

Life is a struggle to procure, store, spend, and return energy.

We will now see how cells do the trick of procuring, spending, and returning energy. The primary source of energy for life on earth is sunlight (Fig 6.4).* Plant cells take in solar energy and store it in the energy-rich carbohydrates by converting atmospheric CO_2 and H_2O, and this process is accompanied by a concomitant release of O_2 into the surroundings. The plant cells need this energy to do their cellular work (such as homeostasis, growth, and defence), and to do this, they break down the stored carbohydrates little by little. This process of breaking of carbohydrates requires O_2 for the oxidation of glucose (plant cells also need O_2 to break down carbohydrates), which again liberates CO_2 and H_2O back into the atmosphere. So the cycle of CO_2 and O_2 is completed (Fig 6.4). Thus, plants are self-sufficient and can survive on their own on the earth, provided they are given sufficient sunlight.

Thus, we may say that it is by endothermic process that cells store energy, and it is by exothermic process that they release energy. The leading endothermic reaction involved in biology is *photosynthesis*, and the leading exothermic reaction is *cellular respiration*. In general, plants perform both photosynthesis and cellular respiration in a cyclic way and are thus self-sufficient ergonomically by themselves (Fig 6.4). However, animals can perform only cellular respiration. They do not perform photosynthesis, but they get energy from preformed sugars from plants or other animals. While some of the animals can use plant carbohydrates for their energy needs (e.g. *herbivores* such as cattle and rabbit), a few cannot use plants, but they use other animals for their stored food (e.g. *carnivores* such as lion and tiger). But most of the animals are versatile; they are *omnivores*—i.e. they use both plants and animals for their energy needs (e.g. man and dog).

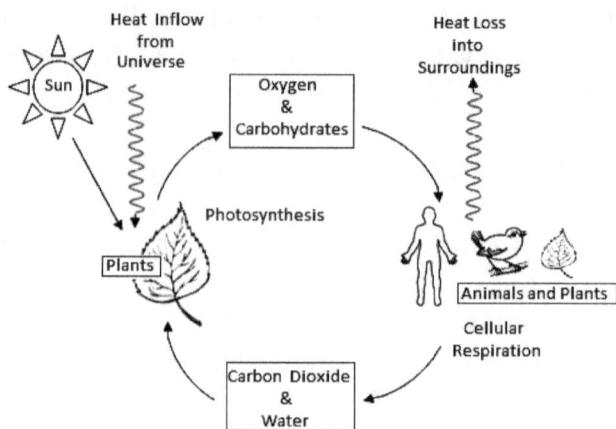

Fig 6.4: Flow of energy in life on the earth.

We will not go into any more details of biology, but herein we will be dealing briefly with only the thermodynamic meaning of energy exchanges in life processes. We will start by answering the following basic questions: We know that sunlight is stored in a molecule of glucose, but where exactly is this energy stored in a molecule of glucose? How is this stored energy of glucose released and delivered to the cells?

Before going further into discussion, the reader must understand one important fact: most of the atoms in nature do not exist as independent units but associate themselves with other atoms of the same kind (H_2, O_2, etc.) or more commonly with atoms of other elements to form molecules (H_2O, CO_2, etc.). Atoms in a molecule associate between themselves by the way of *chemical bonds*. It must be stressed here that energy is *released* into the surroundings whenever bonds form between atoms. And energy is *absorbed* from the surroundings whenever molecules dissociate by breaking down these bonds (see similar analogy in Ch 3-E).

Making bonds releases energy. Breaking bonds requires energy.

Here we will introduce the concept of *bond energy*. The reader has to know the correct meaning of bond energy in order to understand the following discussion. We know that there exists some amount of energy between the atoms in a molecule, which is called bond energy (see Ch 3-E for a similar entity called nuclear binding energy). However, bond energy does *not* mean energy stored in chemical bonds in a molecule. It rather indicates the amount of energy needed to *break* the bonds between atoms; the reader will appreciate this statement better as he/she reads on.

Bond energy is the energy needed to break bonds in a molecule.

Consider this: when 'strong bonds' are formed between atoms in a molecule, more energy is liberated out into the surroundings. And thus, more external energy is required in order to break them apart, and when 'weak bonds' are formed, less energy is liberated, and so less energy is required to break them. Moreover, in general, the more the number of bonds in a molecule, the more energy is needed to break it. For example, a C–C bond has a bond energy of 80 kcal/mol, whereas C=C carries a bond energy of 145 kcal/mol, and for a C≡C bond, it is 200 kcal/mol. Now we will make an important interpretation. By the above analogy, we can say that the 'energy content' of a C–C molecule is *greater* than the energy content of a C≡C molecule (because during the formation of a C–C bond, less energy is liberated, and during C≡C formation, more energy is liberated), and so *less* energy is needed to break the former than the latter.

Now apply this to chemical reactions. If the total energy content of the reactants is higher than the total energy of the products, the difference in energy is released into the surroundings. This is an exothermic reaction, which is favourable in nature. If the reactant's energy content is less than the product's energy content, then energy is absorbed into the system. This is an endothermic reaction, which is unfavourable. For example, formation of water from hydrogen and oxygen is exothermic and spontaneous ($2H_2 + O_2 = 2H_2O + 135\ kcal$), whereas decomposition of water into hydrogen and oxygen needs the same amount of energy to drive in that direction, and so it is endothermic and non-spontaneous. We will now see life's processes in this view.

In photosynthesis, when CO_2 and H_2O molecules combine to form CHO–CHO–CHO molecules (carbohydrates), the energy content of the reactants is less than the product's energy content (because both CO_2 and H_2O molecules have high bond energy, and CHO–CHO–CHO molecule has low bond energy). Thus, in effect, we can say that 'a lot of energy' is stored in the carbohydrates (meaning that less energy is needed to break bonds in carbohydrate). Thus, the formation of carbohydrates is an endothermic reaction, which will not happen unless some energy is supplied to the system from the surroundings, and so sunlight is needed to drive this process, and this we call photosynthesis.

We will analyze this further. In order to form carbohydrates from carbon dioxide and water, the CO_2 and H_2O molecules have to first split up. But this breaking up needs energy (remember: 'making bonds releases energy; breaking bonds requires energy'). The bonds in CO_2 and H_2O molecules are high-affinity, high-energy bonds; therefore, they require high amounts of energy, and this energy is provided by the sunlight. The atoms now recombine (step by step in a series of reactions in the chloroplasts) to form C–H bonds, which are relatively low-energy bonds, so they will not liberate much energy. Thus, in this process, a majority of solar energy is still trapped in the carbohydrates, or we may say that during photosynthesis, a lot of sunlight is stored in the sugars. This is the essence of photosynthesis (we will not go into its complex details). But then finally, we will now look into the change in G value during photosynthesis. In photosynthesis, the ΔH is increased, and the ΔS is decreased, and the resulting ΔG value is increased (+ 686 kcal). A positive ΔG value indicates a negentropic process (Fig 6.3).

We will now see how this stored energy is released on demand, which can then be utilized to perform various kinds of work of the cells, such as homeostasis, locomotion, and defence. Cellular respiration is the process by which glucose is broken down into CO_2 and H_2O to release energy, which is just the reverse process of photosynthesis. And obviously, this is an exothermic process. But then how exactly is this stored energy released at the molecular level? We have seen that when a molecule breaks, it absorbs energy, and so when a carbohydrate molecule splits into CO_2 and H_2O, it should absorb energy rather than release energy! But how does the energy-rich glucose liberate energy?

Since the chemical bonds in carbohydrates are low-energy bonds, they require small amounts of energy to break them, but the broken atoms of carbon, hydrogen, and oxygen are highly unstable, and they recombine (step by step in a series of reactions in the mitochondria) into carbon dioxide and water. CO_2 and H_2O are molecules with very high bond energy, and thus, a lot of energy is released during their formation. Thus, it is the formation of these final products of carbon dioxide and water which actually liberates energy, and this energy, of course, is the original solar energy that was trapped in carbohydrates. The heat released, however, cannot be used directly by the cells but must be first converted into certain high-energy intermediate compounds, such as *adenosine triphosphate* (*ATP*, the 'energy currency' of the cells). The formation of ATP actually helps in regulating the energy release in a controlled manner so that energy can be utilized at specific sites of a cell. By this way, much of the released energy is harnessed by the cell (but not wasted as heat dissipated into the surroundings).

We will now look into the change in G value during cellular respiration. Here the ΔH is decreased, and the ΔS is increased, and the resulting ΔG value is decreased (–686 kcal). A negative ΔG value indicates an entropic process (Fig 6.3).

An Overview of the Thermodynamics of Life: All life processes, at the gross level, take in energy from the surroundings to build up order within them. And this order gradually disintegrates into disorder in the form of cellular respiration to release energy to do cellular work. Thus, the birth and growth of an organism (be it a bacterium, plant, or animal) are characterized by the building up of negentropy in the form of order and growth, which ultimately leads to decay and death of the organism, which results in higher entropy in nature. In short, life builds up negentropy transiently, does work, and finally dies by causing more entropy in nature.

Consequently, we can say that life is essentially a negentropic process which is driven against the odds of nature by non-spontaneous processes of preparation and storage of food, and it finally leads to the inevitable and spontaneous process of release of energy back into the surroundings by decay and death, which represents entropy. Each morsel of food we take builds up order and drives life to sustenance, and with each cycle of respiration, disorder increases and drives life towards senescence. This is the inexorable *thermodynamic cycle of life*.

> *Birth and growth in life denote negentropy;*
> *decay and death denote entropy.*

Having discussed the phenomenon of life in this perspective, we will now go into the energy transactions of the universe at large and see how thermodynamics affects cosmological events.

SECTION E
Thermodynamics and The Arrow of Time

The one vexing question which we haven't asked so far is this: *why* does nature show this great reluctance to reverse the thermodynamic processes of entropy? The prosaic answer to this would be that 'it is because of the very nature of matter and energy in the universe'! However, we may endeavour to answer this question in more analytic terms, hence the following discussion.

We will first ask an array of seemingly devious questions to begin our quest: What are the cosmological implications of the thermodynamic principles we have discussed so far? How do we apply the rules of thermodynamics to the events of the universe, such as the black holes and the big bang? And on the other extreme, how could we apply the rules of thermodynamics to the smallest of the smallest particles in the quantum realm? Are the events of the quantum world ruled by the principles of thermodynamics? Here is a brief account of thermodynamic concepts in relation to these matters. In fact, the essential purpose of this chapter is to understand the thermodynamic implications in the realm of both macroscopic and microscopic worlds, and the principles which guide these affairs are enshrined in the concept of 'arrow of time'. We will now briefly study the various aspects of the arrow of time hereunder.

The concepts presented in this section, once understood, would become tolerably simple and even entertaining. By the end of this chapter, the reader would comprehend not only the role of the second law in all common events that happen around us but also its role in the cosmic dimensions—so that he/she would be able to theoretically connect the causality of shattering of a coffee cup to the birth of the cosmos billions of years ago!

Arrow of Time of the Universe: The most important implication of thermodynamic principles on modern cosmology perhaps is that of the delineation of the *arrow of time of the universe*. This arrow of time would, as we will see now, not only give us an understanding of many phenomena of the present-day universe, but it would also indicate the future and fate of the universe. This arrow of time would appear as a relentless progression of events in nature in a set direction which would take all other events of the universe in its gigantic stride. To understand this grand arrow of time in a better manner, we can discuss it in three major divisions—a thermodynamic arrow of time, a cosmological arrow of time, and a psychological arrow of time. We will first study the thermodynamic and cosmological arrows.

What is this thermodynamic arrow of time? All through the discussion in this chapter, we have seen that nature always prefers entropy, and as time proceeds, disorder increases in any thermodynamic system. This naturally implies that a thermodynamic system proceeds in only one direction, and that is in the direction of entropy, which we conventionally call the 'forward' direction. Thus, we can say that as time moves forward, disorder increases. We will illustrate this with an example. A coffee cup falls to floor and shatters, and this increasing entropy takes time in the forward direction, but even when we attempt to reverse this disorder by reassembling the broken pieces, time still moves in a forward direction and increases disorder in the universe in a general direction, as we have demonstrated adequately in Sec-B! This unidirectional flow of time is called the *thermodynamic arrow of time*.

Now consider this: most of the universal laws we have studied hitherto are time-symmetric (e.g. Newton's laws of motion, gravitational laws, and laws of electromagnetism) in the sense that they behave exactly in the same manner even if the time is reversed. In contrast, this thermodynamic arrow of time bespeaks of an inherent *asymmetry* of nature because as time proceeds in a forward direction, there is an increase in disorder, and this cannot be reversed when time is reversed. And in fact, for this reason, it may even be stated that the reversal of time is forbidden by the second law of thermodynamics. After all, we have, in our experience, always seen cups shattering into pieces, but have we ever seen the broken pieces assembling into a cup again just like that? We have already seen many examples in the above sections, and all of them seemingly indicate that such a thermodynamic arrow drives all the physical and chemical phenomena in the world around us. Why is this asymmetry happening in nature? We will set aside this question for the time being and study the arrow of time in the macrocosm and microcosm.

Now we will try to understand thermodynamics at cosmological dimensions. We have seen in Ch 5 that the universe has started in an infinitesimally gargantuan explosion called the big bang, and ever since the big bang, the universe has relentlessly been expanding. Thermodynamically speaking, this cosmic expansion means that rarefaction of matter increases as the universe expands—or in other words, we can say that disorder (or rarefaction) keeps increasing as cosmic time proceeds forward. By extrapolating the present-day disorder (or entropy) in backwards direction up to the initial condition of the universe, we may say that the universe has begun in an infinitely low-entropy state at the beginning of time. In fact, it may be said that at the time of the big bang, there existed an exquisite degree of initial order. And consequently, it can be said that ever since the big bang, rarefaction has progressed as time moved forward. In other words, this cosmic expansion has provided us with a direction, and therefore, it may be said that all the progression of events that occur in the cosmos are dictated by the state of entropy of the universe at a given time. This arrow of time of the universe is named the *cosmological arrow of time*, which appears to drive all the events in the universe in its giant celestial stride. It may finally be stated that both thermodynamic and cosmological arrows of time move in the forward direction.

Thermodynamic and cosmological arrows of
time always move forward with time.

These arrows of time can actually be shown to lead us into some intriguing aspects about the past, present, and future of our universe. But before proceeding further, we will have to examine whether an arrow of time exists in the microscopic world of atoms.

Thermodynamics in the Quantum World: Until late nineteenth century, the existence of atoms as physical entities was generally not well recognized by the scientists. In 1827, Robert Brown made a puzzling observation that pollen grains on the surface of water show some tiny erratic movements (the so-called *Brownian motion*). However, the cause of these erratic movements was unknown until, in 1905, Einstein explained the basis of Brownian motion in a paper describing them as a result of statistical probability of motion of discrete atoms striking the pollen. And in fact, this explanation stood out as the first empirical evidence for the existence of atoms! In the later years, many developments took place in the quantum realm which showed that not only the movements of atoms but the position of all subatomic particles are also really dependent on statistical probabilities (Ch 2-C).

Now consider this: imagine a large tank of water. In the context of erratic molecular movement, it may appear to be superlatively improbable for us to presume that a certain group of water molecules, spontaneously and without any inducement, move in file in one particular direction to solidify into an ice cube. However, such a probability, statistically speaking, may be improbable, but such an event is not impossible. It may take, say, some trillions of years to happen, but still it is not an impossible affair! Or think of this randomness in this way—if we examine a sample of gas in any thermodynamic state, we cannot predict the direction of movement of any of its constituent molecules; they move haphazardly all over all the time. But then we have seen in the thermodynamic and cosmological arrows of time that the events of the universe relentlessly move in the forward direction. Thus, these arrows of time should, in fact, direct all the atoms in the tank of water or the molecules of gas to move away from each other in a set direction, but this, surely, is not happening. So it can be said that in the microworld of atoms, this sort of time arrow seems to carry no meaning. Thus, we may conclude that the microscopic movements in the quantum world are erratic and haphazard.

The arrow of time is absent in the quantum world.

However, there have been some theoretical attempts to apply time symmetry (i.e. direction) to the wave functions of matter waves (Ch 2-C) so that the probability waves in a wave packet show a consistent tendency to collapse into higher energy states (the so-called *quantum arrow of time*). But there has been no consensus among scientists in this matter. Therefore, we may finally state that the arrow of time is found missing in the case of the microscopic world of atoms and molecules.

Whatever the underlying quantum uncertainties in the microscopic world are, our macroscopic world is predictable because as the size and mass of participating objects in an event increase, the uncertainty dwindles down, and the happenings become tolerably more certain. We can still throw a dart at the bullseye! This predictability is the reason why human civilization has relied upon the worldly events to advance and develop into a better race from the Bronze Age. For example, our Bronze Age relatives have been extracting metal by digging ores successfully. If predictability were not the case with digging, the ancient man, having observed the utter randomness of the events of digging (as the dug-out soil may fall back again magically), would have chucked his ancient spade away and retire to the nearest shady bush for a nap instead of toiling for the ore! This predictability has been the strength and beauty of all physical laws that are operative in nature, which make our world really liveable. We will leave it at that and go back to our discussion on other aspects of the arrow of time.

Psychological Arrow of Time: Yet another arrow of time, called the *psychological arrow of time*, is described by many authors. This is the passage of time from the standpoint of a human being. Humans perceive time as a continuous stream of events with a vague differentiation into the past (the known), the present (the now), and the future (the unknown), and this demarcation seems to exist both at the mental and physical level. The mental effects of time are reflected on the memory we store—the past is that which we have experienced and is mostly remembered, and the future is that which we have not experienced and is not mostly aware. This human mental characteristic of memory that runs in a particular direction from past to future is called the *psychological arrow of time*. The physical effects of time are also evident. For example, our bodies show all features of biological (anatomical and physiological) ageing as time passes ahead. After all, no creature becomes younger as it ages!

Now consider this: it is our scientific understanding that our physical bodies (and perhaps our minds too) are governed by strict physical laws of nature. But then we have already seen that most of these physical laws in general are time symmetrical. They behave in a perfectly similar manner irrespective of whether time moves forwards or backwards. And thus, when a system is symmetrical, time flows forwards and backwards equally, and consequently, that system would not show any discrimination between the past and the future. So now the question becomes obvious: from where do we acquire the capability of storing past memories, and why are we incapable of 'storing' the future events? Or in short, why do we not 'remember' the future? But then all along in the above discussion on the arrows of time, we have seen that the phenomenon of entropy inexorably takes the events of the universe in a forward direction. And so is it possible that the human mind is also governed by the phenomenon of entropy, which explains the human characteristic of memorizing the past events by increasing the order of arrangement of information in the brain (see below) and forgetting it by entropy and disorder. If this is so, then how exactly is entropy related to the psychological arrow? To analyze this, we have to first study the phenomenon of memory (in the general terms of physics).

Because memory is nothing but stored information, we will now study the problem of storage of information as memory and see how this would lead us into the mires of the psychological arrow of time. The study of storage of memory is showcased in a relatively recent scientific theory called the information theory. First of all, we will see what we mean by 'information'. Information is a mechanism by which one agent influences the other agent (these two agents put together constitute a 'system'). The agents involved in a system could be a living being or a non-living object. This needs some explanation. First consider living subjects. We, the humans, use several information modalities by which we can communicate and influence agents around us—*viz.* tell a piece of news to a friend, sing a song to a large audience, write a message to a distant friend, recite a number into phone, do gesticulations, or send a telephonic signal or a telegraphic signal, an email, a radio signal, and so on. While passing information from one person to the other, all these transactions invariably generate some sort of change in the molecular order of the receiving agent (perhaps somewhere in the receiver's brain)—i.e. the molecular order changes from an initial setting to a later (modified) setting. For example, if your fiancée listens to your song and tries to memorize it, this process would cause a change in the molecular order of certain areas in her brain somewhere!

Now consider how a non-living object transfers information. It should be realized that information of every event that occurs in the universe may also be somehow stored, at least transiently, in nature. To envisage this idea, consider this: a piece of wood may be considered to have certain molecular order initially, and when it is stricken by an axe, it changes its molecular order. This means that the agent of the axe has left a signature on the molecular architecture of that piece of wood! And theoretically speaking, all these changes can be retraced and reproduced at a future date by reversing the direction of the causative agent (i.e. theoretically)! This sort of storage of information in nature has puzzled philosophers in the past. For instance, take this historical question put forward by the philosopher George Berkeley: 'If a tree falls in a forest and no one is around to hear it, does it make sound?' It has been said that since no intelligent being is present to record the event, this does not generate any 'sound' so to speak. In fact, the occurrence of that event, or even the very existence of that tree, has been philosophically debated. Philosophy set aside, we may argue that this event of a falling tree has certainly generated some molecular disorder in the surroundings resulting in certain molecular vibrations. This set of vibrations were certainly received by the surrounding structures ('agents'), such as other trees, the ground, and the nearby rocks, and these agents would have 'registered' these changes and stored them as some sort of 'molecular signatures' of the event in them (even when the tympanic membranes of a human or animal were not present during the event). Thus, we may say that sound *was* generated in the event of a falling tree, but a human being was simply unavailable to record the event. Nevertheless, the information was 'recorded' in nature (and certainly this fallen tree has changed the architecture of a rock by breaking it, or it has changed the direction of growth of a few saplings underneath, which would otherwise have grown straight!). This clearly shows that information of all events of nature is transmitted and recorded at all times regardless of human existence (or interference). Or should we become a little witty for the nonce and say, 'If our walls could talk, they would have a lot to say!'

Passage of information is no human prerogative;
it happens in all events of nature.

A non-living object may store information for more extended periods of time. For example, gramophone records or compact discs or hard disks or any other such device can store information for long periods by changing the atomic/molecular state in them, and the stored information is retrieved by retracing the change.

All in a nutshell, this shows that all agents 'communicate' with their surroundings all the time by causing changes in some way! Thus, we may say that all the objects in nature have a past and a future. But then how can we study this stored information analytically and scientifically? This is where the role of information theory comes in.

In 1948, Claude Shannon, the father of the Information Age, unified all the communication signals in mathematical terms and showed that all signals could be broken down into fundamental 'conceptual chunks' which could be encoded into a series of basic signals consisting of 'binary digits' (or 'bits'). In fact, this analytical approach to information has launched a new branch of science called the *information theory*. Information theory deals with problems such as the speed and accuracy by which we can pass information with minimum or no errors. Shannon experimented with the information-encoding machines and studied their outputs and laid down certain indomitable principles. Imagine a system (or a machine) encoding a sequence of events delivered to it. Information theory gives us an idea of the amount of information stored in that system by the way of a simple statement that 'the more random the system is, the more information it will contain'. A measure of the average randomness (or uncertainty) of a system is called entropy (this 'information entropy' is analogous to the thermodynamic entropy as the reader may notice as he/she reads on). It has been shown mathematically that a system's entropy has the maximum value when all the outcomes of that system are equally likely (i.e. maximally unpredictable), and entropy decreases as the predictability of outcomes increases. In other words, it can be said that entropy of information is maximum when uncertainty of the system is maximum and its predictability is low; on the other hand, entropy is minimum when the system operates with more certainty and high predictability (this discussion may confuse a general reader. The mistake is certainly on the part of the author for not being elaborate; however, the reader may get all assistance from certain masterly sources in the net). Now the reader may draw up an analogy between this sort of information entropy and the order–disorder we have seen in the above sections. When the order of a system is decreased, its predictability becomes low, and entropy of stored information increases. In other words, increased disorder increases entropy (and of course, on the contrary, increased order decreases entropy).

We will not go into the details of information theory here any more, but we could state that this conceptual understanding of stored information has resulted in a modern technological renaissance in the sense that it has led researchers to 'digitize' information and conveniently feed it into machines in the form of binary codes, and these concepts have become the foundation principles of all our modern gadgets. And by using these basic principles of information theory, we could now be able to transmit information more efficiently for more distances with minimum loss.

In conclusion, we may say that the events of storage and passage of information in nature are entropy-based events, and it may be said that even abstract actions like thoughts and memory are entropy based. A computer, for example, raises its order as it stores information in its microchips (Ch 2-E), but at the same time, disorder happens in its surroundings in the form of consumption of electricity (as discussed in Sec-B). In the same way, the human brain also encodes information in its neurons, which raises some molecular order in the brain, but this negentropy is countered by an increase in disorder in nature by the way of burning lots of glucose and liberating CO_2 (the human brain consumes about 120 grams of glucose per day to feed the neurons, and the generated heat is taken off by the cooling blood flow across the brain and is later dissipated into the surroundings). Therefore, we may conclude finally that the functioning of our computers and our brains offers some solid theoretical grounds in support of the psychological arrow of time! Or in short, we may say that:

The psychological arrow of time follows thermodynamic principles.

Now in summary, we have got an overall big picture of the arrow of time. It appears that the events of the universe are orchestrated in accordance with the passage of time in the forward direction as guided by the thermodynamic, cosmological, and psychological arrows of time. But then two intriguing questions still bother us. One is that we do not know why the passage of time coincides with all these above gross arrows of time in order to take the universe in its giant stride. And why should the fundamental principles operating the microscopic matter appear to defy this universal principle but are directed by the principle of uncertainty? Clearly, there is more to learn from nature, and we have to come up with some entirely new conceptual understanding in order to decipher this grand design of nature (Ch 12)!

Heat Death of the Universe: We have come to the end of the chapter, but we have not answered the question as discussed above: *why* should this thermodynamic arrow of time exist at all? *Why* should Mother Nature go to prankish levels to see that the net effect of any event in nature results in disorder? If we presume that the universe has started in a highly ordered state, as indicated above, then it means that this order would decrease with the progression of time—a fact we have documented adequately all along in this chapter. But then if you extend this phenomenon of increasing disorder to its logical conclusion, it would mean that someday in the distant future, all the usable energy in the universe may be converted into 'entropic heat'. In other words, at the end, we would have no usable energy at all. In other words, the universe has reached a state of *maximum entropy*. This state is called the heat death of the universe. The reader may see that this state is akin to the big freeze we have discussed in Ch 5-I.

Heat death of the universe results when all energy becomes unusable.

Thus, in this state of maximum entropy, no heat transfer is possible any longer. In fact, a state of perfect thermodynamic equilibrium exists among the objects in space. We have seen all along in the above discussion that any reaction in nature is possible only if it can increase the entropy of the universe. And in this case of maximum entropy, there is no scope for any further increase in entropy, so no heat exchange happens, and thus, all reactions (or events) in the universe come to a standstill. And because there is no heat flow, all the phenomena of motion, work, and entropy are impossible in nature, and consequently, no change is possible in the universe!

However, this maximum entropy has a paradoxical consequence. As the universe extends towards the heat death, all the matter would be gradually converted into energy, and perhaps, the only things that are left over in the universe are the electromagnetic radiation and the fundamental particles. Any transaction between energy and these elementary particles could only result in a change in the form of negentropy (not entropy because maximum change has already happened), and when reactions proceed this way, this would result in the building up of order, and once again, this would result in a collapse of the universe into a *big crunch* (Ch 5-I). This is a mind-boggling conundrum with a never-ending deadlock!

We will rewrite these concepts and address all these issues once again in a simple cosmological model presented in Ch 12.

This completes our general survey of the universe at the macrocosm and the microcosm. With this theoretical background, we will move on to Part II of *Riding on a Ray of Light* and examine a new theoretical model which redefines the older principles in a new perspective.

PART II

THE NEGENTROPIC MODEL

CHAPTER 7

The Mechanism of Motion
And the Meaning of Inertia and Entropy

Overview

In Part I, we have learnt about the basic principles and laws that are operative in the universe. From here on, we will undertake a study of some of the important physical principles of nature and review them in a new perspective which would ultimately uncover certain profound truths. In Ch 7, we will study the fundamental property of motion in the universe, which leads us insidiously into a discussion on the fundamental forces, which, in turn, leads us into a mechanism of generation of light, as we see in Ch 8. In Ch 9, we will look into the structure of the photon, which would soon take us on to a model by which matter is created. Gravity, one of the most essential features of matter, is discussed in Ch 10. In Ch 11, we will discuss two primordial entities of nature that are operative in the background of our universe. In Ch 12, we will set out to see how this new understanding works out to describe a model of the universe!

At first, in Ch 7, we will conduct a short analysis of motion in nature. We have seen in Ch 1 that motion is the most cardinal feature of the universe—*ignorato motu, ignorator natura* (if you ignore motion, you have ignored nature) as Galileo had proclaimed centuries ago. The discussion on motion leads us to dig up a special relationship between uniform motion, speed of light, and inertia, allowing us to gently tweak Newton's first law of motion but of course keeping its essence intact. In the subsequent sections, we will study the concept of non-uniform motion, which defines a relationship between force, momentum, energy, work, and entropy in a novel way and arrive at a 'universal cycle' which may be applied universally to all the events of the universe. In the next section, entropy and inertia are discussed on a common platform to show their deep relationship, and importantly, this study also illustrates their common cosmological meaning. At the end of the chapter, several examples are laid out which bespeak of the importance of the universal cycle in all the events of nature. The reader may notice that throughout the chapter, no new theories are proposed; rather, the known principles of nature are assembled and understood in a unique perspective.

Now off we go into the exciting world of motion!

Section A
Motion: An Introduction

Almost everything we have learnt in the previous chapters is something related to the movement of objects* in space. The theory of relativity, for example, has chiefly dealt with properties of moving objects in the cosmos at large, whereas quantum mechanics has dealt with moving subatomic particles in micro-space. And all the natural phenomena in the universe we have learnt so far also involve motion as a central factor. Be it the theory of expanding universe or the theory of uncertainty principle or the theory of thermodynamics, the underlying principle is that of motion. And also, as we will be discussing in this chapter, at the heart of all the nature's fundamental phenomena, such as force, energy, and work, there is the essential principle of motion. Thus, we can reiterate that

The phenomenon of motion is the cardinal feature of our universe.

---

*   The term *object* here refers to all material bodies in the universe ranging from stars and planets in the cosmos to large boulders to small pebbles or to the smallest subatomic particles in quantum space.

We will now undertake a study of the basic properties of an object in motion and arrive at some important conclusions.

There are essentially two types of motion in the universe— uniform motion and non-uniform motion.* Non-uniform motion, in turn, is of two types: accelerated motion (or non-uniform linear motion) or curved motion (or rotational motion). However, in the world around us, we will be observing many types of motion, such as oscillatory motions, harmonic motions, vibratory motions, and irregular motions. But it must be realized that all these movements result from a complex assemblage of these fundamental types of motion, and thus, they can be reduced to one of these basic types of motion.

All moving objects in the universe are governed by a set of laws which are thought to be applicable anywhere in the observable universe. We will first examine Newton's first and second laws to understand uniform and non-uniform motion, and the third law is reserved for a later discussion in Sec-C. *Newton's first law of motion* states that an object tends to be in a state of rest, or continues to be in a state of uniform motion, unless acted upon by an external force. It defines uniform motion and inertia, so it is also called the *law of inertia*. We will look into uniform motion in Sec-B, and the phenomenon of inertia is discussed later. *Newton's second law of motion* states that an object moves by acceleration when an external force is acting upon it, and the relationship between force and acceleration is represented in the formula $F = ma$. It defines non-uniform motion and force. We will look into non-uniform motion in Sec-C. We will now do a brief analysis of uniform motion in the light of our current understanding of the nature of energy and matter, and at the same time, we will have some new insights in the nature of uniform motion.

---

*   Uniform motion is also termed *constant motion, perpetual motion,* or *inertial motion* depending on the context being discussed. There is also a confusing term called uniform circular motion where the object travels at a constant rate in a circular path, but this is a form of non-uniform motion because here the object does not travel in a straight line but keeps changing its direction continuously (see below for details).

Section B
Uniform Motion

Newton's first law may be analysed in two parts. The first part of Newton's first law says that an object may continue to be in a stationary state for an indefinite period of time until it is set in motion by the application of an external force. This is due to a natural property of the universe called the *inertia of rest*. The inertia of rest is the natural tendency of any stationary object in the universe. The second part of first law says that if an object is in uniform motion in a straight line, it tends to continue in uniform motion for an indefinite period of time until some force acts upon it. This is due to a natural property of the universe called the *inertia of motion*. Inertia of motion is the natural tendency of any object in motion in the universe. It can be said, as per the present-day understanding of nature, that both these states of an object are really one and the same. They actually represent two sides of the same coin, as we will see in our discussion below.

First we will see what exactly is uniform motion. An object is said to be in uniform motion when the object travels in a *straight line* in a given *direction* and at a *constant speed*, and this motion is *eternal* (or *timeless*). The object continues to be in this state *until* it is acted upon by a force, and with the application of force, it becomes non-uniform motion. In uniform motion, the object covers equal distances in equal intervals of time, however small or large these intervals may be. For example, if a spaceship in uniform motion in space covers 200 m/s in the first second, it tends to cover the same distance subsequently in the second, third, fourth seconds, and so on indefinitely.

Uniform motion is at a constant rate, on a straight
line, with one direction, and is timeless.

An object can be considered to be in non-uniform motion when (1) the object moves with linear acceleration and increases its velocity or (2) moves with linear deceleration and decreases its velocity or (3) deviates from the straight path, changes its direction, and takes a curved course so that the object moves in a circular motion (see Sec-B for details). For non-uniform motion to take place, an external force must act on the object. Conversely, for an object to be in uniform motion, it does not require any force (or energy) to maintain its uniform speed across the space for any length of time. Uniform motion is a timeless journey across space without any energy expenditure. This is due to inertia of motion, which is the innate property of nature. In fact, the term *inertia* denotes 'laziness' or 'reluctance', and it denotes the laziness of an object to change its existing state!

Before proceeding further, the general reader is advised to review the introductory part of special relativity in Ch 1-A. In the theory of relativity, we have seen that an object cannot be in an absolutely stationary state in the universe. All objects in the universe are in a perpetual state of relative motion. Hence, inertia of rest is an idealized situation which cannot be achieved in the universe except in theory. In the same way, uniform motion is also an idealized situation and cannot exist in the universe in any conceivable way. The following example would illustrate the point. Take the case of a jet plane cruising at a uniform speed of 200 m/s. It will not only encounter resistance from the surrounding air in the form of friction, but its speed is also affected by the earth's gravity, and both these factors may ultimately slow down the course of the jet. Thus, we can say that, though the jet maintains uniform motion, this is only an illusion of uniform motion because the jet can only maintain its uniform motion by utilizing the force derived from the expenditure of energy (i.e. by burning its fuel). Considering this, we may say that the jet plane is actually travelling in non-uniform motion as the jet is utilizing force from the fuel to prevent its deceleration. It is, in fact, in a state of *simulated uniform motion*. But then if the jet were to be in 'true' uniform motion, any expenditure of energy would not have been necessary to keep it moving because as per Newton's first law, the jet would be in uniform motion constantly for an indefinite period of time (in which case, of course, the aviation industry would be saving on tons of fuel and would be making some astronomical profits!).*

---

* However, the reader may note that for an observer in the jet plane, the jet plane

273

Simulated uniform motion is a form of non-uniform motion.

Now imagine another situation. A spaceship is set in uniform motion while it travels from Earth to Mars. Outer space has no air to cause friction, and so the spaceship can continue its uniform motion, but then it has gravitational force acting upon it. This gravitational force is originating not only from the Earth, Mars, and the Sun but also from some other planets and other distant stars as well (however small this influence may theoretically be; after all, all objects in the universe are under the influence of a blanket of gravity, Ch 4-A). Thus, in theory, there is no escape from gravitational force in the universe, and consequently, we may say that we cannot truly enjoy the benefits of uniform motion in nature. Hence, it can be stated that uniform motion is a hypothetical situation and cannot truly exist in nature.

We will now take the subatomic world and examine uniform motion. Consider a subatomic particle moving in space. An electron may continue its journey in space in uniform motion, but it is under the constant influence of electromagnetic energy. Feynman diagrams (Ch 3: Fig 3.4) show that an electron moves ahead and changes its direction by absorbing a photon, which means that for an electron to move, it needs some force in the form of electromagnetic energy. Thus, it can be said that the electron is really in a state of simulated uniform motion, which is equal to non-uniform motion, as we have seen. Here's another example: a quark (Ch 3-C) moves about within the confines of the nucleus only when it is under the influence of a strong force (in the form of gluon), and thus a quark may be stated to be in a simulated uniform motion. Thus, all subatomic particles are under the constant influence of some field force which makes them move with some velocity in a particular direction. Hence, they can be said to be in a simulated uniform motion (or non-uniform motion). All this means to say that every object in the universe—be it a jet, planet, star, or subatomic particle—is under the constant influence of one of the fundamental forces and so cannot be said to be in a state of absolute uniform motion. In a nutshell:

---

is in uniform motion, and so he/she cannot identify whether the jet is at rest or in uniform motion by any experiment (because the observer himself/herself is travelling with uniform motion in relation to the jet).

Uniform motion in the universe is a myth.

Nevertheless, we have to study this hypothetical uniform motion in great detail because we come across several theoretical situations where we must take uniform motion into account for our proper understanding of the events of the universe. Uniform motion is the key with which we can unlock a treasure of cosmic secrets, as we will be seeing. Thus, we have to work with an idealized setting wherein an object moves with a constant velocity in a perfect straight line for timeless periods. This idealized setting is called the *inertial frame*—i.e. a theoretical frame of reference in which the object in question moves in absolute uniform motion.

We will now arrive at a conclusion. From the above discussion, it is understood that an object which keeps moving in uniform motion, in an inertial setting, does not require any force to keep going. It moves along eternally in space without any expenditure of energy as stated above. This is because *uniform motion is the natural state of motion* for any object in the universe.

Uniform motion is the natural state of an object in the universe.

The statement here may appear paradoxical because despite the fact that uniform motion is hypothetical and unattainable, uniform motion may be considered as the natural state of a moving object in the universe. This means to say that though an object cannot really be able to travel in uniform motion, it has a natural tendency to acquire uniform motion! In other words, *all objects in nature constantly try to attain the unattainable*—or we may say that the objects in nature are attempting to move towards uniform motion without ever achieving it!

Objects in nature attempt to achieve uniform motion without ever achieving it.

We will now study why objects in nature behave in this peculiar way.

The Problem of 'Real' Motion: The basic premise of the first law is that of its eternity. An object at rest continues in that state eternally, and an object in constant motion continues to travel eternally. Uniform motion is a 'timeless' journey. It records no time as long as it continues to move at a constant rate; thus, uniform motion can also be called perpetual motion. Alternatively, we can say that time 'stands still' for an object in uniform motion. One easy way to visualize this feature is to analyze the phenomenon of time dilation. Before going into the discussion, the reader is advised to brush up his/her knowledge on the special theory of relativity and time dilation from Ch 1-C.

We have seen in the special theory that the measurement of the length of a moving object is relative—for a fast-moving object, time tends to slow down, thus making the length of the object appear to contract as measured by an outside observer (see 'light clock experiment' in Ch 1-C for details).[*] Thus, the faster an object moves, the slower its time ticks, and the shorter its length becomes. In other words, for a fast-moving object, the passage of time slows down—i.e. *time dilates*. This relationship between length and time continues until the object attains the speed of light, and once the object attains the speed of light, time stops altogether, and the object's length becomes zero. This fact has been shown in the FitzGerald equation— the length of a moving object is $l = \sqrt{1 - v^2/c^2}$, and if the object's velocity (v) equals the speed of light (c), then the length (l) would become zero. This means that all dimensions disappear for the object at the speed of light. The object simply vanishes as it attains the speed of light (note that if one space dimension becomes zero, the other two space dimensions also become zero because for an object with no length, there is no meaning in attributing any breadth or depth to that object, and consequently, no mass can be attributed to it). Thus, we can say that an object at light's speed behaves like light—or becomes light itself—because it loses all its dimensions and becomes massless.

---

[*]   The reader must realize that the term *observer* does not apply only to a sane and intelligent being or agent, but it also applies to any other gadget making an observation. This was made clear in Ch1-F, and we will discuss this issue further in Ch 11-C.

An object at light's speed becomes light.

Why is this time dilation happening in nature? General theory of relativity gives us an explanation (Ch 1-E). Before going into the details, we will first see how a human being records movement in nature. Consider this simple reality: we will *not* take 'time dimension' into consideration at all when we record motion of an object in nature. Take, for instance, a car moving along the road. In order to record the car's motion, we will consider it in three dimensions—its length, breadth, and depth. But its movement (i.e. distance travelled in a period of time) we record is actually our interpretation of a series of still images in a sequential manner. A series of still images (shown, say, at about twenty-four frames per second) would cheat our brains into a perception of motion (and the reader may realize that this is the basis of our motion picture!). In summary, it can be said that to record motion, we will not take time dimension into consideration for our perception.

For us to take time dimension into consideration, we must first be able to define the 'real' motion. But then what do we mean by the term *real motion*? Once again, we will take our thought experiment of a man on cruise in the Pacific Ocean (Ch 1-A) and deduce the real motion in the universe (look into Ch 1-A for details). A man walking on board of a cruise at a certain pace is actually walking in relation to the cruise, but once we take ocean as the frame of reference, the man's pace picks up its speed. When we take the earth's motion as reference, his motion further gains speed, and when we take larger and larger references, such as our solar system, our galaxy, and so on, and the man gains tremendous speeds until it finally becomes infinite. This endless chain of relative motion—a phenomenon of 'cycles of movement-over-movement'—links up the initial frame of reference to an infinite frame of reference. We will call this phenomenon of infinite cycles of movement-over-movement as *epicyclic kinetic motion* (in order to make a lengthy description short and handy, borrowing the term *epicyclic* from Ptolemy). Thus, the 'real' movement of an object in the universe is the epicyclic motion which is too fast for us to appreciate.

This indicates that the 'motion' we perceive in our 'ordinary' world is just a notion conceived by us which, in turn, is collected from a series of images changing in position in space. However, this 'false' motion serves all our practical purposes on the earth very well indeed! Imagine for a while what could be the upshot if we start appreciating the real motion of an object in our world (even as we take only the modest speed of the earth's motion around the sun at 108,000 km/hr). In order to just board a moving train, you would be attempting to move at the pace of a rocket, and then you would soon be ramming fast into your fellow passenger. Or as another example, if a tiger were to appreciate the real motion of a deer, then it would have been an impossible task for it to pounce upon it with any precision! However, movements on the earth are tolerably slow for us, and we, the earthlings, are quite comfortable with our 3D assessment of the world even though the measurements which we make in this 3D setting are grossly inaccurate and approximate. But then this *is* the reality of the world we live in, and we get along with it all right! Anyway, being sufficiently educated in the theory of relativity as we are, we now understand that this 'worldly thing' is not the absolute truth, and we realize, after all, that the 3D world is only an illusion of our perception!

Therefore, we can conclude that the motion of an object in 3D is not at all absolute. It represents only a relative motion of an object in the universe. But then what is this 'real' or 'absolute' motion? We will now examine the moving object in a 4D setting to understand absolute motion.

Absolute Motion and the Speed of Light: From the above analysis, we can deduce that in order to perceive the real (or *absolute*) motion of an object, we have to start considering the object in an additional time dimension along with the three space dimensions. This means that we have to take into consideration a composite spacetime 4D structure called the *spacetime continuum* (Ch 1-E). Einstein in his general relativity showed that all objects, irrespective of their state of motion, travel with the speed of light when viewed in 4D spacetime. This understanding has some profound implications, as we are going to see. But we will first have a simple understanding of how this phenomenon becomes possible, and then we will see its implications.

Imagine an object in space moving along a ray of light. Let us draw a coordinate system upon the object with the space dimensions of its length, breadth, and depth representing the three coordinates—x, y, and z—and the path of object along the light being the hypotenuse of this coordinate system. Now imagine for a while that the object is moving at the speed of light, c, so that the distance the object covers in time, t, can now be represented as ct. In this situation, ct represents the hypotenuse of the coordinate system (Fig 7.1a). We can find the value of ct by the following Pythagorean equation: $x^2 + y^2 + z^2 = (ct)^2$ or $ct = \sqrt{x^2 + y^2 + z^2}$. This square root of the value of ct is called the *spacetime interval*, and this represents the motion of an object in *both* space and time at a given time.

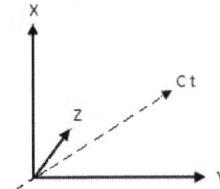

Fig 7.1: Spacetime interval.

Now consider this: according to the relativity theory, any object can be fixed and said to be stationary, and the rest of its surroundings can be considered as moving in its frame of reference. We will now fix the object and consider it as stationary, i.e. the object is now fixed at the origin of the coordinate system. In this stationary situation, the values of x, y, and z in the above equation of spacetime interval would become zero (since an object cannot move in relation to itself). But then we can see that even if the value of the coordinate system becomes zero, the value of the distance in spacetime interval (i.e. ct) does not become zero; rather, ct remains the same. This means that even if the object is absolutely stationary, its spacetime travel continues to be at the speed of light! In other words, a stationary object can be said to be theoretically moving at the speed of light when considered in a 4D setting.

In a 4D spacetime continuum, a stationary
object moves with the speed of light.

We will now consider another situation: an object in space moving in uniform motion at speeds *lesser* than the speed of light. All the objects travelling at speeds lesser than the speed of light are travelling in a *relativistic motion**—i.e. their length and other dimensions are relative to the speed at which they are travelling in space because of time dilation as seen above. Now we will draw a spacetime diagram and extrapolate space dimension on the *x*-axis and time dimension on the *y*-axis (Fig 7.2). First consider this: when an object is moving at the speed of light, it can be said to be moving exclusively in space dimension because speed of light is the maximum velocity that can be attained by a moving object, so we can now say that the object has zero movement along the time dimension, and therefore, the object's elapsed time can be said to have become 'timeless'. In other words, time for the object has stood still. Or we can say that an object at the speed of light moves without time—eternally.

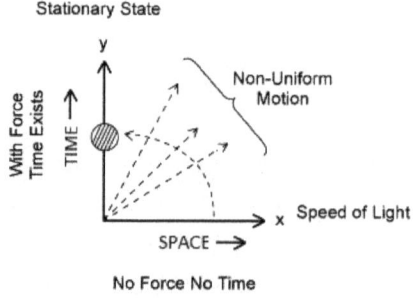

Fig 7.2: **Spacetime diagram.**

And now we will see what happens if the object moves with speeds lesser than the speed of light. For example, assume that a jet plane is moving with a uniform motion of 200 m/s; this is a relativistic motion of the jet. But how can we ascertain its absolute motion? How can we say that the jet moves at the speed of light in 4D? Looking at the spacetime diagram in Fig 7.2, we can say that as the jet's speed decreases, it starts moving less in the space dimension

---

* Generally, the term *relativistic motion* is reserved for objects travelling at nearer the speed of light because at our usual earthly speeds, the lengths of objects have no practical bearing on motion at all. However, in this discussion, the term *relativistic motion* means an object moving at any speed lesser than the speed of light.

but more in the time dimension, which means that it shows varying degrees of non-uniform motion. And an object which is stationary may be shown to travel exclusively along the *y*-axis. However, it can be shown that if an object is moving at speeds lesser than the speed of light, the *combined* speed of that object's motion in space dimension and its motion in time dimension would be equal to the speed of light. In other words, uniform motion at lesser speeds can now be viewed in the spacetime continuum, and they can be said to be really travelling at the speed of light.

By the above discussion, we may say that an object travelling in the time dimension (nearer to, or along, the *y*-axis) has its time measurable, and thus we may say that a car travels at 103 km/h or a Maglev train travels at 610 km/h, or a jet moves at 725 km/h, and so on. These speeds can be 'precisely' ascertained only when we consider their velocities in a 3D setting, but when measured in a 4D setting (i.e. in the spacetime continuum), all objects travel by the absolute speed of light (and time becomes immeasurable), and thus their movement becomes timeless! Therefore, when the object attains the speed of light, it becomes redundant to state that it is travelling at 299,792,458 m/s as this is the maximum speed that can theoretically be attained, and so the object behaves just like light—or in other words, an object at light's speed becomes light, as we have already seen. Thus, as long as the object moves at the speed of light, time is not measurable; it becomes timeless. But when it moves at lesser speeds, time appears in our calculations, and so the speeds become measurable.

Hence, we can say that the cardinal feature of uniform motion is the timelessness of motion, and time stands still for the object once it travels with uniform motion. Therefore, we can say that when an object moves with certain velocity (below the speed of light), even when it is moving at some constant speed (i.e. uniform velocity), we cannot say that this object is moving in 'uniform motion' because the movement in this case is not timeless; the object's movement can be measured (say, as a jet moves by 725 km/h constantly for certain period of time). In other words, we can say that uniform motion is possible only if an object travels at the speed of light. But then the object becomes light at the speed of light!

Uniform motion is possible only with light.

It is interesting to note here that we, the earthlings, who look at everything through our 3D minds, would like to say that light has taken 13.7 billion years to travel from the big bang to our present time. However, for anybody who could have travelled along with light at the speed of light (e.g. a '4D alien') would say that light has taken zero time to travel all along from the big bang to now. And consequently, a four-dimensional creature, if such a thing exists in nature, would see the light to be 'omnipresent' in the universe!

Thus, we may finally say that uniform motion is possible only with light, or conversely, we may say that any object, even though it is moving with a constant velocity, cannot be said to travel by uniform motion but can be said to be moving with simulated uniform motion. Or in other words, basing on the above discussion, we may state that uniform motion is possible only in a 4D spacetime but is impossible in a 3D setting. Or we may say that uniform motion is impossible in relativistic motion. In other words, it may be said that at the speed of light, the relativistic motion is abolished. And to conclude, we may say that uniform motion is a myth because any object moving at the speed of light would soon disappear into a ray of light.

Newton's First Law Revisited: Having discussed uniform motion in this perspective, we will first recapitulate the first law in its old form: an object tends to continue in a state of rest for an indefinite period of time (inertia of rest) or continue to be in a state of uniform motion for an indefinite period of time (inertia of motion) unless acted upon by an external force. However, in the above discussion, we have seen that an object can be said to be travelling at the speed of light in both the situations—at rest and in motion. In other words, we can say that the inertia of rest and uniform motion are one and the same.

Inertia of rest and inertia of motion are two sides of the same coin.

Now we may redefine Newton's first law in the following simple way—of course, keeping its essence intact:

Every object in nature tends to be in a state of uniform motion at the speed of light eternally unless acted upon by force (in which case the uniform motion becomes non-uniform motion).

We will discuss more about inertia in Sec-E, but for now, we will move on to examine non-uniform motion.

SECTION C
Non-Uniform Motion

Having studied uniform motion, we will now examine non-uniform motion and study its properties. Non-uniform motion is understandably complex because various other players, such as force, momentum, energy, and work, become involved in the discussion. However, these seemingly complex concepts can be simplified if we look at them in their right perspective, as we will see below. Here, in this section, the phenomena of force and momentum are studied, and in the next section (Sec-D), energy and work are discussed.

We have studied uniform motion and made some remarkable observations, but then non-uniform motion radically differs from uniform motion in many ways. Newton's second law defines non-uniform motion and force by stating that when an external force is applied upon an object, it sets in non-uniform motion, and this accelerated motion is directly proportional to the force applied on the object and is inversely proportional to its mass. It is encompassed in an equation: $F = ma$, i.e. if force is doubled, the acceleration gets doubled, and if the mass is doubled, the acceleration is halved.

Let us first illustrate the various types of non-uniform motion that can exist in nature. A spaceship, for example, moving in a straight line at 10 km/s and keeping on increasing its velocity at a constant rate of 2 km/s every second is said to be in accelerated motion (or *constant linear acceleration*), and when its velocity decreases at a constant rate, it is said to be in decelerated motion (or *constant linear deceleration*). If the object moves by acceleration and deceleration for varying periods of time, it is said to be in *non-constant linear motion*. If the object takes a circular path around an axis, it is said to be in a state of non-uniform motion because it keeps changing its direction constantly by taking a curved course. When the object is moving along in this curved path at a constant rate, it is said to be in *constant circular motion,* and when the rotating object changes its speed, it is called *accelerated/decelerated circular motion.*

Each of the above combination of movements is an example of non-uniform motion. What differentiates uniform motion from non-uniform motion chiefly is that whereas uniform motion needs no force to keep going and the motion here is timeless, non-uniform motion requires a constant force to keep its acceleration and the motion here is time-bound. When we stop applying force, acceleration ceases. In fact, an object in non-uniform motion has a natural tendency to end up ultimately in uniform motion (upon cessation of force) because uniform motion is the natural state of an object in the universe, as we have seen in Sec-B. Thus, all objects in non-uniform motion have a natural tendency to culminate in uniform motion after the cessation of force.

Non-uniform motion has a natural tendency
to culminate in uniform motion.

What is this force which changes the object's pathway from uniform to non-uniform motion? We will now discuss the essentials of non-uniform motion and force by a systematic analysis, which, by the way, leads us to the myriad and intriguing concepts of momentum, energy, and work.

Non-Uniform Motion and the Concept of Change: It can be said that the phenomenon of 'change' (in the position of object in space) is central to the idea of motion. Velocity is defined as the *rate of change* of a position of a moving object, and acceleration is defined as the *rate of change of velocity.* Since no object in the universe is at rest, as we have seen in the phenomenon of epicyclic kinetics, we may say that the phenomenon of change is inevitable in nature, and thus, we shall unequivocally agree with the Heraclitus's aphorism: *nothing is permanent but change.* We have discussed this phenomenon of 'change' in Ch 6-A in connection with energy, but we will now study the phenomenon of change in another perspective.

Consider this: an object in uniform motion travels timelessly, and so its movement can be said to be infinite. But then when an object's movement is timeless, it can also be considered to be changeless in its position because the concept of change in the position of an object in space does not arise without time. After all, timelessness is changelessness. Hence:

Uniform motion is a changeless motion.

Thus, an object in uniform motion travels in a changeless manner, and only when the object comes under the influence of an external force does a change in its disposition occur. And because no change is happening with uniform motion, it may be said that no energy is spent during uniform motion, and hence, no work can be considered to be associated with uniform motion (see discussion below for a proper understanding of energy and work). Hence, we may say that the phenomenon of change is associated only with non-uniform motion.

Non-uniform motion has four basic manifestations—force, momentum, energy, and work. And these four phenomena may be studied in two sets: one is that of force/momentum (which are vector quantities), and the other is that of energy/work (which are scalar quantities). The reason for this segregation into two sets becomes apparent when we examine momentum, energy, and work in their cause-and-effect relationship. Consider this: we may state that force/momentum represents the *cause* of change in the position of an object in space, and energy/work represents the *effect* of that change. Or to put it in another way, force results in momentum, and the effect of this is the generation of energy, which does a specified work in the universe (the reader may appreciate this cause-and-effect relationship better as he/she moves on). First we will undertake a study of force and momentum and understand them in their proper perspective.

Force/Momentum: Our first step in this discussion is to study the relationship between force and momentum. Force causes an object to move, and any moving object in nature has momentum associated with it simply because momentum (p) is nothing but the product of mass and velocity (i.e. $p = mv$). Velocity has a direction, and so is a vector quantity, and so is momentum. Now consider this: force is defined as mass multiplied by acceleration ($F = ma$), but acceleration is the rate of change in velocity ($a = v/t$); thus, we can say that $F = mv/t$. But then mv is momentum, and so $F = p/t$, i.e. *force is the rate of change of momentum*. In other words, force imparted on an object at a given moment constitutes momentum (i.e. $p = Ft$). Hence, momentum can be said to be the fundamental property of motion; hence, it may be considered as the *cause* of movement (in the cause-and-effect relationship), and so we have grouped it with force.

Now we will discuss an essential property of the universe called the *conservation of momentum* and understand its relationship with non-uniform motion. The law of conservation of momentum says that the total amount of momentum in a given system is constant. This principle can be understood better if we consider *Newton's third law of motion*, which simply states that for every 'action', there is an equal and opposite 'reaction'. In fact, it can be said that Newton's third law is a direct consequence of the principle of conservation of momentum. We will now examine momentum in order to know about this action and reaction.

The third law, in essence, states that two bodies interacting with each other exert equal and opposite force upon each other. For example, as you are standing on the ground, you will be exerting a certain amount of force on the ground, but at the same time, the ground (i.e. the earth as a whole) will also be exerting an equal and opposite force on you but in the opposite direction. One of these forces is called the *action* and the other the *reaction* (it does not matter which way we call it because both of them are equal and reciprocal). All the movements we see in nature can be thought of as actions, and all of them have their counterpart reactions. We can find many examples in nature. When a ball hits the floor, it bounces back with equal and opposite force; if not for Newton's third law, the ball would be stuck to the wall forever! Here's another example. If you are able to walk effectively on the ground, it is also because of Newton's third law. As your foot pushes the ground with some force, the ground also pushes back your foot with equal and opposite force, which enables you to move ahead! However, sometimes in an event, the reaction to an action is not quite obvious. This is because the momentum of reaction gets dissipated into the surroundings owing to the quality of the objects involved in the event, such as the state of the media involved (e.g. solids or liquids), pliability of the surfaces of the objects (soft or rigid), their breakability (hard or brittle), and many other such features. For example, a stone fallen from height on to a soft ground does not generally bounce back with full effect but rather retires to the ground with a dull thud because the momentum of the reaction is redistributed into the surroundings to cause several smaller reactions such as a dent in the ground or throwing up of mud particles into the air or vibrations of the air molecules (to cause a thud). However, the sum of these resultant forces must always be equal to the force of impact—i.e. the sum momentum of the action and reaction in any event is a constant because momentum in a given system is strictly conserved. Or stated otherwise, the total momentum of a system (or an event) should always be *zero* (because we may also designate action and reaction with 'positive' and 'negative' signs as they are acting in opposite directions, and thus, they cancel out each other).

But the reader may note that all actions in nature are examples of non-uniform motion. Uniform motion has no mass to show its effect of force, and hence, no action can be attributed to uniform motion. And all actions in nature are invariably accompanied by equal and opposite reactions. There cannot be a force in nature which is not accompanied by an opposite force, and this is the fundamental principle of conservation of momentum.

There cannot be a force in nature which is not
accompanied by an opposite force.

Now consider this: what exactly constitutes action, and what is this counterpart reaction in the examples supplied above? Action in the above examples is the acceleration generated by the gravitational force of the earth, and the resultant opposite force of reaction is the inertia. We have already seen in the equivalence principle (Ch 1-D) that gravity and inertia represent the same force acting in opposite directions, and it can be said that action and reaction are one and the same, thus saving the conservation principle!

But then there are other fundamental forces (other than gravity) which generate non-uniform motion in the universe. As we have seen in Sec-B, that electromagnetic energy puts an electron in acceleration (or simulated uniform motion), or strong force puts a quark into acceleration, etc. Now what are the principles that guide these events? How do we account for their actions and reactions? Momentum of a given event must be a constant whatever is the type of force involved (or in other words, the total momentum of any given system must be zero). We will see how this becomes possible in the case of momentum created by other fundamental forces.

Take, for example, the case of a bomb explosion. The exploded particles in an explosion are dispersed in a random fashion in all directions. The force behind this explosion is the electromagnetic energy in the form of immense heat generated within the bomb. Each of these ejected particles has its own momentum caused by the EM force, but then the total momentum of the event (i.e. the bomb explosion) must be zero because momentum should be conserved. Conservation of momentum in this case becomes possible owing to the fact that if you consider an ejected particle moving in a particular direction as the action (e.g. right direction or upwards direction and so on), then another ejected particle moving in an opposite direction becomes the reaction (i.e. left direction or downward direction and so on). And then if we add up both the vectors, the result would be a zero. Therefore, every particle in an explosion has a counterpart particle moving in the opposite direction, and thus, the total momentum is conserved. Here's another example. Take a bullet fired from a gun. The EM force generated in the gun will not only push the bullet in the forward direction, but the released energy also pushes the gun itself in the backward direction.

Even in the case of subatomic particles, the principle of conservation of momentum is not violated. If you look at a Feynman's diagram, for example, it depicts a subatomic particle as gaining momentum after absorbing a photon (or a gluon, in the case of strong force), which causes a change in the direction of the particle; this is nothing but action–reaction equivalence (or 'force–inertia equivalence'). Thus, we can say that for any force that kicks a particle in one direction, its opposite inertia tends to move its counterpart particle in the opposite direction with equal force. But then what exactly is this inertia that is opposing the EM force? We will deal with inertia in Sec-E where its meaning becomes clear.

Proceeding with our discussion on momentum, it can be said that a resting body has zero momentum because its velocity is zero ($p = mv$), and by the same analogy, it can also be said that a massless object has zero momentum. The maximum momentum that can be theoretically achieved by any object (or particle) in the universe is by attaining the maximum possible speed, i.e. the speed of light ($p = mc$). But then we have already seen in Sec-A that every object in uniform motion can be considered as travelling at the speed of light, and thus, no force can be said to be acting on such an object while it is travelling in uniform motion. In other words, it can be said that momentum simply does not exist for a particle in uniform motion. Or we can say that force and momentum appear only when the object travels at relativistic (3D) speeds, i.e. non-uniform motion. Let us put this in a clear perspective. It appears that force and momentum are absent in light because photons travel at their maximum velocity, and their mass is zero.

Light has no force or momentum.

Therefore, force and momentum become operative only at speeds lesser than the speed of light. Or it may also be stated that force appears to impede the speed of light.

Force impedes the speed of light.

However, the reader may note that this statement can be appreciated better in Ch 8 where we will be discussing force in a different perspective. This principle also has an important bearing in the formation of matter as discussed in Ch 9. For now, we will proceed with non-uniform motion.

The Paradox: But then when we say that light has no momentum, here comes in a paradox. Consider this: light (i.e. EM force) is the source of energy. Or simply put, light carries energy, and this energy causes changes in the form of work in nature. Moreover, as we have seen in Ch 2-A, light (or EM radiation) is considered a fundamental force.

And also, we have seen that all our common energy transactions are mediated by the EM radiation in the form of heat, light, chemical energy, etc. (Ch 6-A), and the resulting exchange of energy is the essential property of EM radiation. But then if light has no force or momentum associated with it, then how can we say that it provides energy and does work? Or putting it this way, how can EM radiation cause change when change itself is not an essential property of light? The reader may also consider the following interesting phenomenon: the sun is estimated to be losing 4 million tons of mass each second by the way of generating light energy by nuclear fusion (Ch 4-D). But then when photons have no mass of their own, then where is this mass going?

Light appears to have no force and momentum,
but it has energy and does work.

We will first have an important historical insight into this paradox. Though photons are massless, it has been mathematically shown that they do have certain momentum of their own. We have here one explanation. In 1922, Arthur Compton has theoretically showed that when a photon with a certain wavelength hits an electron, both the electron and photon get scattered (following the principle of conservation of momentum), but after collision, the photon increases its wavelength (called the *Compton effect*). This means that the photon behaved as though it has mass and momentum! Compton argued that the photon indeed has a tiny momentum associated with it even though it has no mass. He mathematically combined the equations of energy of mass ($E = mc^2$) and energy of photon (i.e. $E = hf$, where h is the Planck constant and f is the frequency of EM radiation) from which the following equation can be derived: $m = hf/c^2$. We can plug this into $p = mc$ (the maximum possible momentum) and arrive at the equation, $p = hf/c$. But then the relationship between the wavelength, frequency, and speed of light is already known as $c = \lambda f$ (where λ is the wavelength, Ch 2-B), and plugging c in the equation $p = hf/c$, we get the momentum of photon, which is equal to the Planck constant divided by its wavelength, i.e. $p_{photon} = h/\lambda$.

However, we can see that this calculated momentum carried by an EM wave is only a mathematical derivation, but it has no proper theoretical explanation. Moreover, consider this. Though it is claimed that no mass appears in the above equation (h/λ), it must be realized that Planck constant itself is obtained just by dividing the energy of a quantum of electromagnetic radiation by its frequency (Ch 2-C), and its value is $6.6260695729 \times 10^{-34}$ kg m^2/s. This value, in fact, denotes that we are considering mass for the calculation, which means that the relativistic mass of a moving particle has a bearing on the equation. In other words, considering Planck's constant, we may say that the momentum of photon is represented by the smallest possible mass with the smallest possible momentum. But then the relativity theory has firmly asserted that the mass of a photon would become *zero* at the speed of light, thus putting these two concepts at theoretical loggerheads!

When all is said and done, it is an unimpeachable fact that light carries energy and mass with it, and we need a concrete theoretical explanation for this phenomenon. The reader may see that this paradox gets resolved with a conceptual understanding of the structure of a light ray as presented in Ch 8 and Ch 9. And we will take up this issue once again in Ch 10-B to categorically describe a mechanism by which EM radiation carries force, momentum, and energy. For now, we will concentrate on energy and work in the universe, which would consequentially lead us into other myriad affairs of thermodynamics, as we will see.

Section D
Energy, Work, and The Universal Cycle

In the following discussion, we will examine energy and work* at the most fundamental level and find their relationship with uniform and non-uniform motion. Before proceeding further, the reader may go back and revise the concepts presented in Ch 6-A regarding energy and work, which would come in handy in our discussion here.

---

* The reader, however, must not consider the term *work* in our ordinary sense of work done. Work here simply means a measure of the movement of objects (or particles) in space as a result of the force applied to them.

So far, we have seen that force *causes* movement, which imparts momentum on to the objects, and that force and momentum are vectors. The *effect* of force and momentum is shown to be a change in the disposition of matter, and we have seen in Ch 6 that the ability to cause this change is measured in terms of energy, and a unit of this change is considered as the work done ($W=F.d$). And we know that energy and work are scalars. We will now look into the phenomena of energy and work in a more fundamental manner so that we can understand them in a new perspective.

We will first examine energy and work with the help of a few examples. A train expends energy to do work while it moves. In order to get this work done, the train has to move its wheels, and for this to happen, the fuel has to burn, liberating its stored energy. If we closely examine the exact mechanism of movement of the train, we may be able to say that first the energy that is stored in the fuel is released into the surroundings, which is manifested as an increase in the velocity of the constituent molecules/atoms of the fuel as an initial step. This increased motion, in turn, moves the molecules of water (forming steam), and ultimately, this moves the wheels in a stepwise way, thereby moving the train. In other words, the energy that was stored in the fuel as potential energy (in the form of chemical energy, see below) is converted into kinetic energy, which is manifested as heat (Ch 6-B). Or putting it in another way, we may say that this minuscule potential energy of the fuel molecules, when converted into kinetic energy, is transferred step by step to cause greater degrees of change and movement, and this is manifested as greater and greater magnitudes of momentum, which ultimately enables movement of the mighty locomotive. And more importantly, we have also observed in Ch 6 that in the process of doing work, some amount of energy is lost (or 'wasted') into the surroundings as entropy.

Thus, in summary, we can say that while doing work in the form of locomotion, some amount of force (that is stored as energy) in the fuel is utilized to do work, and the rest of the force is liberated out into the surroundings as entropy in the form of heat. What is the exact mechanism by which these thermodynamic events take place? What is the relationship between energy and work, and why should entropy occur at all? By the end of this chapter, we would be able to answer the real purpose of entropy logically, but before proceeding further, we will take up some more examples to make clear the ubiquitous nature of this sequence of events of potential energy transforming into work and entropy.

When you lift a boulder of rock, you do work. When you give a lecture to your students, you do work. Even when you are quietly standing on the ground, you will be continuously working against the gravity to keep yourself from falling (this is work done in the form of stretching a few muscles and relaxing a few others). All these events are possible because the stored potential energy of the muscles in the form of glycogen is converted into kinetic energy of the protein molecules of the muscles, which is finally manifested as greater and greater degrees of change in the form of contraction of millions of molecules of the muscle (as in the case of locomotion of train), and that allows you to do work. During this process, some amount of energy is also lost into the surroundings as heat, i.e. entropy. Here are other examples. The various physiological processes of heart beating, breathing, digesting, etc. are all also examples of work done. Underlying all the actions of the cells in the body, there is an incessant chain of metabolic reactions which keep them alive (Ch 6-D), and all the metabolic reactions represent the same general sequence of events of converting potential energy of larger molecules into kinetic energy of smaller molecules and subsequently into work and entropy.

Yet another example of work done is the generation of electricity in a power station. The potential energy stored in the fuel, be it coal or diesel or uranium, or the potential energy stored in the body of water above as in the case of hydroelectricity (we will be looking into the stored potential energy below) is converted into kinetic energy and is released (through the turbines) as electrical energy, which does our household work. And dutifully, some amount of energy is lost during this process into the surroundings as heat.

We will now analyze the phenomena of potential energy, kinetic energy, work, and entropy and arrive at some important conclusions.

Potential Energy: Classically, potential energy is described as the energy possessed by an object by virtue of its position—i.e. the magnitude of the potential energy is determined by its relative position in space (see below for explanation). Kinetic energy, in contrast, is the energy possessed by an object by virtue of its motion. First we will discuss potential energy, and later we will look into the evasive kinetic energy.

At the outset, we may state that force is stored in an object as energy, and when liberated, it does work in the universe. The following analysis clarifies this statement. Potential energy represents the stored force in an object (or a particle), which is capable of initiating change in the relative position of objects in space. Potential energy, in fact, is the force which is held suspended 'in reserve' in an object. And this potential force has a natural tendency to be converted into kinetic energy as nature tends to minimize potential energy and prefers kinetic energy, as we will see below.

Nature abhors potential energy and favours kinetic energy.

Thus, an object with potential energy can really be considered as a transient custodian of the force that was imparted on to it, and this object (or particle) shows a predilection to convert it into kinetic energy to do work (i.e. to cause change) in the universe.

Potential energy exists in nature in different forms—gravitational, mechanical, chemical, electromagnetic, and nuclear (the reader can observe here that all the four fundamental forces of nature are represented in one of these different forms). Potential energy, in any one of the above forms, has a potential to transform into a different form of energy to do work in the universe. We will briefly look into these different forms of potential energy.

Gravitational potential energy is all pervasive—it occurs on the earth, in between the planets, in between the stars, in galaxies, and so on. In fact, as was envisioned by Newton, it exists between all the objects in the universe. On the earth, it can be said that gravitational potential energy represents the energy stored in an object by virtue of its height from above the ground—the higher an object is placed above ground, the greater is its potential energy. When we have lifted a stone and placed it above on a board, we have actually transferred our energy to the stone (to do work against gravity), and this energy is now stored as potential energy. And when it is released, this energy is transformed into kinetic energy, and the stone falls on to the ground where it assumes a state of least potential energy—i.e. the stone tends to slip off and fall more readily than it would stay on board. After all, nature has an inexorable tendency to attain a state of least potential energy. Imagine a ball resting on the floor of a valley – since it is not moving, no energy is expended on it. To make it roll up the valley, we have to strike it so that some mechanical energy is transferred to the ball. The ball now acquires kinetic energy and moves up to reach the height of the valley, and then the ball stops for a brief moment before it rolls back to the floor. Now consider this: what has happened to the energy while the ball has stopped for the briefest second at the height of the valley? The ball's kinetic energy has not vanished but is converted into potential energy, which has allowed the ball to stay still for some time. And as the ball starts rolling back, the potential energy is gradually converted back into kinetic energy. The magnitude of this potential energy can be calculated by using the formula $PE = mgh$, where m stands for mass, g for the rate of acceleration due to gravitational force on the surface of the earth (which we know is 9.8 m/s/s), and h for height from the ground. As the ball moves up and down, the total energy remains constant but periodically switches between kinetic and potential energies. The ball ultimately stops rolling because it encounters friction from the walls of the valley and resistance from the air, which gradually converts some of the energy into heat and is absorbed by the walls and the air, and this heat is dissipated into the surroundings. It may be imagined that if there were no friction/resistance, the ball would tend to swing forever! But then it is inevitable that in all these types of work where gravitational energy is converted into mechanical energy, there will be some loss of energy into the surroundings in the form of entropy.

Mechanical potential energy is also prevalent in nature. Here are a few examples. A clockwork functions by transferring mechanical potential energy to move the chain of wheels, a piston of an airgun transfers mechanical potential energy to the ball, and all our muscle movements are also based on mechanical energy, as we have seen above. Elastic potential energy is a well-studied form of mechanical energy. For example, when you pull a rubber string, the energy of your muscles is transferred and stored in the string as potential energy, and when released, it does work in the form of movement. Of course, in all these actions, entropy inevitably happens by the way of some heat loss.

Chemical potential energy is the energy stored in chemicals, which, when liberated as thermal energy, does some work. For example, food contains chemical potential energy, which, when consumed and broken down, liberates kinetic energy in the form of increased molecular movement, which, in turn, allows the cells of an organism to do work. All cells survive this way by doing cellular work (Ch 6-C). Chemical energy may also take a different form. For example, a battery cell may convert chemical energy into electrical energy to power a light bulb, or chemical energy may be converted into mechanical energy to move a motor. However, in all these actions, entropy invariably happens by the way of heat loss.

Electric potential energy arises when two charges are close enough to each other to exert either an attractive or repulsive force. For example, when two positive charges are brought close to each other, their repulsive force pushes them apart. However, when they are brought together against resistance, they develop an electric potential energy, and this energy does work when released. The closer the charges are to each other, the higher the potential energy stored and the more work done when released. It is needless to stress that, ultimately, when electricity does work, it results in entropy.

Nuclear potential energy is the energy stored up in the nucleus of an atom, which can be released when the nucleus changes its configuration by the way of various nuclear reactions (Ch 3-D). We will look into nuclear reactions in Ch 9-C.

Kinetic Energy and the 'Natural Sequence' of Events: We will now examine kinetic energy. Whereas potential energy is the energy conferred to an object by its position, kinetic energy is the energy conferred by its motion. We have seen that potential energy is the stored energy of an object which, when liberated, does work. But it must be realized that for the work to be done, potential energy has to be first converted into kinetic energy. Thus, work and kinetic energy share a common footing. In fact, it can be shown that work is actually measured in terms of kinetic energy, as we will see.

Almost all activities we see around us stand as examples of kinetic energy and work. Water stored above ground has potential energy, which is converted into kinetic energy as it falls down and does work by breaking the stones underneath (or by moving the turbines to generate electricity—here the gravitational energy is converted into electrical energy). A plane ready to take off has potential energy; as it moves, it loses potential energy, gains kinetic energy, and work is done. A meteoroid falling to the ground has immense kinetic energy, which does its work by creating a huge crater on the ground. An electron shot from an electron gun has more kinetic energy, and when it hits a fluorescent screen, it does work by dissipating its energy in the form of light and heat. However, it's imperative to add that entropy is an inevitable part of these instances of conversion of kinetic energy into work.

Thus, we can summarize that all moving objects in nature follow a *natural sequence* of events from cause to effect, which can be depicted in the following 'linear' pathway:

Force→Momentum→Potential energy→Kinetic energy→Work→Entropy

We will now examine the relationship between work and kinetic energy. This can be easily understood if we know how we calculate work. Work done in the universe is measured as the magnitude of displacement of an object that has taken place because of the force applied on the object. In other words, work is force multiplied by the distance travelled: $W = F.d$. However, because $F = ma$, this equation can be rewritten as $W = m.a.d$. But then acceleration is velocity over time, v/t, and the distance travelled is calculated by the formula $\frac{1}{2}at^2$. Plugging in these details into the above formula, $m.a.d$, we can arrive at the famous equation by which we can calculate the work done by a moving object: $\frac{1}{2}mv^2$. And the reader may notice that this is also the equation to calculate the kinetic energy in nature!

In fact, the net work done by a system is calculated as the difference in the kinetic energy of the object at the start and at the finish: $W_{net} = \frac{1}{2}mv_f^2 - \frac{1}{2}mv_i^2 = \Delta KE$.[*] In other words, the work done

---

[*] In this formula, we have done calculations for 'translational kinetic energy' (i.e. work done by an object during motion in straight paths), i.e. in the equation $\frac{1}{2}mv^2$, m here actually represents 'translational inertia', which, in simple

on an object by a net force equals the change in kinetic energy of the object, and this relationship between work and kinetic energy is called the *work–energy theorem*. Therefore, the units of measurement of both work and kinetic energy are the same—it is newton-metre (which is nothing but 'F.d'—i.e. force in newtons and distance in metres), or in short, this unit is called a *joule*.

Thus far, we have seen the inexorable relationship between energy, work, and entropy. But now we will see how we can rearrange these concepts into a meaningful outcome which bespeaks of a universal truth.

The 'Universal Cycle': Before going further, we will first try to understand one important aspect of the final player in the above sequence of events—entropy. We have already seen in Sec-B and C that uniform motion is the natural state of an object in the universe and that all non-uniform motion has a natural tendency to culminate in uniform motion. But then the reader may observe that the phenomenon of entropy may be equated with uniform motion because the lost heat in the linear pathway above may be considered as a form of a ray of electromagnetic energy, and we have seen that a ray of EM light travels in uniform motion at the speed of light. In short, entropy may be thought of as a state of uniform motion.

Entropy, as lost heat, is nothing but light in uniform motion.

Now by looking at the linear pathway of natural sequence of events, the reader may observe that we can join both the ends of this linear pathway by using uniform motion as the 'merger' and make it a circular pathway, i.e. 'cycle'. We will now analyze how this cycle may be arranged (the reader may look at Fig 7.3). Imagine that an object (or a particle) is travelling in uniform motion, and when an external force is imparted on to the object, it causes non-uniform motion. The object now attains some momentum, becomes

---

terms, could be said as mass (see below), and v represents linear velocity. But for 'rotational kinetic energy' (i.e. work done in rotational movements), the equation would be $\frac{1}{2} I\omega^2$, where I represents *moment of inertia* (here the force applied is called *torque* or 'turning force') and ω represents *angular velocity*. We will not go into further details here.

a custodian of potential energy, gains kinetic energy, does work, and ultimately joins back its original natural state of uniform motion by ending up in entropy. Thus, the final result is uniform motion at the speed of light, which appears to be the final goal of this cycle. And this we will be calling the *universal cycle* of nature!

As shown in Fig 7.3, we can depict both 'force' and 'work' as 'offshoots' of this universal cycle, and without them, the cycle will go on and on forever without interruption because only when the agent of force exists would the offshoot of work exist. Or alternatively, we may say that force, once it enters the cycle, gets off the cycle by doing work.

Force enters the universal cycle and exits as work done.

But then in a situation where there is absence of force and work, no change is possible, and the universe would just be a monotonous sheet of uniform motion which pervades the universe at all places and at all times! Thus, in the absence of this force, everything in the universe would have been a *tabula raza*—a clean sheet. However, we will keep away from such an uneventful universe and presume that, time and again, we encounter the incursions of force into the cycle with the resultant work being efficiently done in the universe; otherwise, where are we?

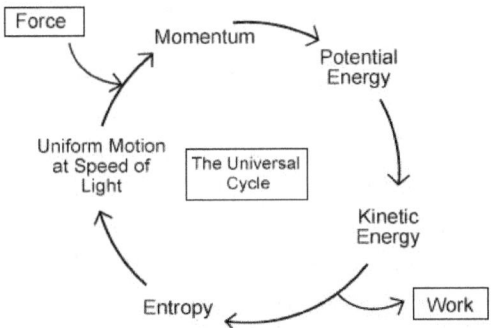

Fig 7.3: The universal cycle.

By this universal cycle, we can arrive at the conclusion that the natural sequence of events encompassing energy and work is, in fact, a recurrent theme of the universe. All the events associated with

RIDING ON A RAY OF LIGHT

non-uniform motion ultimately have to end up in this universal cycle. In other words, this natural state of uniform motion can be represented as a 'bottomless sink' into which all the moving objects must fall in. We have laid out many examples in Sec-F to demonstrate the importance of this universal cycle, and this would actually improve our understanding of nature in the correct perspective (as we will see in Sec-F).

Thus, in summary, we can say that the external force imparted on an object is always represented in non-uniform motion, which does work in the universe, and finally culminates in the natural state of uniform motion.

Force creates non-uniform motion, does work, and dissipates as entropy.

At this juncture, it may be recapitulated that we have discussed about inertia of rest and inertia of motion in Sec-B and demonstrated that they both are the same because they both represent uniform motion at the speed of light. And here we have also seen that entropy is also concerned with uniform motion. And thus, there must be a relationship between entropy and inertia, and what could be this relationship? We will now explore this issue in the following section and draw parallels between entropy and inertia.

Section E
Relationship Between Entropy and Inertia

In Sec-B, we have redefined Newton's first law and stated that inertia of rest and inertia of motion are one and the same and ascertained that all objects tend to be in uniform motion at the speed of light. In Sec-D, we have learnt about another universal tendency in the form of entropy, which is observed as an end result of all thermodynamic events in nature and which also culminates in uniform motion. Now what is the relationship between these two universal events of inertia and entropy?

Here in this section, we will discuss inertia and entropy in clear terms, and the following discussion would actually lead us to arrive at a profound theoretical conclusion as to the real purpose of inertia and entropy. And by the end of this section, the reader can see the cosmological significance of these entities. First we will begin our discussion with entropy.

The Concept of 'Cosmic Entropy': We may deduce by the discussion in Sec-C and D that all the players in the natural sequence, such as force, energy, and work, are merely manifestations of non-uniform motion (or acceleration). Once the object attains uniform motion, there is no force or momentum, and consequently, there is no energy or work done. By this discussion, we can also deduce that an object always tends to spend away the force imparted on to it preferentially by increasing its kinetic energy rather than keeping still and storing it as potential energy.

Thus, finally, we may say that the force which is converted into kinetic energy ultimately ends up in two ways. First, a portion of it is converted into 'useful' work by causing displacement of the objects (or particles) involved in the system; this is the actual work done in the universe (see Gibbs free energy in Ch 6-B). And second, the remaining portion of kinetic energy is invariably converted into heat, which cannot be used to do any useful work in the form of displacement, and this is the energy lost as entropy. In other words, we can state that nature allows converting kinetic energy into work *only* when a portion of it (in fact, about more than 50% of it! see Ch 6-B) is allowed to transform into entropy. This is nature's trade-off between work and entropy (and this feature is adequately stressed in Ch 6-A and B).

Or we may think about the above discussion in a new perspective, which we have already probed at the end of Sec-D. The phenomenon of entropy represents lost energy, which is nothing but heat in the form of a ray of electromagnetic energy, and this ray of EM light travels in uniform motion at the speed of light (and because of this feature, we have included entropy into the main stream of the universal cycle). When we look at the universal cycle, force enters the cycle as potential energy and gets converted to kinetic energy, which ultimately ends up doing work (in the form of non-uniform motion) and entropy (in the form of uniform motion). The inevitability of entropy bespeaks of one important fact—the phenomenon of entropy is nature's attempt to attain uniform motion. In other words, force can convert to work only when some portion of it is utilized to become heat in the form of electromagnetic radiation which travels at the speed of light. Or to put it in simpler words, the phenomenon of entropy is an essential necessity of nature!

Entropy is a necessity of nature.

Having seen the inevitability of entropy in all thermodynamic events and having also learnt that entropy is nothing but the natural state of uniform motion at the speed of light, we can arrive at the necessary conclusion that the essential purpose of all the events in nature by the way of spending energy and doing work is to ultimately end up in this natural state, i.e. entropy, which, again, is nothing but uniform motion at the speed of light.

The essential purpose of all thermodynamic
events is to end up in entropy.

But then if the essential purpose of events is to end up in the natural state of entropy, then why does nature *not* allow *all* the energy involved in an event to convert into entropy? The answer to this question becomes obvious in Sec-F below and in Ch 8. But for now, we will settle with this important statement: nature always *attempts* to convert *all* the force involved in any thermodynamic event into entropy, failing which, nature tends to trade off not only by making entropy an inevitable accompaniment of all thermodynamic events but also by taking away as much energy as possible in the form of entropy! This is the real meaning of entropy in nature, and this is the reason for its stubborn persistence in all thermodynamic events!

Failing complete conversion, nature converts
majority of force into entropy.

The reader may note here that all the thermodynamic events that take place throughout the universe dutifully follow this principle. In other words, we may say that entropy is a universal principle. This universal tendency of all thermodynamic events in nature to end up in entropy, in fact, appears to set up a purpose and direction to this gigantic universe, as we have already seen in the thermodynamic and cosmological arrows of time in Ch 6-E. This gross universal tendency of events in this grand cosmos to culminate in as much of entropy as possible can be termed as *cosmic entropy*.

All thermodynamic events in the universe end up in 'cosmic entropy'.

Having discussed entropy in this broad perspective, we will start our discussion now on inertia.

The Concept of 'Cosmic Inertia': We have already redefined Newton's first law in Sec-A by stating that inertia of rest and inertia of motion are one and the same, and all objects tend to attain uniform motion at the speed of light. By this definition, we may state that inertia is the universal tendency of an object to attain uniform motion.

Bertrand Russell, the philosopher and mathematician, had introduced the concept of the *law of cosmic laziness*,* which actually reflects a profound universal truth. We will now examine this phenomenon in a simple manner and see how it leads us to the concept of 'cosmic inertia'. To start with, we will recapitulate the spacetime diagram presented in Fig 7.2. We can see that all objects which travel exclusively along the space dimension (i.e. along the *x*-axis) travel at uniform motion with the speed of light, and this can be said to be a 'timeless' travel in space because the object is not at all travelling along the time dimension (i.e. *y*-axis). Conversely, an object which is travelling exclusively along the time dimension can be said to be stationary, i.e. at rest. And an object which travels at a velocity below the speed of light—i.e. as it travels in between the *x*- and *y*-axes—may be said to travel by non-uniform motion. But then as per Newton's first law, an object which travels exclusively along the space dimension does not need any force to keep its uniform motion at the speed of light, which, in turn, clearly means that non-uniform motion is possible only with the application of force.

---

* This phenomenon of law of cosmic laziness is discussed expertly by the physicist Martin Gardner in his book *Relativity Simply Explained* (pp. 88–90).

However, we have seen that all objects in non-uniform motion have a tendency to ultimately end up in uniform motion.

By the above discussion, we may say that all objects in nature always have a tendency to take a path which is more in space dimension (x-axis) and less in the time dimension (y-axis). This natural preference of an object in the universe to show reluctance to travel by time dimension (or in other words, its natural preference *not* to deviate from space dimension) is called the law of cosmic laziness. This cosmic laziness causes objects to move more through space whereby they may ultimately cease to travel by time dimension at all. This universal tendency of nature to achieve cosmic laziness may be considered to be the real meaning of inertia, and in fact, we may actually call it by the apt term *cosmic inertia*.

We may also discuss this phenomenon of law of cosmic laziness (= cosmic inertia) in another perspective. In Ch 1-E, we have discussed about geodesics and stated that all objects moving in a spacetime distortion of other heavier objects (e.g. earth moving in the spacetime distortion created by the sun) tend to take the straightest possible path, and that is the path of geodesic (the reader may refer to Ch 1-E for details). This tendency of an object to take the straightest and shortest* possible path between two points on a curved surface can also be viewed as the manifestation of the law of cosmic laziness (or cosmic inertia). Now we will collate our renewed understanding of cosmic entropy and cosmic inertia into a single framework and see how they are related.

Cosmic Inertia and Cosmic Entropy in One Framework: Thus, it has become evident to us that there are two sets of events in the universe which tend to culminate in the natural state of uniform motion— one is the cosmic inertia, and the other is the cosmic entropy. All the *thermodynamic events* that happen in nature (each morsel of food

---

* Several authors have discussed on the matter of 'longest' or 'shortest' distance of a geodesic, but we may say that this nuance is pointless because the preference of nature is for all objects to travel by just uniform motion in a straight path at the speed of light, which is 'tightly and surely' fixed at 299,792,458 meters per second—not a bit less, not a bit more by space or time—so that it doesn't really matter if it is shortest or longest! Similarly, on a curved surface, the geodesic simply takes the straightest possible path; it is neither short nor long.

we take, each ounce of fuel we ignite, each cinder of wood we burn) invariably generate entropy as a part of an attempt to join the cosmic entropy to culminate in the natural state of uniform motion.

In the same way, all the *physical processes* in nature (be it a ball rolling up and down a valley, a train speeding on its tracks, or a feather falling to the ground) invariably experience inertia (as a reaction to the force applied) as part of an attempt to join the cosmic inertia of uniform motion at the speed of light. Thus, it can be said that the inevitability of entropy and the inevitability of inertia are the consequences of the same natural principle.

> *Inevitability of entropy and inertia are the*
> *consequences of the same natural principle.*

The above universal principle, when we are dealing with a thermodynamic system, we call it entropy, but the same universal principle when dealing with a moving physical system we call inertia. Or putting it in another way, we may say that in a physical system, entropy is disguised as inertia, or in a thermodynamic system, inertia is disguised as entropy. And we are taught of this universal phenomenon as entropy at one place in our textbooks and inertia at some other place, while both of them are serving the same purpose in the universe.

> *In a moving physical system, entropy is disguised as inertia.*
> *In a thermodynamic system, inertia is disguised as entropy.*

Or finally, in a grand way, we can say that inertia and entropy are one and the same.

> *Inertia and entropy are two sides of the same coin.*

With the help of this theoretical framework, we will now go into the implications of universal cycle in our understanding of various events of nature. However, some of the readers may find the above discussion and its conclusion a bit unsettling as it were, but the examples provided in Sec-F would certainly restore their confidence in this fundamental understanding.

Section F
Implications of The Universal Cycle

We have seen at the beginning of this chapter that the phenomenon of motion is the cardinal feature of the universe, and by using step-by-step analysis, we have arrived at the conclusion that all motion in nature culminates in uniform motion at the speed of light, and this phenomenon is depicted in the universal cycle. Or by logical understanding, we may state that this cosmic principle, in fact, holds the 'driving principle', which moves forward all the events in the universe. In other words, we may state that the various kinds of behaviour of moving objects or the thermodynamic activity we observe in the universe can be explained simply by assuming that they are all making an attempt to move towards the goal of this uniform motion at the speed of light. In short, we can say that this universal cycle of nature is the underlying principle which governs all the mechanics of motion in nature.

Universal cycle drives all the mechanics of motion in the universe.

To understand this in its proper perspective, we may take any event in the universe and show that every event of nature ultimately ends up in the universal cycle. But then the reader must realize that our universe is a complex place. Various forces operate together at a given time with varying magnitudes and in different directions, and these complexities of nature may effectively mask this underlying universal principle from our direct observation. However, by a diligent analysis of any event in nature, we may be able to unmask the overlying complexities and show that the underlying basic principle guiding all the events is nothing but the universal cycle culminating in uniform motion.

We will now examine a few events of the universe in this perspective which exemplify this basic principle. By the way, in the following discussion, we will also arrive at some profound truths of nature which explain a variety of other phenomena of nature in a logical way.

Event 1—An Ideal Situation: At first, we will consider a thought experiment. Imagine a bullet being fired into a space where, hypothetically speaking, there is absolutely no ambient force existing in it. The space is absolutely devoid of all forces, including the energy of empty space (perhaps beyond the 'edge' of the universe). Now we can say that the only force that acts on the bullet is the force of thrust derived from the energy of gunfire (in the form of electromagnetic force). This external force acting upon the bullet gives it acceleration and keeps it moving. As the bullet moves ahead, it loses its potential energy, but at the same time, it gains kinetic energy—or in effect, the bullet keeps increasing its velocity. And it keeps moving in a straight line as there is no additional force acting upon it to divulge its path, and now this bullet may be said to be in a state of linear acceleration (Fig 7.4). But then what would be the ultimate fate of the bullet? We can say that it keeps increasing its velocity by acceleration until it attains the maximum permissible velocity—i.e. the speed of light (explanation below).

Fig 7.4: Event 1—an ideal situation.

This is because the bullet which is hitherto stationary in relation to the gun has now started accelerating into a space where no other force is existent. But since the bullet is under the influence of force (i.e. EM force), it has to accelerate; it cannot travel in uniform motion. The reader may argue that after a certain period of time, the bullet may decelerate and bend its course. But for the bullet to decelerate and to stop or to change its direction, it needs some opposite force, and this additional force (or energy) is absent in our hypothetical space, so the bullet has to increase its acceleration incessantly in the forward direction and travel in a straight line. But then how can we ascertain at which point the bullet attains uniform motion at the speed of light? This distance should obviously depend on the initial rate of acceleration at the gun. It would take less time if the initial acceleration is more, and it would take more time if the initial acceleration is less; even a slow-pace carom striker ejected into such a hypothetical space should ultimately attain the speed of light after travelling for a certain distance!

Putting this in another way, we may say that the bullet merges with light because at the speed of light, it loses all its spatial dimensions and its mass. It now travels in the space dimension alone, and time dimension does not exist for it any more, thus making the passage of time immeasurable, as we have seen in Sec-B. We may also state that the bullet has become a ray of light and has merged with the maximum inertia possible, i.e. the state of cosmic inertia.

The bullet becomes a ray of light and merges with cosmic inertia.

However, the reader must realize that this is an ideal setting. No object or event in nature can practically attain this situation, and it is outside our experience. However, we can discuss this ideal situation in practical terms, as shown below, which in fact improves our theoretical understanding of the universe. Thus, we will continue with Event-1 for some more time.

We may also understand Event-1 in terms of the work done in the universe because the bullet was under the influence of a force, and it has travelled a certain distance ($W = F.d$), and this displacement is the work accomplished by the bullet in the universe. And in such a case, the final outcome of the event—i.e. the uniform motion at the speed of light—may also be reckoned as 'entropy' that has resulted out of the work done. We may analyze this ideal setting in the context of work and entropy. In Ch 6-B, we have studied entropy under two headings—'entropy as disorder' and 'entropy as heat loss'. We will see how our bullet in the hypothetical space could have achieved both of these entropy states. Consider this: as the velocity of the bullet kept increasing due to linear acceleration, the relative space between the source (the gun) and the bullet kept increasing constantly (Fig 7.4), and this can be construed as increasing disorder or increasing randomization (or rather we would call it increasing 'rarefaction', which is a better term). And it can be said that this increased rarefaction is in consonance with 'entropy as disorder'.

But then obviously, the bullet has also finally ended up in uniform motion at the speed of light, which is nothing but a ray of light (electromagnetic radiation). And this can be taken as 'entropy as heat loss' (Fig 7.4). Thus, in this ideal situation, we may see that the bullet's work has achieved both of these phenomena. In fact, we can state that both increased disorder and heat loss are one and the same—increased disorder ultimately means increasing the rarefaction to the maximum possible extent so that it reaches the uniform motion at the speed of light! In other words, we may state that 'entropy as disorder' and 'entropy as heat loss' are merely two sides of the same coin.

'Entropy as disorder' and 'entropy as heat loss' are one and the same.

And when we consider the amount of work done in this ideal event, we may say that this is the maximum possible amount of work permissible for any event in the universe. But then we may also say that the work done and the resultant entropy in this ideal situation are also one and the same. Simply put, we cannot differentiate between work and entropy in this ideal situation; this feature becomes clear to the reader as he/she reads on.

In an ideal setting, both work and entropy are indistinguishable.

And then finally, in this ideal situation, the total amount of energy that was imparted on to the bullet is ultimately converted into entropy. We may compare this to our ordinary situations. In any thermodynamic event, out of the total amount of input of energy, some amount of it is converted into work, and some is converted into heat loss. This makes us perceive both work and entropy separately. For example, if you strike a rock with hammer, the rock will not disappear in a flash of light, but rather, some of the energy is utilized to break down the rock into disorderly pieces (i.e. work), and some of it is dissipated as heat and light (i.e. entropy). In contrast, in this ideal setting, both work and entropy have identical outcomes. In fact, both of them are the same.

In summary, theoretically speaking (as in this idealized situation), we may say that the phenomena of force, energy, work, and entropy are indistinguishable from each other. In fact, in this new perspective, the concept of 'entropy as disorder' itself needs some revision! All the disorder we see in the universe is an attempt (or a drive in the universe or a struggle of nature) of the objects to achieve the ultimate goal of moving at the speed of light and to merge with the universal cycle. In short, it is the drive of nature to convert all force and energy of any event into 'entropy as heat loss'. The disorder that accompanies any event is merely an illusion of our human understanding. This is the deeper meaning of all movement in the universe—the merger with cosmic inertia/entropy!

The implications of this theoretical understanding are astounding. On the one hand, it has a bearing on the existence of matter in the universe (Ch 9), or by a logical extension, it can also predicate on the very existence of the universe itself (Ch 12), and we will discuss these issues in the appropriate chapters.

Event 2—A Down-to-the-Earth Situation: Now we will look into a practical situation. Imagine a cannonball being fired on the earth which takes a curved trajectory and finally hits the ground. The cannonball is initially under the influence of force created by the electromagnetic energy of the gunfire. The cannonball when left alone, with no other force interfering with it, travels straight, reaches uniform motion, and merges with the universal cycle as our bullet has done in Event-1. However, on the earth, we have a plethora of forces to complicate its trajectory. For example, we have air to give friction, and we have gravity to pull it down. Initially, the cannonball moves forward, utilizing the EM energy provided by the gunfire. And it accelerates for some time to a certain distance, and then, encountering air resistance, it decelerates and slows down. All the while, the potential energy stored in the cannonball keeps converting into kinetic energy, which keeps it in motion. But then the kinetic energy also has to do an additional work of moving the air molecules (ignoring the weak gravity for the time being), and this is appreciated as mechanical energy in the form of friction. This frictional force, offered chiefly by the air molecules, prevents further acceleration of the cannonball. However, during this conversion of kinetic energy into mechanical energy, some of the kinetic energy is lost as heat and is dissipated into the surroundings by the way of the heated air molecules (Fig 7.5). The heated air molecules move with increased velocities and thus become more disordered. This heat and disorder, as the reader can see, are the representatives of the 'entropy of heat loss' and 'entropy of disorder', which are inevitable accompaniments of any work done in the universe.

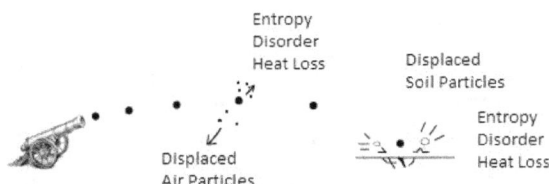

Fig 7.5: Event 2 showing work and entropy.

The cannonball continues its journey but becomes sufficiently decelerated to come now under the influence of gravity; thus, it changes its trajectory and starts falling down to the earth. As it descends to the earth, it undergoes acceleration (due to the gravitational acceleration, g). The cannonball now falls on to the ground with an impact which results in displacement of chunks of soil, creating cracks in the ground, accompanied by soil particles flying up into the air, producing a thud. At each step in the above trajectory, the cannonball has done some work in the universe, and each step is invariably accompanied by the generation of heat and disorder (Fig 7.5). Thus, the kinetic energy of the cannonball is finally converted, step by step, into a variety of different sorts of energies during its transit from the cannon to the ground. And there is loss of heat and an associated displacement and rarefaction of matter in the trajectory.

By looking at these events, we can notice certain essential differences between Event-1 and Event-2. In Event-1, though the kinetic energy is converted into work and entropy, both of these end results are indistinguishable. Hence, energy in Event-1 has ended up doing *no* 'useful' or 'measurable' work in the universe (see below). Consequently, the total amount of force that was imparted on the bullet has merged with the universal cycle and transformed into uniform motion.

In an ideal setting, the total amount of imparted
force is converted into entropy.

In Event-2, in contrast, the kinetic energy of the cannonball has ended up in several forms of energy in its staged journey. In other words, the electromagnetic energy is converted into various forms of mechanical energy and has done some 'useful' work (i.e. the energy which could be measured as work). And thus, in Event-2, we can say that some of the imparted force on the object has resulted in a change in the disposition of matter in the form of work, and some of it has merged with the universal cycle of inertia as entropy.

At this juncture, we may make a conclusion: though Event-1 is hypothetical, the end result of this thought experiment is of paramount importance because all objects and all events of the universe have a natural tendency to merge with the universal cycle of inertia. They all tend to do no measurable (or 'useful') work in the universe, but they tend to end up in uniform motion at the speed of light. However, as noted above, this is only an ideal situation which is unattainable in nature. But then once this ideal status is attained, objects will no longer be objects, but they simply vanish into streaks of light!

We have arrived at a similar conclusion in Sec-B: 'Uniform motion is the natural state of motion in the universe, and all the objects are attempting to move towards uniform motion without ever achieving it.' These two conclusions are parallel—all objects (or particles of matter) tend to exist in an ideal situation where there is no interference of force, and in this state, the objects achieve uniform motion with the speed of light. In other words, given the right conditions (i.e. with no force interfering with movement of an object in space, as in Event-1), the objects have a natural tendency to slip into uniform motion without doing any demonstrable work in nature. Speaking the other way, we can say that all objects tend to 'cast off' the imparted force on them and end up in uniform motion. The profound significance of this concept becomes apparent in the next chapter, Ch 8.

Now look at the trade-off played by nature. The above ideal situation being hypothetical and unachievable, nature attempts a trade-off—if a total conversion of force into entropy is not possible in a system (as in Event-1), then the nature makes a compromise and allows at least a portion of force to be converted into entropy, and then only it allows the rest of the force to be utilized to do work (as in Event-2). This is the bargain of nature. This is the true meaning of entropy; this is the reason for its inevitable appearance in all thermodynamic events of nature.

> *Nature trades off by converting at least a portion*
> *of imparted force into entropy.*

But then in Sec-E, we have seen that entropy and inertia are equivalent, and they fundamentally mean the same thing. And thus, we may say that entropy/inertia is the essential purpose of all thermodynamic events. It is the goal of any thermodynamic event in the universe, and this signifies the essence of cosmic entropy or cosmic inertia.

> *Cosmic inertia/entropy is the essential purpose of the universe.*

Event 3—Nuclear Reactions: Now we will briefly look into the world of subatomic particles and see how nature contrives to achieve inertia/entropy even in the case of the most powerful reactions of nature, the nuclear reactions. The phenomenon of *nuclear fission* (Ch 3-E) can easily be accounted for in this cosmic purpose. Nuclear fission is a spontaneous process wherein large unstable nuclei are broken down into smaller nuclei, and a lot of nuclear energy is liberated during it course, which portends the occurrence of both 'entropy as disorder' and 'entropy as heat loss', respectively.

However, in the case of *nuclear fusion* also, the inevitability of cosmic inertia/entropy can be accounted for. Nuclear fusion involves not only the formation of heavier nuclei, which has increased order in nature butit also involvesdisorder and heat loss (in excess of the order it has caused)! This becomes obvious if we look at the proton–proton fusion (Ch 4-D) and notice that during the chain of events, so many smaller particles (such as electrons, positrons, and neutrinos) are liberated ('entropy as disorder') along with the liberation of massive amounts of energy in the form of gamma rays ('entropy as heat loss').

All these features of nuclear fission and fusion bespeak of the fundamental nature of the universe in promoting 'entropy as disorder' and 'entropy as heat loss' by the way of achieving the ultimate cosmic purpose! However, the reader may also notice that nuclear fusion is *not* an essentially spontaneous process (because it primarily increases order), but it happens in nature in the presence of great gravitational pressures and intense temperatures (as that happens inside the stars, Ch 4-C), but even so, the final result is an increase in disorder and heat loss as nature would always prefer!

Having studied motion and learnt of its cosmic secrets, we will now use these concepts to explore the true nature of light and a mechanism by which light is generated in nature in Ch 8.

☼ Carry-Over Snippets—Chapter 7 ☼

☼ **Uniform motion is a myth**. *All uniform motion we observe in nature is* **simulated uniform motion** *(which is a form of non-uniform motion).*

☼ *Any object moving at the speed of light becomes a ray of light!*

☼ **Uniform motion re-examined** *and* **Newton's first law of motion redefined:** *every object in nature tends to be in a state of uniform motion at the speed of light eternally unless acted upon by force.*

☼ **Non-uniform motion** *has a natural tendency to culminate in uniform motion. All objects tend to achieve the natural state of uniform motion without ever achieving it.*

☼ *Inertia of rest and inertia of motion are two sides of the same coin.*

☼ **Universal cycle of nature** *depicts the natural sequence of events starting from uniform motion to force to momentum to potential energy to kinetic energy to work and ending finally in uniform motion in the form of entropy. All events or objects in nature exist as a part of this cycle, all of which ultimately tend to culminate in it—universal cycle drives all the events of the universe.*

☼ In all the events we observe in nature, force enters the universal cycle and exits as a work offshoot. However, an **ideal universal cycle** has no work offshoot. In an ideal universal cycle, work and entropy are indistinguishable.

☼ **Entropy and inertia are one and the same.** In a moving physical system, entropy is disguised as inertia; in a thermodynamic system, inertia is disguised as entropy. The inevitability of entropy and inertia in nature is the consequence of the same natural principle. Entropy and inertia are cosmological necessities.

☼ **Entropy as disorder** and **entropy as heat loss** are one and the same.

☼ **Cosmic inertia/entropy** is the essential purpose of nature.

CHAPTER 8

Generation of Light
And the Negentropic Model

Overview

The precise meaning of the phenomenon of light, despite science progressing to lofty heights, has remained an enigma. Einstein, at the end of his life, had conceded that 'all these fifty years of pondering have not brought me any closer to answering the question of what are light quanta!' And even today, the theoretical basis of many of the properties of light remains unexplained.

In this chapter, we will discuss the mechanism of the generation of light in nature. At first we will put forth an array of questions which leads us to examine the true relationship between force, light, and matter. Having understood the central role of force, we will define force in a new perspective, which allows us to embark on the study of all the four fundamental forces, highlighting the need for a new classification of the fundamental forces. We will then make a thorough examination of the nature of light and realize that light may be regarded in two distinct forms. One is the 'cosmic light' which measures no time and spreads across the universe infinitely, and the other is our 'local light' which travels with a finite velocity and lights up our living rooms and heats up our daily bread! A further analysis of these two sides of light shows us the involvement of force in the generation of light in nature.

In this chapter, a versatile model of the generation of light called the *photonic negentropic model* is presented wherein the central concept of 'negentropic–entropic equivalence' explains almost all the unique characteristics of light categorically. This model also elucidates the precise mechanism by which a ray of light is generated at its source in nature.

Now we will have a sparkling discussion on the generation of light!

Section A
Interconnections of Force, Light, and Matter

To begin with, we will examine the relationship between force, light, and matter in a new perspective. We will first recapitulate the paradox we have encountered in Ch 7-C. When we have depicted light as an 'electromagnetic force', it naturally implies that light is a force of some sort, and, in fact, light empowers the world around us in the form of 'electromagnetic energy' and does all our work, as we have already seen. But then we have also seen in Ch 7-C that force no longer becomes operative once an object reaches the speed of light—force and momentum appear only at speeds lesser than the speed of light. Considering the above two paradoxical statements, we will ask an array of following questions: Is there force really hidden in light? If so, where is this force incorporated in a photon, the unit of light? Why should force, momentum, and mass appear to be somehow 'concealed' in a photon so that they do not overtly show up? And moreover, even though photons impart force, momentum, energy, and mass to the atoms (as, for example, in the case of boiling of water where molecules absorb EM energy and show it up as kinetic energy, or in the case of photoelectric effect where atoms absorb massless photons and emit electrons, or in the case of matter gaining mass upon heating), why do our experiments agree well with the concept of photons being massless?

Where is force hidden in light?

We have also another set of intriguing problems regarding light and force. We have also seen in Ch 7-C and F that once an object (or a matter particle) reaches the speed of light, the object acquires the properties of light—or simply, the object becomes a ray of light. This means that a matter particle assumes the form of a photon once it attains the speed of light. Since we have already seen in Ch 7-C that only at subluminal speeds do force, matter, and mass appear, we may speak the other way and conclude that force converts a photon into a matter particle (because without force, the matter particle attains the status of a photon).

Force converts a photon into matter particle.

And lastly, we have another direct association between light and force: the phenomenon of mass–energy equivalence principle (depicted by the equation $E = mc^2$) shows us that a ray of light may be converted into a particle of mass. Or putting this the other way, a particle of matter may annihilate into rays of light. Since mass can appear only at subluminal speeds, we may interpret the above phenomenon in simple terms: light, when it is slowed down, creates mass!

Light, when slowed down, creates matter particle.

The reader may find the above discussion a trifle confounding, but the situation clarifies itself greatly when we refine the above complex questions and statements into the following two simple questions: Where is force hidden in a ray of light? What is the mechanism by which force converts light into matter? We will undertake a study of these issues in this chapter and the next, but before answering these questions, we may just sum up the above discussion to arrive at one important conclusion: force is the pivotal agent that is somehow involved both in the generation of light and in the creation of matter.

Force is the pivotal agent in the formation of light and matter.

In this chapter, we will see the relationship between light and force and study a mechanism by which light is generated, and by the way, we will also see why light has all its unique properties. In Ch 9, we will examine the relationship between light, force, and matter and see how force plays a critical role in the formation of matter.

However, the reader may see that the first step to begin with our quest is to conduct a thorough study of the pivotal agent that is responsible for the generation of light and matter—i.e. force. In classical physics, we have studied force in its four forms—electromagnetic, gravitational, strong, and weak interactions. But then in modern theoretical physics, electromagnetic and weak forces are merged together into electroweak force, giving us quantum electrodynamics (Ch 3-F), and subsequently, the strong force is also brought into the quantum fold, giving us the concept of quantum chromodynamics (Ch 3-F). Thus, we may say that all the three non-gravitational forces are amalgamated and understood as one in the quantum realm.

Even the gravitational force, as we have seen in general relativity (Ch 1-E), is no longer considered a force in its classical way. Rather, it is considered as the manifestation of spacetime warpage. So we can see that the overall 'big picture' of the four fundamental forces is now shattered, and thus, we may say that the concept of 'four' fundamental forces of nature is only of historical interest and is no longer tenable in the modern setting. This bold realization would really prompt us to go ahead and redefine the fundamental definition of force in newer terms. In this context, we may say that because force appears to have some intricate relationship with light and matter, it would not be very demanding to state that the chief goal of this new definition of force would be to establish a connection between force, light, and matter.

All because of the above reasons, we will undertake the study of force in this chapter in a renewed way so as to get a proper understanding of the nature of light and matter. We will start with a discussion on the fundamental forces as they are classically discussed, and later we will arrive at a new perspective which gives us a novel classification of the forces of nature, and that would be our real starting point in our search for a mechanism of the generation of light and matter in the universe.

SECTION B
A New Classification of Fundamental Forces

The four fundamental forces of nature are diverse in their manifestations. They may act on a long range or short range, they may be strong or weak, they may be charged or neutral, they may act on some particles and may not act on others, and so on. In short, there is no defined way by which we can unify them all in a single file. However, all these fundamental forces share one common characteristic—they all cause a change in the disposition of matter on which they act. This means that when a force particle is exchanged between two matter particles, the matter particles *change* their relative orientation in space and move from their respective positions. In short, all forces cause a change in the disposition of matter in space. We will now see that the four fundamental forces can be reclassified in a different way in accordance with the type of change they induce.

Force is an agent which causes 'change' in nature.

We have seen in the preceding chapters that there have been vain attempts by the scientists to unify the fundamental forces into a single force. In fact, defining this unified force (Ch 3-F) has been the cherished goal of many leading thinkers and scientists of the past. However, we will now examine these forces from a different point of view and arrive at a new classification which turns out to be theoretically illuminating and practically useful for our understanding of nature. Before going into the classification, we will first categorically recapitulate the essential features of these four forces of nature in a different perspective, and this description would indeed let the reader appreciate the need for a new classification. We will start with gravitational force first.

Gravitational force. Gravitational force is the fundamental property of matter – all matter attracts all other matter. Gravity is a force in the long range. Its influence ranges in astronomical scales, and in fact, it can be said that its range is infinite. For example, the earth's gravity spreads for more than 400,000 km (which is why our moon is tethered to it), but then, theortically speaking, the earth's gravity is never zero, rather it spreads infinitely into outer space (though diminutively). In fact, gravitational force is ubiquitous in nature, which pervades across the known universe and spreads across the universe like a blanket. However, gravity is an extremely weak force.

One special feature of gravitational force is that it is always attractive. The law of gravitation states that the strength of gravitation is directly proportional to the mass of the objects in question and inversely proportional to the square of the distance between them. In other words, this law states that gravitational field force of an object increases with the amount of matter in it— more the matter, more is its field force. And equally importantly, this law states that the gravitational field force dampens as it spreads out in distance from the object.

Gravity is attractive, and its strength is dependent on mass and distance.

The strength of its force of attraction is depicted by a constant called the *universal gravitational constant* (G). It is 'universal' because it is thought to be constant throughout the universe. Gravitational constant has a tiny strength with a value of $6.673 \times 10^{-11} \text{m}^3 \text{ kg}^{-1} \text{ s}^{-2}$ (Ch 2-F). What this number actually denotes is that an object of 1 kg mass exerts a force of 0.00000000006673 newtons on another object of 1 kg at 1-metre distance (from centre to centre of both objects). So miniscule is the gravitational force that no wonder it is the weakest force in nature. Consider this to understand the nanoscopic nature of the strength of gravity. The gravitational force of the entire mass of our massive earth on a cup of coffee can be easily countered to lift it up by the tiniest electromagnetic force of our hand! Gravitational force is theoretically suggested to be transmitted by *gravitons*, which are spin-2 particles, and they are considered to be massless and to travel by the speed of light.

Now consider this: we have already seen that force is an agent which causes change. The gravitational force being always attractive, the change that this force brings about between two objects under its influence is to cause them to come nearer. Or putting this in another way, it can be said that when objects attract and come nearer, the 'order' in nature has gone up. Thus, we may say that gravity always tends to increase the order in nature. For example, the planets in the solar system are under the influence of the sun's gravity so that they revolve in defined orbits around the sun, and thus the solar system is considered 'orderly'. In the absence of gravity, planets drift apart and move randomly, and the solar system becomes 'disorderly'. Similarly, all the objects that circle round the earth, such as the moon, our satellites, and ISS, can be said to be moving in an orderly fashion and revolving in predictable paths because of the earth's gravity, without which there would be a lot of chaos. As a matter of fact, the whole of the universe may be said to be 'orderly' in the sense that it has uniform density of matter throughout its extent, and it is homogeneous and isotropic (Ch 4-A). This orderliness can be said to be due to the attractive force of gravity (without which all the constituents of the universe fly away). Thus, we may say that without gravity, there would be disorder. Gravity binds together objects to build up order.

But then we have seen, in Ch 6-B, that there is a term coined by Schrödinger called *negentropy*, which is used to describe the building up of order in nature. Negentropy is the opposite phenomenon of entropy. Whereas the term *entropy* signifies disorder, *negentropy* signifies order. With this idea in the background, we can say that gravitational force tends to negate entropy and works in its opposite direction to bring about order. Thus, we may call gravity a *negentropic force*. To get a better idea of this feature, the reader has to read further on!

Gravity builds up order; it is a negentropic force.

Strong force. Now we will examine the characteristic features of strong force. It is a force in the short range. Though strong force is the most powerful of all the fundamental forces, it can act only on the matter particles within the ultrashort span of nucleus of an atom (the reader may brush up his/her knowledge on strong interaction from Ch 3-C). Strong force is attractive; not only does it bind protons and neutrons together to form a defined structure of atomic nucleus, it also binds quarks together to form nucleons (protons/neutrons). The strength of strong force dwindles down as it spreads further away from the source. Thus, its force is really strong when operating in between the quarks to make them protons/neutrons, but it diminishes rapidly as it binds protons and neutrons, and eventually, this force becomes negligible outside the limits of the nucleus. Or in other words, it remains strong only for a tiniest distance from the quarks. In fact, the *gluon*, its force carrier, can travel only for about 1 femtometre (10^{-15} meters) in space, which shows that strong force is very strong at the level of quarks, but its strength progressively diminishes as it spreads across the nucleus. Gluons are massless spin-1 particles.

Now consider this: strong interaction is essentially an attractive force, without which the quarks and the protons and neutrons would be flying across the space in a haphazard manner. In other words, strong force brings matter particles together, and consequently, it can be said that it has a tendency to increase order in nature. Hence, strong interaction can also be considered a negentropic force.

Gravitational force and strong force are negentropic forces.

We have seen in the beginning of Sec-B that there is one indomitable characteristic feature of force—that is, its effect to cause *change* in nature. And we have demonstrated that gravitational force and strong force do cause change but generally by displacing matter particles (from stars to planets to boulders to subatomic particles) towards each other. Now we will examine the other two forces in this perspective.

Electromagnetic force. EM force (or EM radiations) is also a force in the long range. It extends in astronomical scales, and it tends to travel for infinite distances (the reader may brush up his/her knowledge on electromagnetic force from Ch 2-B, which may come in handy in our discussion here). Light, in fact, can be considered to span across the entire universe from the big bang to perhaps the 'end' of the universe (Ch 1-E and Sec-C). Thus, EM force spreads across the universe infinitely, and it can be said that there could be no place in the universe where we do not find the EM force in some form or the other (except, perhaps, inside the black holes, Ch 4-C, Ch

12-D). Consider these three points which bespeak of the ubiquitous nature of EM force: (1) There is no place in the universe which has a temperature of absolute zero, which means that at least a minuscule amount of EM energy must exist at each and every point of space in the universe. (2) We have seen that all through the interplanetary and intergalactic space, matter is distributed in the form of molecules or atoms or ions (Ch 4-B), and we have seen that matter and EM energy are interchangeable, so we may say that energy exists everywhere. (3) And finally, we have seen that the quantum empty space (the vacuum) is not an inert space but is laden with tremendous amounts of hidden energy that pops in and out of the empty space constantly (Ch 3-H). Thus, it can be said that electromagnetic interaction in the universe is all-pervading, just like the gravitational force.

We have seen that electromagnetic force can be theoretically shown to be composed of electric and magnetic field forces (which are charged forces), but then EM force itself is *not* considered as a charged force (Ch 2-B). It is an effectively neutral force because equal quantities of both the positive and negative charges cancel each other out (Fig 2.1). EM force is carried by photons, which, of course, travel by the speed of light. Photons are massless spin-1 particles.

However, unlike the gravitational force, electromagnetic force, in general, is a repulsive force.* The following three cases are supplied here to illustrate this feature.

Case 1. We may say that when we supply energy (i.e. EM radiation) to any object in nature, the matter particles in the object gain kinetic energy, thereby increasing the temperature of the object, and in consequence, the matter particles move apart from each other, causing an expansion of the object. This expansion of matter may be construed as increased 'disorder' (or we may call it by a better term, *rarefaction*) in nature. Thus, EM radiation may be considered as a disruptive force. For example, a mountain held together by the cohesive effect of gravity, when supplied with EM energy (such

---

*   The reader may note here that 'electric force' or 'magnetic force' may be attractive or repulsive (like charges/poles are repulsive; opposite charges/ poles are attractive), but 'electromagnetic' force as a fundamental force is chiefly repulsive (as discussed here).

as from the super-hot lava), explodes and breaks up into pieces. Many more examples are supplied below, which would highlight this feature of the disruptive nature of EM force.

Case 2. We have learnt in Ch 6-D that atoms, when they combine to form molecules, liberate electromagnetic energy; and molecules, when they dissociate into individual atoms, absorb energy. This indirectly means that EM energy, when supplied to molecules, causes them to dissociate into atoms, thereby increasing disorder in nature; and EM energy, when taken out, causes cohesion of atoms into molecules. In fact, at very high temperatures, the atoms themselves dissociate into a mixture of isolated protons and electrons, which is a fourth state of matter called *plasma* (Ch 3-D). This dissociation of matter particles may be reckoned as increased rarefaction in nature, which showcases the fact that electromagnetic force is a repulsive force.

Case 3. When atoms absorb photons, the electrons in their shells gain momentum and shift to higher orbitals. This increased spatial separation between the electrons and nucleus may also be construed as a form of repulsion or an increase in rarefaction. In fact, when sufficient energy is pumped into the atomic orbits, the electrons may even fly away from the atom into the surrounding free space, and this can again be interpreted as increased disorder in nature. Conversely, when atoms lose energy (i.e. when they eject EM energy in the form of photons), the electron tends to fall back into the inner orbitals. This may be construed as a form of decrease in rarefaction (because this loss of energy has resulted in a decrease in the spatial separation between electrons and nucleus). All these various phenomena indicate that EM force is a disruptive force.

Now we will make a point: considering that electromagnetic force has a tendency to cause repulsion of matter particles, thereby causing increased disorder in nature, it can be said that EM force represents *entropic force* (in contrast to the negentropic force which causes increased order as described above).

Electromagnetic force is an entropic force.

325

Weak force. Weak nuclear force, like strong force, also works in the short range (Ch 3-C). It affects matter within the span of the nucleus of an atom. The weak force is represented by three particles: two charged force carriers (W^+, W^-) and a neutral force carrier (Z). The weak interactions are responsible for the decay of massive particles into fundamental particles (as we will see in Ch 9-G). It is this force which is mainly responsible for the cause of radioactive decay of the heavy or unstable nuclei of the radioisotopes (Ch 3-D), and it is also responsible for the transmutation of proton into neutron in proton–proton nuclear fusion reactions that powers the stars (Ch 4-D). As the reader may see, all these processes generate smaller particles from bigger particles and so can be considered as essentially disruptive (Ch 9-C). Moreover, we have already seen that the weak and the EM force are unified into an electroweak force (Ch 3-E), and thus, it may be assumed that they share a common property of causing increased rarefaction in nature. And so it can be concluded that weak force represents an entropic force.

Electromagnetic and weak interactions are entropic forces.

To summarize, we can now say that we have two groups of forces—one group of chiefly repulsive forces and the other of chiefly attractive forces. The EM and weak forces, being repulsive, are called entropic forces, and they generally increase disorder in the universe. The gravitational and strong forces, being attractive, are called negentropic forces, and they usually decrease disorder in the universe. However, the following clarification would allow the reader to understand the real significance of negentropic and entropic forces.

The Real Nature of Negentropic and Entropic Forces: The real significance of this classification must be understood in a proper perspective. It is very important to note here that the terms *entropic force* and *negentropic force* are *not* merely descriptive phenomena applied to these four fundamental forces. Rather, entropic and negentropic forces are two important phenomena which can be applied to any general events in nature. The significance of this statement is appreciated better in the subsequent discussion in this chapter and the next. But for now, the following explanation is sufficient: force, in general, can be defined as an agent which causes change in nature. The phenomena of entropic and negentropic forces represent only the *direction* in which this change can happen in nature – a force resulting in a change which increases rarefaction in nature is called an entropic force, and a force which results in a change which increases order in nature is called negentropic force. Thus, in fact, these forces represent the natural tendency of events in nature to happen in a particular direction as the reader will see in the later sections.

Negentropic and entropic forces only indicate the direction of change.

Having viewed the fundamental forces in this perspective, we will now examine the interaction of light and force in a different perspective, and this discussion would certainly emphasize the importance of our new classification of forces.

SECTION C
Force and 'Finite Light'

So far, we have discussed about the new classification of fundamental interactions into entropic and negentropic forces. In Sec-A, we have seen that force, light, and matter are related. In this section, we will examine specifically how force and light are very intimately related to each other. We will start first by studying the nature of light in space, and later we will see how force is fundamentally related to the generation and propagation of light.

The Unique Features of Light: Light (or electromagnetic radiation) is a unique phenomenon. It behaves differently in different contexts. In our ordinary optical experiments, light demonstrates such commonplace properties as reflection, refraction, polarization, absorption, etc. But light exhibits certain uncanny and universal features such as the following: light has the maximum attainable speed in nature, and as such, it can be taken as a universal constant (as shown in the special theory of relativity). Light is also shown to spread across the universe infinitely, timelessly, and in an all-pervading way with no beginning or ending when considered in a four-dimensional perspective, and also it is shown that all objects, when travelling at the speed of light, attain the status of light (as shown in general relativity).

Now consider this: how do the visible light rays that are generated from our flashlight theoretically differ from Einstein's all-pervading, infinite 4D light? Think that by no stretch of the imagination, we can say that the light ray of our flashlight has no point of origin, nor can we say that the flashlight has existed eternally *before* it was generated from the flashlight!

Our ordinary light shows up reflection, refraction, absorption, and other features.

The cosmic light is all-pervading and infinite with no point of origin.

Also, in this context, consider another set of questions: What is the exact mechanism by which one source of light generates a low-frequency radiation, such as infrared rays (e.g. a warm steel ball), whereas another source of light emits a medium-frequency radiation (e.g. a hot tungsten filament or a burning candle), and yet another gadget generates high-frequency radiation (e.g. an X-ray machine)? In other words, what is the mechanism by which different sources of light generate EM waves of vastly varying wavelengths ranging from the most powerful gamma rays (with a wavelength of about nanometres, 10^{-12}, Ch 2-Table1) to the least powerful radio waves (wavelength of some kilometres, 10^5)? And finally, what is light made up of, and what is the structure of a photon? Of course, we know answers to some of these questions (see below), but here in the following sections, we have some comprehensive (and refreshingly new) answers to these questions. These issues are presented here by discussing them at the most fundamental level in a step-by-step analysis, which would highlight some of the most profound concepts of nature.

To understand the nature of light and to answer these intriguing questions, we will first study light in two different ways: light in a four-dimensional setting and light in a three-dimensional setting. The reader may argue that these two views are merely two different views of the same phenomenon, and thus we may assume that light can be understood in either a 3D setting or a 4D setting, as we have seen in the preceding chapters (Ch 1 and Ch 7). However, in this section, we will see that there exists an important relationship (or actually an important difference!) between these two views which carries a significant theoretical bearing on our understanding of the universe. We will first recapitulate the general characteristics of light in both these settings.

The 4D View: We have seen in Ch 1-E and Ch 7-B that light, when viewed in a 4D continuum, spreads across space endlessly and timelessly. 4D is an absolute setting wherein the three spatial dimensions fuse with time dimension and, consequently, light behaves as an infinite phenomenon. And thus, in a 4D setting, no finite velocity can be attributed to light so that the meaning of motion and time is lost here. Furthermore, we have seen in Ch 1-E that when we describe the spatial orientation of any object in the universe in a four-dimensional spacetime (which is also called *4D continuum*), all the measurements become invariant (i.e. the same for all observers) irrespective of their own relative motion. In fact, for anybody who could have travelled along with a ray of light at the speed of light in the fourth dimension, it would appear that light has taken zero time to get from the big bang to the present time as already discussed. In other words, we may say that in 4D, there is no beginning or ending of the universe, and thus, it spreads across the universe limitlessly and timelessly. The 4D view of the universe is a monotonous spread of light for an infinite extent of space and time. Because timelessness is changelessness (Ch 7-C), we may say that the 4D universe is a 'changeless' universe. And finally, we can say that in such a 4D universe, neither space nor time can be defined—i.e. it is a 'no-space-no-time universe'. This situation may also be called a state of 'spacelessness and timelessness'!

> *4D light is a state of spacelessness and*
> *timelessness—a 'no-space-no-time' entity.*

However, we can see that this *infinite light* (as we will be calling it from now on) is not the 'light' we see day in and day out; this 'theoretical' light is not the light that would take precisely 8.3 minutes to travel from the sun to the earth; this is not the light we have experimented with in the Michelson–Morley experiment (Ch 1-B); this is not the light that has bent its course during Eddington's solar eclipse (Ch 1-F); this is not the light that has caused electrons to eject from metals in Einstein's photoelectric effect (Ch 2-C) – and finally, this is certainly not the light (or EM energy) that would warm us up in winter or cook up our food in our microwave ovens or generate spark to ignite gas in our cars. This, in short, is definitely not the light of our mundane experience! This 'theoretical' light will not do any 'useful' work for us in the universe. But then of course, all these 'practical' properties of light can be conveniently explained if we consider light in the 3D perspective (see below).

Consequently, this infinite light of the 4D continuum loses all its status as a fundamental force, and it cannot be attributed with any of the effects of force, such as momentum and energy. In short, when we view light in a 4D setting, we cannot consider it as an 'electromagnetic force' at all, and thus, it can no longer provide energy in the thermodynamic events of the universe or perform work of any sort in nature (Ch 7-E).

4D infinite light cannot be qualified as electromagnetic force.

Thus, we may say that when we understand the universe in a 4D perspective, it is only a hypothetical universe, and in this universe, the very existence of matter and motion becomes impossible— simply because all objects disappear into rays of light! In short, the universe in a 4D setting merges with a no-space-no-time continuum.

4D universe is a hypothetical universe.

However, we have to make a deep study of this hypothetical universe for one important reason—this four-dimensional continuum is the state which would provide us the clue to unravel the mystery of the fundamental nature of force, light, and matter as is presented in this chapter and the next.

The 3D View: We will now examine light in a 3D setting (but before proceeding with the discussion, the reader may look into Ch 2-B: Fig 2.2). In this 3D view, light appears to travel with a certain *finite* velocity, which means to say that in this 3D view, 'time' and 'distance' have appeared for us to measure. And thus, this has enabled us to precisely record a measurement of the speed of light at 299,792,458 'metres' (a unit of distance, i.e. space) per 'second' (a unit of time). In other words, we may say that 'space' and 'time' make their appearance once we study light in a 3D perspective.

3D is a 'space-and-time entity'.

We may call this phenomenon as the *finite light* because light in a 3D setting travels by a finite velocity. However, we have seen that light in this setting travels by a fixed velocity, which is the maximum attainable speed in nature. And hence, no material object (i.e. object with mass) in nature can neither attain nor surpass its velocity. But then it has to be realized that though light in a 3D setting travels by a finite velocity, it also travels for infinite distances. This means that when we shine a ray of flashlight, for example, into the vast empty space, though we may be able to ascertain a point of origin for the light (at the head of our flashlight), we may not be able to pinpoint an end to this fixed light because, theoretically, it would continue to travel in space for indefinite periods. This means that the finite 3D light, like infinite 4D light, also spreads across the universe infinitely.

3D light travels by finite velocity but for infinite distances.

However, this finite light differs from infinite light in many other ways: it can either be studied as a wave form or particulate form, signifying wave–particle duality (Ch 2-B); it can be studied with features such as wavelength, frequency, and amplitude; it has an electric force field and a magnetic force field oriented perpendicular to each other and running along the direction of its travel; and the finite light may take a bent course under the influence of strong gravity. And none of the above properties belong to 4D light. More importantly, because of these properties, this 3D finite light can now be considered as 'electromagnetic force', and thus, it can impart all the effects of a force (such as momentum, energy, and work) to the material objects that it encounters in nature. Or in other words, we can say that this 3D finite light can now participate in all the thermodynamic activities of the universe.

Now consider this: we have seen that in a 4D setting, space and time dimensions are welded together into 'spacetime', and in fact, we have seen that it presents as a 'no-space-no-time entity'. In 3D setting, the space and time dimensions are segregated into separate entities. However, here is an important catch: though the space and time dimensions are segregated, they have a 'constant relationship', which means that for the finite light to travel a span of 299,792,458 metres, it takes exactly 1 second, no less no more! In other words, a photon (be it a gamma ray photon or a radio wave photon) always travels by this constant relationship and so has a fixed speed. This is actually what we mean by 3D space-and-time entity.

But then what would be the crucial difference between 4D infinite light and 3D finite light? What is that which makes us to understand them in so radically different perspectives? What transforms 4D spacetime into 3D space-and-time? Below we will do a brief analysis to get the answer.

The Crucial Difference between 4D and 3D Light: To understand what converts 4D continuum into 3D light, we will first examine how motion of light takes place in the four-dimensional setting. Because the 4D continuum is a no-space-no-time entity and because the infinite light takes an absolutely straight line, it can be construed that the infinite light has no waves or particles. In fact, in a universe where there is no space or time, the idea of 'motion' itself is not a valid concept. However, we can assume hypothetically that in a 4D continuum, the infinite light takes an absolutely straight line for an indefinite period of time, and thus we can say that the infinite light travels by an *absolute uniform motion*. But then we have already seen in Ch 7-B that absolute uniform motion is only an imaginary entity, and thus, we must realize that the 4D universe, infinite light and uniform motion are only hypothetical entities and are unattainable in nature.

With this understanding, we will look at the characteristics of motion in the case of 3D finite light. The following discussion underlines the uniqueness of 3D light motion, which also showcases the crucial difference between 3D and 4D light. Because finite light tends to travel at a constant speed of 299,792,458 metres per second, every second, endlessly, it may be said that 3D light travels by uniform motion. However, consider this argument: 3D light travels by a fixed span of 299,792,458 metres, but absolute uniform motion is a timeless, dimensionless entity. But then 3D light travels by a fixed time of 1 second! Considering this finite travel, we may say that 3D light does *not* travel by *true* or *absolute* uniform motion – rather, we would say that the 3D finite light only *appears* to travel by uniform motion! In Ch 7-B, we have called this sort of motion as *simulated uniform motion*, which is nothing but uniform motion under the influence of force (and so we may really call it as non-uniform motion). Thus, it can be said that 3D light really travels by simulated uniform motion.

Uniform motion of 3D finite light is 'simulated uniform motion'.

But then we have also seen in Ch 7-C that simulated uniform motion is possible only if driven by the addition of some force! So finally, we may now be able to say that the uniform motion of 3D light at a constant rate is possible only because it is constantly under the influence of force. With the cessation of force, the 3D finite light would tend to revert back to 4D infinite light, which would then take an absolute straight path without any of the characteristics of 3D light. In short, we may say that the 3D finite light has attained the status of a fundamental force, i.e. the electromagnetic force.

3D light propagates in space under the influence of force.

In summary, we may say that the 4D infinite light cannot demonstrate the effects of 3D light simply because it does not carry force along with it (Table 8.1). Putting it this way, we may say that 3D light carries an agent called force, which would convert the 4D continuum into finite 3D light. Thus, we can assume that the 'addition' of force to 4D infinite light could achieve this feat of segregating space from time into a space-and-time entity. Another useful way of understanding this is that when we are referring to 4D infinite light, we endeavour to 'eliminate' force by eliminating time (i.e. by merging time dimension with the three space dimensions), and when time is eliminated, it precludes light of its other features, such as momentum, energy, and curvature. The reader would get a better picture of this analysis as he/she reads on.

Table 8.1: Differences between Infinite and Finite Light

	4D Infinite Light	3D Finite Light
Space and time	No-space-no-time entity	Space-and-time entity
Extent	Infinite	Infinite
Speed	Limitless (infinite)	Absolute (finite)
Motion	No motion (uniform motion)	Simulated uniform motion
Course	Straight	In waves
Perception	Intangible	Tangible
Thermodynamics	No	Yes
Force	Not present	Present

Now consider this: all the above discussion actually means that some sort of force must be operating at the source of light generation (e.g. our flashlights, tungsten filaments, and X-ray tubes), which would convert 4D spacetime to our ordinary 3D light (we have to wait till Sec-E and F for details of light generation!). However, it may be considered that when we switch on the 'power' of our flashlight, we will be imparting the required 'force' to the 4D spacetime to convert it into a ray of 3D light (details later, Sec-D, E, and F), and when we switch off the power, our flashlight would cease to provide the required force, and so it stops to shine!

But what exactly is the nature of this force which converts 4D light into 3D light? And how does this force act on the 4D infinite light? We have seen in Sec-B above that forces in nature may be of two types—entropic and negentropic. Now consider this question: which one of these two forces are likely to act on the infinite light to generate finite light? The answer to this question becomes obvious in the following sections. In Sec-D, E, and F, we will look into the mechanism by which 3D light is generated and propagated in space, and this discussion would not only indicate the nature of force involved in the generation of light but also highlight the importance of our new classification of forces. We will now embark on a study of the mechanism of the generation of finite light.

SECTION D
Propagation of Finite Light

First we will study the propagation of light (and in Sec-F, we will study its generation for reasons which become obvious as we read on). At the outset, we will recapitulate what we have learnt in Ch 7 regarding uniform motion, and then we will apply the same principles to explain the propagation of 3D finite light and to understand why the finite light behaves as it does with many of its unique ways. We have seen that uniform motion is the most preferred state in nature, but we have also seen that no object in nature can attain uniform motion and that it is only a hypothetical situation. Furthermore, we have seen that all objects in non-uniform motion have a natural tendency to culminate in uniform motion. Thus, as a consequence, we have made a statement in the preceding chapter that 'all objects in nature are constantly attempting to move towards uniform motion without ever achieving it'. We have also learnt that uniform motion is only possible in a 4D continuum but is impossible in a 3D (relativistic) setting. And we have seen in Sec-C that all the objects in a 3D setting (including 3D light) move only by non-uniform motion (i.e. in the form of simulated uniform motion, which is really a form of non-uniform motion).

4D State as a Universal Scaffold: By analysing the above discussion on motion, we can summarily state that non-uniform motion in the case of 3D finite light always has a tendency to culminate in absolute uniform motion in 4D infinite light. Since we know that both absolute uniform motion and 4D continuum are only hypothetical and can never be attained in nature, we can state, finally, that all objects in nature (including finite light) constantly attempt to achieve motion in 4D state without ever achieving it!

3D finite light attempts to achieve 4D state without ever achieving it.

This is the relationship between 3D and 4D light—a relativistic 3D finite light constantly attempting to merge with a hypothetical yet absolute 4D continuum without ever really achieving it. The relationship between 4D and 3D light may also be interpreted in the following way: the 4D spacetime continuum always works in the background of our universe, and the 3D space-and-time entity appears as a result of force acting on 4D spacetime. And this 3D light has a natural tendency to once again merge with the 4D spacetime. In fact, it can be stated that this natural tendency is the drive of the universe (as we have discussed in Ch 7-F).

Here at this juncture, it is very important to realize the real meaning of the term *hypothetical universe*. It does not mean that this sort of 4D universe *does not* exist – it simply means that we, the human beings, are not equipped with suitable faculties to understand and perceive things and events in 4D continuum. This understanding is essential because the tendency of 3D to merge with 4D is real and actually becomes operative in the background in nature at all times (without we ever noticing it), and this universal tendency is the actual drive which propels all the events of the universe. The only perplexing thing is that we, the humans, cannot appreciate this 4D continuum so we keep harping that it is 'hypothetical' and say that 3D 'attempts to achieve 4D without ever achieving it'. The reader will get a better appreciation of this aspect in Ch 11.

The above statement can also be rewritten this way: a 'relativistic' and 'true' state of 3D light takes its origin from an 'absolute' and 'hypothetical' 4D continuum by the introduction of an agent called force. And once again, this 3D state continuously strives to return to the original 4D status by eliminating the force imparted upon it. This struggle of 3D to 4D status is the drive which moves all the events we observe in the universe. In fact, it is shown in the subsequent sections that all the thermodynamic events of the universe are driven by this fundamental principle. In view of the above understanding, we can imagine our universe as being composed of a 'background 4D continuum' with a '3D space-and-time grid' superimposed upon it (Fig 8.1), and this space-and-time grid can be imagined to attempt continuously to merge with the spacetime continuum. In other words, we can say that 4D state forms a *universal scaffold* which spreads limitlessly in the background over which the 3D world gets laid down by the action of force (Fig 8.1). And all the events we observe in the universe—from the nuclear reactions in the stars to the turbulence of the electrons in atoms to the movements of the living cells—are merely the events of the 3D world which play as a never-ending drama, the ultimate goal of which is to end up in an unachievable 4D spacetime continuum! The significance of this statement becomes apparent as we move along in our discussion in this chapter and the next.

All events of 3D world are driven to merge with universal scaffold.

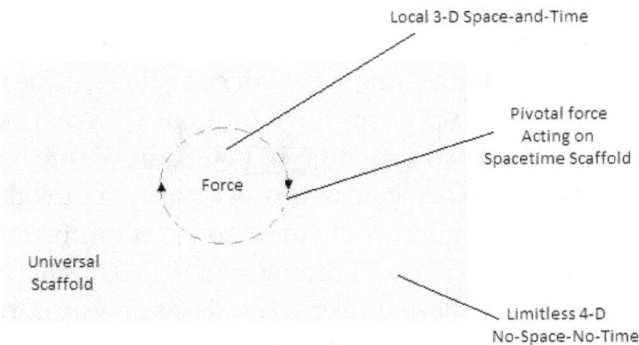

Fig 8.1: The 4D universal scaffold with 3D space-and-time grid.

Universal Cycle *sans* Work: The above discussion, as the reader can see, has finally boiled down to two important entities that operate in nature to generate light—one is the 'force', and the other is the 'spacetime continuum'. Or as we have seen above, force converts a 4D continuum into a 3D space-and-time (the nature of this agent of force itself will be discussed later on). Now we will rephrase the above concept into a leading question: what is the mechanism by which force converts 4D spacetime continuum into 3D finite light? Or in short, we may ask, how does force act on spacetime at the most fundamental level? We will now discuss the mechanism by which transformation of infinite light to finite light takes place.

We will run a brief thought experiment. Imagine a hypothetical '4D light ray' (in fact, there is no 4D light ray in nature; it's only an infinitely spread continuum) travelling in spacetime continuum endlessly. This ray continues to go on forever with uniform motion in an absolute straight line unless a force is applied to it. But once a force is applied on to spacetime, it changes the course of its journey to take a curved path. The hitherto uniform motion of the light ray, because of the effect of force, has slowed down, took a curved path, and became non-uniform motion (Fig 8.2). However, we have already seen in Ch 7-C that non-uniform motion has a tendency to culminate in uniform motion. With this knowledge, we will see what happens to the light ray under the effect of force.

The infinite 4D light under the influence of force bends its course and becomes non-uniform motion, and because of its natural tendency to revert back to uniform motion, this spacetime curvature would soon tend to take a straight path once again—i.e. the spacetime curvature again ends up in uniform motion. We have seen in Ch 7-D that this feat of transformation of non-uniform motion back to uniform motion can be depicted as a 'universal cycle'. In other words, force sets in a cycle of non-uniform motion and uniform motion in a cyclic fashion (Fig 8.2). And it can be said that 3D light propagates forward by repeating these universal cycles of non-uniform motion and uniform motion for an indefinite period of time.

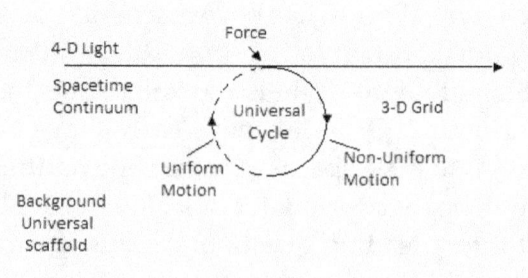

Fig 8.2: Universal cycle and 'no work-maximum entropy'.

Now the reader may notice an incongruity. The universal cycle, which was generated with the application of a certain magnitude of force at the source of light, must 'die down' after a single cycle or, perhaps, after a few finite number of cycles depending on the magnitude of the applied force! But we know that all light rays, once generated from any source, have an infinite extent—i.e. light rays travel indefinitely in space. For example, the light generated from a distant exploding supernova (Ch 4-C) continues to travel on and on even after the original event ceases millions of years before it reaches the earth!

In other words, light from any source tends to propagate forward by cycle after cycle, which goes on and on forever infinitely! Why does 3D light propagate as an endless ray of light even after the cessation of force? Moreover, a light ray travels on and on in space with a finite and fixed velocity without any 'dampening effect'. In other words, a light ray always travels with the same magnitude for indefinite periods of time. Why should a light ray behave this way?

To understand this, we must see that there is an essential difference between the universal cycle we have seen in Ch 7-D and the universal cycle we have depicted here in Fig 8.2. The universal cycle described here is the *most ideal cycle* that could ever exist in nature. This is because the force that is imparted in this cycle ends up doing *no work*. This feature can be understood with some explanation. We have seen in Ch 6-B that in any thermodynamic event, some amount of energy is expended to execute work related to the event, and the rest of the energy is lost as entropy into the surroundings. And in Fig 7.3 (in Ch 7-D), we have depicted this work done, leaving the cycle as a 'work offshoot', and entropy is depicted as being continued in the cycle. However, here in this case of an ideal universal cycle (depicted in the process of transformation of 4D infinite light to 3D finite light), there is no work done. The force used to bend the 4D spacetime is completely expended to create uniform motion, which joins back the 4D continuum. And we have already seen in Event 1 in Ch 7-F (Fig 7.4) that in an ideal setting, both work and entropy are indistinguishable. In other words, it can be said that in this ideal universal cycle, the entire force that is imparted on to it is converted into entropy, and there is no 'loss' of force (or energy) in each cycle in the form of work offshoots, and for this reason, these universal cycles can go on and on forever.

3D light travels indefinitely because it has
ideal cycles with no work offshoots.

And thus, we may say that *all* the force which was applied on the infinite light is converted into entropy and thence to uniform motion. And we have seen in Ch 7-E that this state of entropy is nothing but cosmic entropy, which is the maximum entropy that is possible in nature. In other words, in this ideal situation of 3D finite light, the total amount of force that is applied is utilized to pass the universal cycle to attain cosmic entropy. In short, when force acts on 4D spacetime, it culminates in cosmic entropy, which would present to us as a ray of light!

This 'no-work-maximum-entropy' condition allows the 3D light ray to go on and on endlessly (Fig 8.2) because force, once applied (on the 4D infinite scaffold) at the source of light (such as a flashlight), is not spent as work but is converted into entropy *in toto*. This is the reason the finite light continues its journey in space forever with an instant application of force at the outset (as with any source of light).

A light ray travels indefinitely because of 'no-work-maximum-entropy'.

Now we will move on to Sec-E wherein we will look into the nature of force that would initiate the generation of 3D light.

SECTION E
The Photonic Negentropic Model

So far, we have discussed about the propagation of finite light. We will now examine a working model of generation of light in nature called the *photonic negentropic model*.

Before going into the actual mechanism of generation of light, we will first investigate into one of light's most important properties—the property of wave motion. Why does light propagate by waves? What are these waves composed of? What determines the wavelength and its energy content? We will now study these intriguing features of light, but we will first outline the essential features of these light waves (the reader may review Ch 2-B for more information).

Electromagnetic radiation is composed of alternating waves of crests and troughs, and each unit of EM wave consists of a crest followed by a trough (Ch 2: Fig 2.2a). *Wavelength* is the distance between two successive crests/troughs, and *frequency* is the number of waves that occur in a unit of time. Each type of EM radiation is characterized by a specific wavelength and frequency. Each unit of wave is composed of an electric field force and a magnetic field force which are oriented perpendicular to the line of propagation, and these electric and magnetic fields themselves are oriented 90° to each other. This means to say that as the light wave propagates, the electric and magnetic fields oscillate at right angles to each other and in the direction of wave propagation or energy transfer (Ch 2: Fig 2.2b).

The propagation of light can be understood better in terms of 'corkscrew movements'—wherein a corkscrew moves in one direction and the cork particles are displaced outwards to the axis of the movement, i.e. to the right angles to the direction of penetration. This is something akin to the circular arrangement of a helix of a staircase—i.e. a circular helix which is a curve in 3D space. We may depict this feature as in Fig 8.3.

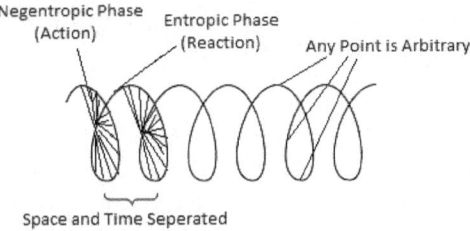

Fig 8.3: Negentropic and entropic phases—and their equivalence.

Now consider this: on the coil of the corkscrew, we cannot precisely point to the peak of a crest (neither can we precisely point to the nadir of a trough) because a wave has no beginning or ending. And so in order to measure the span of each coil on a corkscrew, we may have to select any two arbitrary points encompassing a crest and its adjacent trough. Thus, we may say that a screw is a continuous phenomenon. We may envision EM waves in a similar manner. The reader may also look into the footnote in Ch 2-B for a further insight into waves as a continuous phenomenon. The significance of such a continuous phenomenon is that there are really no peaks or troughs in a wave, and consequently, we cannot 'break' one cycle away from another cycle in any definitive manner—each wave is only an arbitrary unit. To sum up, we may say that the measurement of its wavelength is only arbitrary. And this signifies an important feature of light—it bespeaks of the fundamental property of light being endless or infinite (as noted in Sec-D).

An electromagnetic wave is a continuous and infinite 3D phenomenon.

However, as we have already observed, we can 'quantize' each unit of EM wave by selecting any two points between a crest and trough so that we may also be able to depict a unit of wave as a 'particle' (or a 'wave packet'), i.e. a photon. In fact, as we have seen in Ch 2, light behaves in both ways—as waves and as particles—this feature of 'wave–particle duality' being the essential property of an EM wave. We will study light as a photon in Ch 9-B, but for now, in this chapter, we will continue to deal with its wave property.

Now we will come back to the original question: what is that which determines the wavelength and energy content of a light wave? We will now discuss the mechanism of wave formation to answer this question.

Negentropic–Entropic Equivalence: We have already seen that force is the pivotal agent which converts 4D spacetime into 3D finite light. And we have also seen that the fundamental forces may be classified into two groups—negentropic and entropic forces. Now which one of these two forces is responsible for inducing non-uniform motion in a 4D continuum? In order to understand the exact nature of the force which acts on the 4D spacetime at a given source of EM radiation, we will do the following critical analysis.

We will start with a statement: it can be theoretically ascertained that it is the negentropic force which converts 4D continuum into 3D finite light. And we will follow up this statement with a supporting argument (see Sec F for an elaborate discussion). Consider this: if we apply entropic force to the 4D infinite light, it should cause more entropy in the infinite light, and this is a meaningless concept. The infinite light is already at its maximum entropy, and causing more entropy is not possible. Thus, it can be said that an entropic force will not show any effect whatsoever on the infinite light. On the other hand, negentropic force acts in the opposite direction of entropy. Hence, the application of negentropic force on the infinite light would cause a decrease in entropy in the form of deceleration and curvature of spacetime, resulting in the transformation of uniform motion of 4D light into non-uniform motion of 3D light. Thus, we can say that negentropic force can initiate a *change* in the *changeless* infinite light. Or in other words, we may say that negentropic force impedes the infinite spread of 4D light and causes it to transform into 3D light with a finite speed. And this causes the spacetime to curve and enter into a universal cycle.

Here is an important reminder: by the term *negentropic force*, we shall *not* think of gravitational force or strong force at all as we have pointed out at the end of Sec-B. We must rather think of force as a general tendency to cause change (and we have, in fact, classified as per their tendency to cause change in nature!). The reader understands this sort of negentropic force better as he/she reads on.

Negentropic force can induce change in the changeless 4D continuum.

We can see that these small, localized universal cycles of 3D finite light contain two phases—one, the phase of non-uniform motion caused by the applied negentropic force, and the other, the phase of uniform motion which happens because of the natural tendency of non-uniform motion to revert back to uniform motion. Thus, we may call the former event as *negentropic phase* (represented by non-uniform motion) and the latter event as *entropic phase* (represented by uniform motion) (Fig 8.3). And this cycle repeats itself infinitely because of the 'no work maximum entropy' nature of this universal cycle, as we have already seen. Because no work is done (and therefore no force is expended) in this ideal universal cycle of a 3D light ray, the magnitude of entropic phase is *exactly* equal and opposite to that of the magnitude of the negentropic phase. In other words, as there is no 'loss' of force in the form of work, both the negentropic and entropic phases remain symmetric and equivalent, and this is the reason for light's endless travel.

Now we will put the above discussion in some other perspective. We will call the negentropic force which has caused the non-uniform motion as 'action' and call the entropic phase as 'reaction'. We have seen in Ch 7-C that an action is always accompanied by a reaction. This feature of action-and-reaction is depicted by directed arrows in Fig 8.3. However, the reader may remind himself/herself that we have dealt with a similar viewpoint in Ch 1-D when dealing with general relativity. In the 'Galilean Leaning Tower experiment', we have noticed that the time taken for the fall of an object from a height is independent of its mass because the tug of war between the gravity and inertia is equal—gravity being exactly equal and opposite to inertia. In fact, Einstein had said that gravity and inertia are not only equal forces but that they are one and the same. This remarkable 'equivalence principle' of gravity and inertia has caused Galileo's heavy cannonball to fall to the ground at the same time as his lighter cork ball. Similarly, the negentropic phase in a light wave being exactly equal and opposite to entropic phase, we may say that *all* EM waves, irrespective of their frequency or wavelength—be it the shortest-wave gamma ray or the longest-wave radio wave—travel by a fixed and finite speed. This phenomenon is called the *negentropic–entropic equivalence*.

Negentropic–entropic equivalence explains fixed and finite speed of light.

The reader may see that this negentropic–entropic equivalence is possible *only* because the 3D light propagates by no-work-maximum-entropy condition. If work is done, then the universal cycle is split into work and entropy, and so the cycle is no longer equal on both sides. This interesting affair is discussed in Ch 9.

Here, we will pause for a while and look at the negentropic–entropic equivalence from the correct point of view. Whereas the negentropic phase of an EM wave must be understood as the 'spacetime folding' resulting from the application of force, the entropic phase must be interpreted as the 'spacetime unfolding' as reaction to the folding. In short, the entropic phase must be understood as 'unfolding' of the spacetime, which was hitherto 'folded' under the influence of force. However, it must be understood that the recoiling of entropic phase is the fundamental property of the spacetime continuum, which always tends to remain in an absolute straight course. We will look into the properties of spacetime and force in detail in Ch 11-A.

Negentropic phase is spacetime folding; entropic phase is unfolding.

Redefining Light: With this understanding, we can now redefine an electromagnetic wave. An electromagnetic wave is a continuous chain of *changes* initiated by an agent called force acting upon the hitherto *changeless* spacetime, converting spacetime into an *infinite* chain of universal cycles which travel by a *fixed* and *finite* speed. These events may also be represented as a continuous sequence of 'action–reaction' or 'order–disorder' or 'negentropy–entropy'. In short, a light ray is an endless chain of ideal universal cycles composed of equivalent negentropic–entropic phases.

A light ray is an endless chain of equivalent universal cycles.

We will now go into the details of energy content of an electromagnetic wave.

Energy of EM Wave: What determines the energy content of an EM wave? Or we may rephrase this question this way: what sets in the proportionality factor between frequency and energy, which is known as the Planck constant (Ch 2-C)? Obviously, it is the magnitude of force which was applied initially at the source of light generation (e.g. a candle or a flashlight or an X-ray tube—see Sec-F for full explanation) that determines the frequency and energy content of a light wave—the more the magnitude of force applied, the more the curvature of the negentropic phase of the resultant wave is, and the more the energy content of the wave is. This increased force enhances the acuteness of the curvature and thus shortens the negentropic phase (Fig 8.4). And the corresponding entropic phase, which is equal and opposite to the negentropic phase, also becomes more curved and shorter. Consequently, if the force applied at the source of light is more than the wavelength of the released EM,

345

radiation becomes shorter, the frequency increases, and its energy increases. We have already learnt that this relationship between energy and frequency is depicted in the formula $E = hf$ (Ch 2-C). We will get a complete picture of this feature in Sec-F.

Increased magnitude of force on spacetime
increases frequency/energy of finite light.

Thus, the above discussion leads to the natural implication that the initial force imparted upon the 4D continuum at the source of light determines the curvature of the negentropic phase of the universal cycle in the EM radiation, which, in turn, dictates the wavelength of the radiation (Fig 8.4).

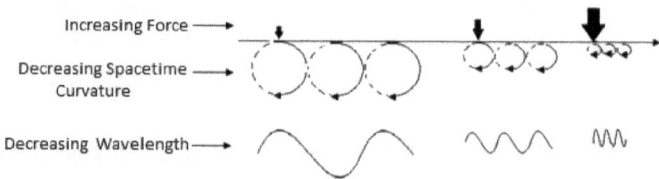

Fig 8.4: Relationship between force, curvature, and wavelength.

But then the reader may argue, here at this point, that the force acting in the above examples of sources of light (such as a flashlight or an X-ray tube) is *not* a negentropic force. It is certainly not gravitational force or strong force, but it is generally electromagnetic energy, which, in fact, is an entropic force! But then the reader would realize, surely, that this force would not be an entropic force either because, as stated above, entropic force could not add more entropy to the already existing maximum entropy of the spacetime continuum. So what is the exact nature of force that is acting on the 4D spacetime at these sources of light? How do we understand this force? We have seen in Sec-B that the real meaning of the new classification of forces into negentropic and entropic forces is not just to divide the four fundamental forces into two divisions, but this new classification encompasses a general tendency of a given force to cause change in a particular direction (either towards the direction of maximum entropy or away from it, as we will see in the following discussion). Thus, in Sec-F, we will look at the nature of this force that acts on 4D spacetime to cause curvature and non-uniform motion to initiate 3D light.

Section F
Formation of Negentropic Phase

In continuation with the photonic negentropic model, we will try to discover the exact nature of force applied on 4D spacetime. We have stated in Sec-E that it is the negentropic force (but not the entropic force), which is the initiating force, which excites the 4D continuum to generate 3D finite light at its source, but we have not supported this view with any evidence. Of course, in order to search for evidence, we will have to examine the source of light in detail, as we will do now.

At first sight, the reader may find it easy to believe that it is the entropic force which induces the generation of EM radiation in nature. More specifically, the reader may state that it is the electromagnetic force which is responsible for the generation of light. Thus, it is something like 'light begetting light'. For example, we may say that 'heat' ignites a candle, giving out more heat and light; or 'electricity' powers an incandescent bulb, giving out more heat and light; or 'chemical energy' may cause a glow in matter (as in the case of glow-worms), and so on. All these initiating forces are examples of electromagnetic force (which is nothing but an entropic force). Moreover, we have never seen any negentropic force (either gravity or strong force) directly initiating the generation of light of any sort in nature!

But then the following discussion demonstrates that the terms *entropic force* and *negentropic force* are *not* merely descriptive phenomena applied to electromagnetic/weak or gravitational/strong forces. Rather, they are phenomena which can be applied to certain general events in nature as pointed out in Sec-B. The property of this initiating force (which induces non-uniform motion and negentropic phase in 4D continuum to generate 3D finite light) can be studied by a careful analysis of the various sources of EM radiation in nature as presented below.

First we will classify the sources of EM radiation—both natural and artificial—and then we will look into a general pattern by which EM radiation of any wavelength (from gamma rays to radio waves) is generated. It will suit our purpose well here to classify the sources of radiation into three types according to the method by which they are generated—viz. *nuclear, atomic,* or *ballistic emissions.* However, it must be stressed that these three divisions are only arbitrary, and they may have many overlapping features as the reader can see below.

In *nuclear emissions,* the subatomic particles of the nucleus create EM radiation, utilizing the force generated inside the nucleus. In other words, here nuclear energy is converted into EM radiation. Generally, this type of emission is a high-energy radiation with very short wavelengths, such as gamma rays and X-rays. Examples are nuclear fission of heavy atoms, radioactive nuclear reactions (Ch 3-D), nuclear fusion reactions in the cores of the stars (Ch 4-D), etc. We will not look more into these nuclear reactions here, but we will examine them in connection with the mechanism of the creation of matter in Ch 9-C.

In *atomic emissions,* the electrons oscillate within their orbits around the atoms to generate EM radiation (exact mechanism described below). This is the commonest and the most useful method of light production for our practical purposes, and there are many varieties of natural and artificial sources of atomic emission in nature. The light generated here may be typically in the medium range, such as UV rays and visible light or infrared radiation, but high-energy X-rays are also possible when the event at the source is energetic enough. Examples of atomic EM emission include various sources of incandescence, such as light bulbs, halogen lamps, and neon lights, and various sources of luminescence and florescence, such as bioluminescence (e.g. glow-worms), phosphorescence, and light-emitting diodes (LED). There are also many other natural sources of atomic emission, such as a burning cinder of wood, a lightning in the sky, or heat waves from a hot body.

In *ballistic emissions,* the source of EM radiation are generally *free-moving* charged particles (mostly electrons) in space when they come under the influence of some strong electric or magnetic field (exact mechanism described below). The generated EM radiation here is characteristically of longer wavelength, such as microwaves or radio waves. The most notable natural example of ballistic emission are the radio wave emissions in a radio galaxy; radio waves are generated when the fast-moving charged particles in the intergalactic space change their course under the influence of huge galactic magnetic fields. Generation of microwaves in an oven stands as an example of artificial source of ballistic emission. However, sometimes, the generation of higher-energy emissions is also possible in ballistic radiation when high-energy particles interact with a very strong medium; one particular form of such radiation is called *bremsstrahlung* (i.e. 'breaking radiation'), which occurs when a fast-moving free electron is suddenly decelerated or stopped by the atomic nucleus (or by some orbital electrons of an atom), which results in a change of course of the electron followed by the production of high-energy X-rays.

Now we will study the actual mechanism of the generation of EM radiation below by taking the example of atomic emission first. Herein, we will concentrate on the exact nature of force at the source of EM generation, which acts on the 4D spacetime to produce 3D light, and see why we should call this initiating force as negentropic force.

Events at the Source of Light: Fundamentally, as we have seen in Ch 6 and 7, any action (or movement) in the universe or displacement of matter in nature can be considered as work done, and this work requires a certain amount of force (or energy) for it to happen. And also, this work done is invariably accompanied by the phenomenon of heat loss in the form of entropy. This is the essence of nature.

Generation of light also entails involvement of work because at the source of the production of light, one form of energy is converted into another form of energy accompanied by displacement of matter and loss of heat (entropy). For example, in the case of atomic emissions, the potential energy of the displaced electron gets converted into light (see below for details); in ballistic emissions, a somewhat similar mechanism is at play (see below); and in nuclear emissions, matter particles are converted into light (Ch 9-C). And we will see that all these mechanisms depict negentropic force in some way! We will now see how exactly this sort of work takes place at the source of light, how matter is displaced, and how entropy is finally achieved in the case of generation of light.

We will now describe a mechanism of generation of light in an incandescent light bulb. We have already seen in Sec-D that the 4D spacetime continuum spreads across the universe infinitely in the form of a universal scaffold. Now imagine the tungsten filament of the light bulb, which is suspended in the vacuum. This vacuum represents a bit of the universal scaffold which surrounds the filament. Thinking this way, it may be said that the filament is the source of force which can be delivered to the 4D spacetime to initiate cycles of negentropic–entropic phases which form a ray of light!

The bulb's filament delivers force on 4D spacetime,
initiating negentropic–entropic cycles.

In consequence, we may say that the sudden application of force on the 4D spacetime scaffold surrounding the filament initiates the spacetime distortion—like a sharp pat on the still waters of a silent tank. This spacetime distortion initiates the conversion of 4D spacetime into 3D light. But then what is the exact mechanism by which the electrical excitement of filament initiates cycles of negentropy–entropy to generate light?

In order to investigate into the mechanism of this conversion, we will examine the events that happen at the filament upon its electrical excitement. We have known that it is the shift of an electron in the atomic shell that leads to the generation of an EM wave. Now we will see why this shifting of electrons generates light.

Electrons are arranged in concentric shells around the nucleus, and the shell that is the nearest to the nucleus has the least amount of energy. And hence it is occupied by the least number of electrons, whereas the outermost shell has the maximum energy and is occupied by the maximum number of electrons. When energy is supplied to the atoms, the electrons in the outermost shell gain energy, become excited, and jump from its 'ground state' (i.e. a level at which the least amount of energy is possible for that shell) to a higher-energy state. Thus, the excited electron moves up and occupies a higher level. In other words, this excited electron can now be said to have 'stored' energy; therefore, it possesses a higher potential energy than in its unexcited state (as in the case of an object on the earth which has a higher potential energy when it is stationed at a higher position above ground, Ch 7-D). We have seen that nature always abhors potential energy, and an object always tends to drift towards a state of least potential energy—so does an electron which always tends to occupy the lowest possible energy state. Thus, an excited electron falls back to its ground state (i.e. to its de-excited state), and in this process of de-excitation, the extra energy that is stored by the atom (during its excitation) is released into the surroundings as a photon with a specific frequency (and energy) that is specific to the atom of each element at a particular temperature (and this is the reason for the specificity of each element's spectral lines). This is basically the mechanism by which light is generated in most of our light sources, such as candles, incandescent bulbs, and LEDs (see below). In the following paragraphs, we will see why we must equate this potential energy with negentropic force and look into a mechanism by which this potential energy of an electron initiates the formation of the negentropic phase. But before going into that, we will also briefly look into the mechanism of ballistic emission, which shares certain common features.

In the case of ballistic emission, the charged particle (mostly an electron) comes under the influence of a strong magnetic or electric field. The electron, which is hitherto moving with a great amount of kinetic energy, is now impeded precipitously by the strong force field so that it keeps losing kinetic energy but starts gaining potential energy. But then nature tends to regain its kinetic energy (by losing potential energy), and this process of an electron losing its potential energy not only causes its deflection from its path but also initiates the formation of a negentropic phase, thereby generating a ray of photon. This is the mechanism of ballistic emission of light.

We will deal with the mechanism of nuclear emission in connection with matter formation in Ch 9-C where it becomes more relevant, but the general rules of potential energy being converted into the negentropic phase of EM radiation hold good here also.

Mechanism of Formation of Negentropic Phase: Now we will see a mechanism by which the potential energy initiates spacetime curvature and forms cycles of negentropic-and-entropic phases. In other words, we will see how exactly potential energy initiates an EM wave, and we will also see how potential energy and negentropic force are one and the same. To begin with, we will recapitulate the linear pathway of natural sequence presented in Ch 7-D:

Force àPotential energyàKinetic energyàEntropyàLight

The reader may notice that we have 'removed' *work* from the list because, as we have seen in Sec-D, no work is done in this idealized form, and we have 'added' *light* because, as we have seen in Ch 7-E, entropy is nothing but uniform motion at the speed of light. But now we will further simplify this linear diagram as shown diagrammatically in Fig 8.5. At the right-hand side is the maximum entropy (or cosmic inertia/entropy as shown in Ch 7-E) that can be attained in nature, and at the left-hand side is the maximum potential energy. Or in even more simplistic terms, we may say that on the right-hand side, there exists no force at all because it is in a state of cosmic inertia/entropy, and the hypothetical object attains uniform motion at the speed of light, which is devoid of any force. And at the same time, we may say that the left-hand side has a state of burgeoning potential energy when the object reaches a state of absolute rest. The reader may see that both the ends of this linear diagram are only conjectural situations, the state of uniform motion and the state of absolute rest being only hypothetical.

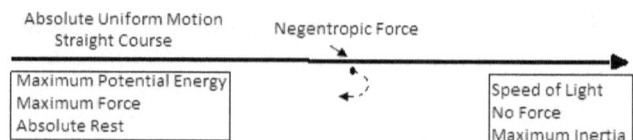

Fig 8.5: Linear diagram showing natural sequence and reversal of time.

Putting the above discussion the other way, we may say that, in nature, force starts to appear only when an object travels at speeds lesser than the speed of light!

Force appears only when an object travels at subluminal speeds.

Or speaking otherwise, we may also say that all motion at speeds lesser than the speed of light is non-uniform motion. And it is curved because at subluminal speeds, time becomes shorter, and any object travelling with time must travel relativistically with non-uniform motion, and so the object bends its course (whereas at the speed of light, the object attains uniform motion and travels straight). A curved course may, in fact, be extrapolated as making a retrograde or backwards journey (Fig 8.5). Or in other words, we may state that an object moving at lesser speeds would move backwards.

At subluminal speeds, time becomes shorter, its
course is bent, and its journey is backwards.

Now we will make another point. We have also seen in event 1 of Ch 7-F that 'entropy as disorder' has reached its maximum at the speed of light, and in fact, it has merged with 'entropy as heat loss' so that both these events become indistinguishable in the ideal situation. Thus, we may say that at uniform motion and at the speed of light, the 'disorder' (read it as 'rarefaction', a preferable term indeed!) is at the maximum. This discussion finally boils down to a simple statement as follows: as force starts to decrease, rarefaction increases, and entropy maximizes.

As force decreases, rarefaction increases, and entropy maximizes.

Furthermore, it may be construed that an object moving in a straight line is increasing disorder as time proceeds (see Ch 7: Fig 7.6), and in the same way, an object taking a curved course may be construed as increasing order. Now speaking otherwise, we may say that any force that acts on spacetime would increase order (Fig 8.5). Thus, we may say that the potential force which is acting on the spacetime results in its warpage, causing a curvature. This curved course, in turn, increases order. And thus, the potential energy that we have seen being applied on to the spacetime scaffold in the case of atomic emission is a negentropic force exerted by the electron as it jumps from higher orbital to a lower orbital (Fig 8.6).

This negentropic force causes curvature of spacetime, leading to the creation of a negentropic phase.

Negentropic phase is made up of potential energy.

And it can now be said that the potential energy (i.e. the negentropic force), as it keeps increasing in magnitude, makes the curvature more and more acute, making the wavelength of generated light shorter and shorter.

Here the reader may consider this explanation: a larger curvature of spacetime is nearer to a straight line than a shorter curvature. And thus, a shorter curvature of spacetime is capable of storing greater potential energy than longer curvatures, which is in full agreement with the fact that a gamma ray photon, which has a greater curvature, could hold more energy than a radio wave photon, which has a lesser curvature! And this is also in full agreement with the Planck formula, $E = hf$, which shows us that if the force applied at the source of light is more, the wavelength of the released EM radiation is shorter, the frequency increases, and its energy increases.

As potential energy increases, curvature increases, and wavelength decreases.

However, this negentropic phase is not really favoured by nature as we have already seen, and thus, the spacetime reverts back to generate an equal and opposite entropic phase. And because of the no-work-maximum-entropy principle, cycles keep continuing to generate a ray of light at the filament!

We will now summarize the photonic negentropic model by explicitly describing the exact mechanism of the generation of a light ray in the case of a common source of light such as our household light bulb.

What Makes the Bulb Glow? Consider an incandescent bulb emitting visible light (Fig 8.6). The tungsten filament in the bulb may be envisaged to be suspended in vacuum in the bulb (in reality, in the modern bulbs, filaments are no longer suspended in vacuum but are surrounded by an inert gas for technical reasons into which we will not go!). Thus, the tungsten wire may be said to be surrounded by nothing else other than a general spread of spacetime scaffold around it. Now the electrical energy that traverses the wire excites the metal atoms in it, which bounces the outer electrons into higher orbitals, making them gain more potential energy. De-excitation causes the electrons to fall back to lower levels, and this potential energy (or negentropic force) kicks in a change in the hitherto changeless spacetime scaffold. This sudden application of force imparts a curvature in the hitherto characterless spacetime surrounding the filament, and this initiates the generation of a negentropic phase in the spacetime, which, in turn, results in a corresponding entropic phase. These tiny universal cycles, each with an equivalent negentropic and entropic phase, propagate by repeating themselves on and on, making the light ray travel for infinite distances, and this we perceive as a ray of light. In simple words, each wriggle of the electron in space can be said to generate a ray of light at its source (Fig 8.6). If the momentum of the wriggle is more, then it generates high-energy light rays, and if the momentum is less, then it generates less powerful rays (and if the energy of the wriggle is not sufficient enough, then the effects may die down without any light generation). Needless to say, these cycles cease to exist once the power is switched off because the electrons stop wriggling at the source, but the photons which are already generated keep going in space for infinite distances—as starlight of dead stars keep reaching us even now (unless, of course, they are absorbed by the intervening matter!).

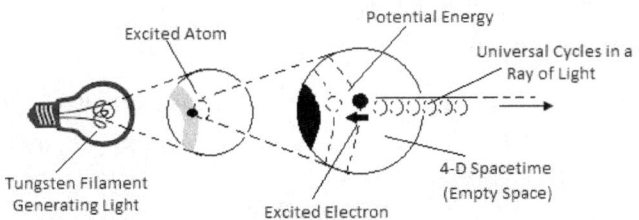

Fig 8.6: Generation of light by an incandescent bulb.

Having answered several questions pertaining to light in this chapter, such as its finite speed, its infinite extent, its wavy nature, and its frequency–energy relationship, we are now left out with one intriguing property of light which needs proper explanation – this is the problem of the mass of a photon. Consider this: when two particles of matter with certain definite mass annihilate, they disintegrate totally into massless photons (Ch 9-C); or conversely, when we heat up a metal bar (i.e. when we supply massless photons), its mass increases (Ch 1-C). Where is this mass hidden in photons?

We have seen in the above sections that force is incorporated into a light ray in the form of a negentropic phase. This force is certainly the 'precursor' of the energy that is injected into the atoms, and we know by mass–energy equivalence that energy can be read as mass. But then if force, energy, and mass were present in the light, why does a photon pretend to be 'massless'? And how does this force/energy get transferred to matter? Where exactly is this energy/mass hidden in a ray of light? We will examine these issues in Ch-9.

☼ *Carry-Over Points—Chapter 8* ☼

- ☼ **Force** *is defined as an agent which causes change in nature. Without force, it is only a changeless 4D spacetime continuum. Force is the pivotal agent in the formation of light and matter.*

- ☼ **Fundamental forces** *are classified as* **entropic** *and* **negentropic forces.** *Entropic forces (such as EM force and weak force) cause disorder in nature; negentropic forces (such as gravity and strong force) cause order.*

- ☼ *4D* **light** *is infinite and has cosmic properties; it is a* **no-space-no-time entity.**

- ☼ *3-D* **light** *is finite and has physical and thermodynamic properties; it is a* **space-and-time entity.** *Force causes conversion of 4D light into 3D light.*

- ☼ *3D light is an electromagnetic force. It travels by* **simulated uniform motion** *at a fixed rate. 3D light constantly strives to achieve 4D status without ever achieving it.*

- ☼ *Force initiates curling up of spacetime, thereby causing a* **negentropic phase.** *Spacetime responds by uncurling, thereby creating an* **entropic phase.** **Negentropic—entropic equivalence** *is the* **sine qua non** *of a light ray.*

☼ *A light ray travels for infinite distances with fixed velocity because of the* **no-work-maximum-entropy** *condition of its ideal universal cycles. Light is redefined as an endless array of ideal universal cycles.*

☼ *Increased magnitude of force applied over spacetime increases spacetime curvature, which, in turn, increases the* **frequency** *of light.*

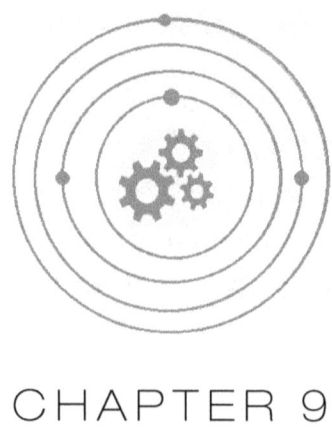

CHAPTER 9

The Anatomy of Atom
And the Fermionic Negentropic Model

Overview

T his chapter describes a model of creation of matter in nature called the *fermionic negentropic model*, which is indeed an extended version of the photonic negentropic model we have studied in Ch 8. We have seen in Ch 1-C that, as per the mass–energy equivalence, mass and light interchange, but we do not know how exactly a ray of light is converted into matter or the precise mechanism by which a bit of matter annihilates into light. However, it may be guessed that the first step to understand this transformation mechanism is to understand the precise structure of the particle of light, i.e. a photon (which is a boson), and the particle of matter (which is a fermion), and we will undertake this task in this chapter.

In the first section of this chapter, we will sum up the properties of matter, such as mass, motion, and charge, and study the intrinsic property of spin in some detail, and this discussion takes us into the world of bosons and fermions. Next, the structure of a photon is delineated explicitly based on the principles laid out in the photonic negentropic model. In the succeeding sections, we will see how the negentropic–entropic equivalence (the stronghold of a light wave, as we have seen in Ch 8) gets dissociated, which, in turn, results in the formation of the fundamental building block of matter, the quark. The fundamental aspects of mass and motion of quarks and gluons are discussed, and this would explain various hitherto unexplained phenomena in quantum mechanics, such as asymptotic freedom and quantum confinement.

Subsequent sections show the correct meaning of charge which has defied scientists until now. Next, the creation of various particles, such as electrons and neutrinos, is discussed. A brief outline of the structure of an atom is discussed at the end.

We will now proceed with a gritty discussion on matter particles!

SECTION A
The Nature of Matter

In the preceding chapter, we have studied light and its properties, and we have seen a mechanism by which electromagnetic energy is generated by the interaction of force and spacetime. However, we have analysed the relationship between force, light, and matter in Ch 8-A and arrived at the conclusion that matter is created from light. Anyway, special relativity has already shown us, by the famous equation $E = mc^2$, that a tiny bit of matter can be created from massive amounts of energy. We will now look into a mechanism by which matter is actually created in nature. But before going into these details, we will have to first examine the essential properties that distinguish matter from energy.

One of the most important properties of matter is the phenomenon of *mass*. As we have seen in the preceding chapters, every matter particle (be it a small subatomic particle or a large planet) has certain mass, and thus, mass may be stated as the most important characteristic feature of matter. Another unstoppable characteristic feature of matter is *motion*. In fact, it is the 'relative motion' that is characteristic of matter because all matter particles move eternally in space in relation to one another (whereas 'absolute motion' is the characteristic of 4D light, which is unattainable by any object in nature, Ch 7-B). To sum up, we may say that these two qualities of matter—mass and motion—qualify universe for its very existence because a universe without features of mass and movement is no universe at all but is merely a *tabula rasa,* a clean, characterless sheet of light of infinite expanse without any lumps and bumps!

Spatial orientation is another property of a matter particle. All matter particles can be depicted in space with certain spatial orientation, which may be represented in a three-dimensional pattern. In addition to these characteristics, some matter particles may have a *charge* on them (e.g. electrons and protons), and some of them may be neutral (e.g. neutrinos and neutrons). Even some matter particles like quarks may have a different sort of charge on them, such as colour charge. And moreover, all matter particles have an exotic property called the *spin,* the character of which exquisitely differentiates matter particles from energy particles. We will be studying this property exclusively in the following paragraphs.

We will now summarize the properties of matter. By considering the above discussion, we may classify properties of matter into two categories—*intrinsic* and *extrinsic properties.* Mass and charge are the intrinsic properties of matter. However, spatial orientation (or we may call it relative position in space) may be considered as an extrinsic property. But then what about motion?

Motion of a matter particle can be considered as both intrinsic and extrinsic property; the following discussion would highlight the point. We have already studied the motion of matter on a large scale in the theory of relativity (Ch 1) and ascertained that all motion in the universe is relative. But now we will examine the motion of a subatomic particle. Imagine a free-moving electron in space as shot, for example, from an electron gun. It may be said to possess *linear momentum,* which is an extrinsic property, because as we have seen in Ch 7, the electron is actually set in motion by an external force which propels it forward with certain speed. However, consider an electron orbiting around the nucleus in an atom. It has two kinds of movement—*orbital angular momentum* and *intrinsic angular momentum.* Orbital angular momentum is the momentum imparted on to it by the electromagnetic field force existing in that atomic orbital, and thus it is an extrinsic property. And this may be likened to the planetary *revolution* around the sun.

Intrinsic angular momentum (otherwise called the *spin* in particle physics) is of tremendous interest to us for our study here in this section. Spin may be likened to planetary *rotation* around itself, and this may be considered as an intrinsic property of a matter particle. This fundamental property of spin is both intriguing and illuminating in the sense that it sheds light on the formation of matter, as we will see in the subsequent sections. Now we will undertake a study of the basic concept of spin here. The reader may look into Ch 3-C for an introductory discussion on spin.

Spin and Spinor: The term *spin** in the quantum realm does not actually mean that the particles physically spin like tops, but it's a mathematical expression of the property acquired by a particle while it is moving in space. Consider this: owing to their mass and movement, many matter particles exhibit an angular momentum which, in turn, is manifested as a magnetic field (just as the earth generates a magnetic field as it rotates), and this magnetic field thus created is quantitatively expressed in terms of spin. And thus, in consequence, it can be said that it is mathematically similar to the intrinsic angular momentum of a particle. Spin of all particles in nature (both force and matter particles) can be quantized either in *integer* units of 0, 1, 2 . . . times reduced Planck constant or in *fractionated* units such as 1/2, 3/2, 5/2 . . . times reduced Planck constant. For example, a photon, which has a spin of 1, actually means that a photon has spin value that is equal to Planck constant divided by *2 pi* (i.e. *reduced Planck constant, $\hbar$*).

All particles which have an integer spin are called *bosons*— such as photons, gluons (which are spin-1 particles), and gravitons (which are thought to be spin-2 particles). As we can see, all force particles are bosons. All particles with a fractional spin are called *fermions*, such as electrons, muons, and quarks (all of which are spin-1/2 particles). And this property of fractionated spin has been the cornerstone of all the matter particles that make up our world as discussed below.

---

* Technically, spin may also be equated to the *wave function* of that particle. There are two kinds of wave functions: *symmetric* and *antisymmetric*. Bosons are symmetric, i.e. if you flip a particle around the centre, the wave patterns remain the same, and fermions are antisymmetric, i.e. if you flip it around the centre, they will not be the same.

We may understand the concept of spin in another perspective. Spin may be imagined as a property of matter particle informing us about what it looks like in different spatial orientations. A *spin-0* particle is like a dot—it looks the same in any orientation; Higgs boson (Ch 3-G) is an example of spin-0 particle. A *spin-1* particle may be imagined to look like an arrowhead—in order to look exactly the same, we have to turn the arrowhead to one full 360° turn (Fig 9.1). Photons and gluons are examples of spin-1 particles. A *spin-2* particle would be like an image with two opposing arrowheads. We have to turn 180° (360 ÷ 2) to make it look the same (gravitons are supposed to be spin-2 particles), and for a *spin-3* particle with three radiating arrowheads, we have to turn 120° (i.e. 360 ÷ 3) to make it look the same, and so on.

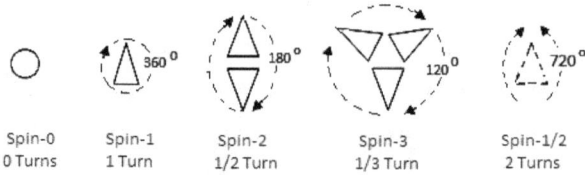

Spin-0   Spin-1   Spin-2   Spin-3   Spin-1/2
0 Turns   1 Turn   1/2 Turn   1/3 Turn   2 Turns

Fig 9.1: Spin and spinor.

Now, imagine a *spin-1/2* particle: by the above analogy, we may say that, in order to return to its original state, we have to rotate it by 720° (360 ÷ 1/2) (Fig 9.1)! In other words, it would take *two turns* to make a spin-1/2 particle look the same! This is called a *spinor* (rhymes with the currency 'dinar'). A spinor is an object which returns to its original state when it is rotated twice, and it is considered a special kind of vector. This is an extraordinary claim—a spin-1/2 particle has some sort of odd configuration which makes it behave uniquely! Spin-1/2 state bespeaks of a behaviour which is impossible for us, human beings, to imagine in our mental eye.* Nevertheless, it must be noted that spin-1/2 is not merely a mathematical expression, but particle physicists have practically documented its existence

---

*   The reader may find it difficult to visualize this spin-1/2 state, but he/she may watch some really enticing videos in *YouTube*, a very simple and interesting demonstration being presented by the physicist Dr. Andrew Thomas in http://tinyurl.com/particlerotation.

experimentally. For example, when a beam of neutrons (which are fermions with 1/2 spin) with a uniform rotation in a magnetic field is rotated by 360°, and when it is combined with another beam with similar orientation, it appears that the two beams cancel out each other, but when one beam is rotated by 720°, the combined beams augment when they interact!

Further experimental evidence came up in support of the half-integer nature of spin by the way of the *Stern–Gerlach experiment*. Otto Stern and Walther Gerlach, in 1921, conducted a famous experiment wherein they have sent a beam of electrons through an inhomogeneous magnetic field and studied their deflection. They were anticipating a smooth column of deflection of electrons on the screen from top down. This is because any charged particle moving in space (like a moving electron) may be expected to behave like a bar magnet (because when it moves, it acquires a magnetic field), and these little magnets (i.e. electrons) may be oriented in many different directions in space at a given column. And when they are deflected, they are scattered differently, making all the possible degrees of deflection—hence a smooth column. But to their surprise, they found that the electrons have split up into only two distinct columns—one up and the other down—suggesting that the electrons can exist only in two states—*spin-up* or *spin-down*—meaning that only two *eigenvalues* (= 'allowed states') are permitted for the spin of electrons. In simple words, this means that all charged subatomic particles (such as electrons) exist in only two states of intrinsic angular momentum—spin-up (i.e. an angular momentum of $+1/2 \, \hbar$) and spin-down (that of $-1/2 \, \hbar$).

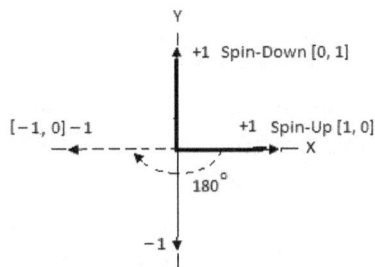

Fig 9.2: Spin-up and spin-down.

We will now approach the spin-1/2-up and spin-1/2-down states by assigning them mathematically as vectors in the quantum state and by depicting them on a graph (Fig 9.2). The spin-up can now be depicted as [1, 0] and spin-down as [0, 1]. We can see that these two vectors are in an orthogonal state—i.e. they are at 90° to each other. Now consider this: if we have to physically convert a spin-up electron in space to spin-down state, we have to rotate it *physically* by 180° (by turning something pointing up into something pointing down); but if you look at the graph, we can say that their rotation in the quantum state (i.e. in mathematical terms) is only 90°. This looks as if the physical state of matter particle is connected to its quantum state by a 2:1 ratio. And consequently, we may say that if we have to return a spin-1/2 particle to its original position (Fig 9.1) by rotating it by 360° in the *quantum state*, we have to *physically* spin the particle by 720°. This means to say that we have to rotate it physically by two turns. This is remarkable—quite bizarre a quantum feature indeed!

For spin-1/2 particle, 720° physical turn equals 360° quantum turn.

However, in order to gain the full significance of this quantum feature of a spin-1/2 particle, we will have to look at the graph once more. If an electron is rotated physically by 360°, its quantum spin would only be 180°, and in this situation, the spin-up vector [1, 0] can be depicted as [−1, 0] (Fig 9.2). However, these are not the coordinates of a spin-up electron, but in fact, they are its coordinates multiplied by −1 (i.e. [1, 0] × −1 = [−1, 0]). This shows that when we rotate an electron by 360°, we end up in a quantum state which is multiplied by −1. Thus, to get back to its original position, the particle has to be turned by two turns, and this is the property of a spinor. Now consider this: if two electrons (or any fermions, or matter particles, for that matter) are exchanged in a physical state, the swapping means a 360° turn physically, and their quantum state is multiplied by −1. And only when you continue to exchange again would the quantum state revert back to its original position—i.e. in order to restore the original state, we have to multiply again by −1(−1 multiplied by −1 gives us +1).

Now consider this: when two identical fermions in space have exchanged their positions, their physical state becomes unchanged.

In other words, we may say that they have swapped their positions by a rotation of 360°, and they may be said to have returned to their original state. But then their quantum state can be considered to have changed (because their coordinates are multiplied by −1). So now we have arrived at two contradictory statements in this situation. The physical state of fermions has not changed, but the quantum state has changed. This is illogical and is unacceptable. But the only number which does not change when multiplied by −1 is 0. However, we have already noted that the spin of a matter particle may be represented by its wave function (Ch 2-C), but then no wave function can exist with a value of zero amplitude. Thus, basing on the above discussion, we may conclude that no two fermions can occupy the same quantum state at a given place and time.

No two fermions can occupy the same quantum state.

Now the reader may see that we have unwittingly arrived at the fundamental idea of the Pauli exclusion principle (Ch 3-D)! The exclusion principle states that no two fermions (or matter particles) can occupy the same place. Speaking the other way, we may state that two fermions in nature *may be identical*, but they *are not indistinguishable*. For example, all electrons in the universe are identical, but each electron in the universe may be distinguished owing to its specific position and location. In other words, each electron is special in the universe owing to the fact that it has a specific location which is distinguishable from all the rest of them!

Fermions may be identical, but they are not *indistinguishable.*

This is the unique property of all matter particles. In contrast, we know that bosons (e.g. all force carriers which are spin-1 particles) do not behave this way! For example, all photons in the nature are not only identical but also indistinguishable (see below). This means that, theoretically speaking, all the photons in the universe can fall into one place at a given time! This is a remarkable feature of bosons! Consequently, bosons do not obey the Pauli exclusion principle.

Bosons are identical and indistinguishable.

This behaviour of fermions is of tremendous significance. It stands out to explain how fermions can build the world as we see today. So happily for us, we are able to stand on a piece of land because of the Pauli exclusion principle. The matter particles of our feet and that of the earth below may consist of all very identical subatomic particles, such as electrons and quarks, but they are not indistinguishable. They exist individually in their locations and do not share their positions with their neighbouring comrades! And this is also the reason why, sardonically, speeding vehicles ram into each other, causing gruesome accidents, but such accidents would not happen in the case of two colliding beams of light which can pass through each other without a hitch! And thus, we may finally say that fermions are the reason why the universe exists. If there were no fermions in the universe where are we? The entire universe would have bundled up into 'nothing'!

By fermions we exist; by fermions the universe exists.

Having discussed bosons and fermions, we will come back to the original question of the mechanism of the creation of matter from light. To start with, we will have to first rephrase our question: what is the mechanism by which a photon, a spin-1 bosonic particle, is converted into a fermion, a spin-1/2 matter particle? To answer this question, we will have to make further investigation into the properties of our prototype boson, the photon, which is the unit of light (and a typical 'massless particle'!). We will now examine the structure of a photon in the light of a new perspective.

SECTION B
The Anatomy of Photon

In Ch 8, we have studied light essentially as a wave. In this section, we will again study light but not as a wave, rather as a spin-1 particle, i.e. a photon. In fact, this discussion would really become the first step in the formation of matter as the reader will see.

Defining Photon: In Ch 8-E and F, we have studied light as a wave and laid the foundation to the photonic negentropic model. However, each unit of light may be described as a quantum particle, and each photon represents the unit of energy contained within the confines of a unit of EM wave (which includes one crest and one trough, Ch 2-B). In simple words, each unit of an EM wave may be represented as a pulse of a photon. Basing on the photonic negentropic model, we may now state that each photon—be it the photon of a gamma ray or a radio wave—is composed of a negentropic phase and an equal and opposite entropic phase. Thus, a photon may be diagrammatically represented as a symmetric unit (such as a circle) with one half representing the negentropic phase (depicted in black) and the other half representing the entropic phase (depicted in white) (Fig 9.3). Now looking at the spin-1 particle in Fig 9.1, the reader could say that the photon can now be depicted as a spin-1 particle which can come back to its original position by a rotation of 360°, and thus, the photon may be said to behave like a boson.

A photon is a spin-1 particle composed of one
negentropic phase and entropic phase.

Now with this understanding, we will try to find out the relationship between the wave form and the photon structure (Fig 9.3). We may imagine that each unit of a wave, from crest to crest, may be depicted as a series of phases of a spin-1 particle (or photon) making one full rotation of 360° across the length of the wave. Or in other words, each phase of the electromagnetic wave may be depicted by an equivalent phase of photon. Imagine this sequence: at the crest of the wave, the photon may be thought of to be oriented exactly perpendicular to the line of transmission, and as the wave progresses, the photon 'rotates spirally' to make a full turn by the end of each wave wherein it is once again oriented perpendicular to the line of transmission.

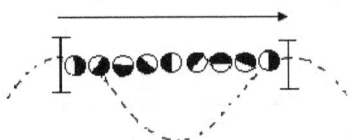

Fig 9.3: Photon depicted as various spin states in a wave.

But then what does each photon 'contain'? We know that each photon carries energy, but how exactly is the energy incorporated in a photon?

The intriguing feature is that each photon—be it a photon of the shortest wavelength such as a gamma ray or a photon of the longest wavelength such as a radio wave—has a constant relationship between its length and its energy. This feature can be envisaged by the following analysis: we have seen in Ch 2-C that the proportionality of energy content (E) of a photon and the frequency of the EM wave (f) are constant by the equation $E = hf$ (h being the Planck constant). However, this equation can be rewritten as $h = E \lambda/c$ (because $c = \lambda f$), and this means that, h and c being constants, the energy content of the photon and its wavelength ($E.\lambda$) are inversely related—as the wavelength decreases, the energy increases, and *vice versa*. This again means that as the force imparted on the spacetime at the generation of light increases, the curvature of the negentropic cycle increases along with an increase in the curvature of the corresponding entropic phase, and this makes the resulting cycle smaller (Ch 8-E; Fig 8.4). In other words, as the magnitude of the force increases, the cycle becomes smaller and smaller (Fig 9.4), making its wavelength shorter and shorter. This naturally implies that the wavelength of light decreases as the magnitude of force at the source of light generation increases, and it increases as the force decreases (as shown in Fig 9.4).

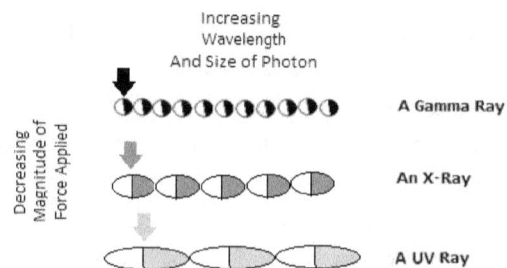

Fig 9.4: Force and wavelength relationship.

There is another important derivation from the above analysis. When we look at the negentropic–entropic equivalence of a photon (Ch 8-E), we may say that a photon is perhaps the only place in the universe where a given unit of force in the form of negentropic phase is effectively and totally countered by the equivalent entropic force so that it may travel by uniform motion – this is possible owing to the 'no-work-maximum-entropy' feature we have studied in Ch 8-D. We may put this idea this way: at any time, if the negentropic phase exactly matches the entropic phase in nature, then it results in the formation of a light ray! In Sec-D below, we will see how this negentropic–entropic equivalence is 'disrupted', which plays an important role in the creation of matter.

Only in a photon the effect of force is annulled
by negentropic–entropic equivalence.

With this idea in the background, we will now go into the formation of matter in the universe. In Sec-C, we will look into the general features of matter creation, and then in Sec-D, we will describe the actual mechanism.

SECTION C
Formation of Matter – General Considerations

The generation of light (or EM energy), as we have seen in Ch 8-F, is a ubiquitous phenomenon. EM energy is generated in many forms in nature, such as light, heat, and electricity, on a continuous basis (think of this—just by switching on your table lamp, you will be generating photons!). On the other hand, generation of matter in nature is not an overtly observed phenomenon in nature, and hence, we will not appreciate matter formation in the same way as we appreciate light generation. Nevertheless, formation of matter is not so rare a phenomenon in nature, as we will see now.

We will begin our study by categorically studying the situations where matter is created in nature. But before proceeding further, we will have to first understand the correct meaning of the term *matter creation* in the context of this chapter. We have already come across two terms connected with matter formation in the universe as described in the preceding chapters—*primordial nucleosynthesis* (Ch 5-E) and *stellar nucleosynthesis* (Ch 4-B).

Primordial nucleosynthesis is a process which is described in connection with the creation of matter at the time of the big bang. During the initial stages of the universe, because of the superlatively high Planck temperatures, matter is thought to exist in a superplasma state wherein the quarks, gluons, and leptons existed independently (*quark–gluon plasma*, Ch 5-G), and as temperatures dropped with the expansion of the universe, the particles recombined together, forming simple atoms of hydrogen and helium. Thus, *primordial nucleosynthesis* is the term applied for the creation of matter in the form of simple atoms from the *already* existing quarks and leptons. However, we are *not* concerned with this phenomenon here in this chapter.

Stellar nucleosynthesis is the formation of matter in the form of heavy elements from simple atoms of hydrogen and helium. This happens in the cores of massive stars owing to the tremendous gravitational pressures and temperatures prevailing in them (Ch 4-B). Neither are we concerned with this phenomenon in this chapter.

Rather, we will deal, here in this chapter, with a more fundamental understanding of the formation of matter. Here we will be studying the mechanism of the formation of the elementary particles in nature. In other words, here we will describe a mechanism by which fermionic matter particles are synthesized in nature. Hence, we will call this mechanism the *fermionic negentropic model* (in contrast to the 'photonic' negentropic model described in Ch 8). As we will see in this chapter, the fermionic negentropic model, apart from answering the most intriguing concepts such as mass and motion at the fundamental level, explains the fundamental meaning of charge, spin, etc.

Now consider this question: do we find any evidence of matter formation in the world around us? Yes, certainly! We have several examples of matter formation both in the natural world and in the man-made setting. Interestingly, almost all these examples are related to the various high-energy interactions that happen in and around the nuclei of atoms. All our 'ordinary' chemical reactions merely exchange matter and energy during their interactions, and the total amount of *matter* and *energy* is always conserved. After all, chemical reactions involve interactions with only the outer rind of electrons of the atom. However, in nuclear reactions, most of the times, as we will see below, matter is *not* a strictly conserved entity, and some of the matter particles involved in the nuclear reactions disintegrate into energy (of course, dutifully following the equation $E = mc^2$). And of course, in many nuclear reactions, some matter particles are also created, at least for transient periods. Thus, it is *only* the energy which is absolutely conserved during nuclear reactions.

In nuclear reactions, matter particles are both destroyed and created.

We will now discuss instances of matter formation in nature. And because almost all these instances happen in relation to the nucleus, these processes may be studied simply as nuclear reactions. Nuclear reactions may be classified into three categories for our purpose of discussion here—*particle decay, particle annihilation,* and *pair production.* Now we will briefly look into some of the examples here, and this discussion would actually leave us at the doorsteps of a general mechanism of matter creation in nature.

Particle Decay: We have already discussed radioactivity in Ch 3-E and found that unstable nuclei of heavy atoms undergo decay by several means, such as alpha decay, beta decay, and nuclear fission. We will consider beta decay and see how, along with decay of matter particles, new matter is created here. Beta decay may be of two types—beta-plus or beta-minus decay.

Beta-plus decay is a process wherein a proton is converted into a neutron. Here, one of the up-quarks of the proton loses its positive charge to the weak force carrier, W^+ boson. And this W^+ boson, being an unstable particle, soon decays into a positron and a neutrino. By this feat, one of the positively charged up-quarks (+2/3) (Ch 3-C) is converted into a negatively charged down-quark (–1/3), and the total charge of the nucleon would then become zero (uud à udd), thus converting a proton into a neutron. The reader may observe here that two new particles are created in the process of beta-plus decay along with generation of a gamma ray—a positron and a neutrino. The reader may note that neither positron nor neutrino is a natural resident of the nucleus, but these particles are created as the nuclear reaction proceeds. The neutrino escapes out into the surroundings and travels to far distances because of its non-interacting nature (Ch 4-D), and the positron combines with one of the ambient electrons and annihilates into gamma rays.

Beta-minus decay is the reverse process wherein a neutron is transmuted into a proton. Here one of the down-quarks loses its negative charge to a W⁻ boson, and this W⁻ boson, being unstable, decays into an electron and an antineutrino, thus converting the negatively charged down-quark into a positively charged up-quark (udd à uud) and thus converting a neutron into a proton. And of course, both electrons and antineutrinos are not the natural residents of the nucleus but are created in the vicinity of the nucleus during beta-minus decay.

However, there are some other essential differences between β+ and β− decay. A radioactive atom can disintegrate by b+ decay (i.e. positron emission) only if the resulting daughter nuclei have a greater binding energy (i.e. a lower total energy, Ch 3-E and Ch 6-D) than the parent nucleus as in the case of certain types of artificially generated radioisotopes (e.g. carbon-11, nitrogen-13, and oxygen-15). However, this is not a spontaneous process, and it needs a powerful field force of a particle accelerator (such as a *cyclotron*) to synthesize such radioisotopes. b+ decay may also happen if tremendous energy is driven into the event as, for example, what happens in the cores of the stars under tremendous gravitational pressures and temperatures (Ch 4-D). This non-spontaneity of decay of a proton into a neutron is because of the fact that a proton is remarkably stable (see below), and it needs large amounts of energy to convert it into a neutron. The process of β− decay, on the other hand, happens spontaneously. This is because a free neutron is inherently unstable (see below), and thus it tends to disintegrate into a proton.

It is worth reiterating here that in these decay processes, the generated matter particles, such as electrons, positrons, or neutrinos, are *not* the normal residents of the nucleus; rather, they are created *de novo* in the vicinity of a nucleus during the nuclear decay processes, which indicates the creation of new matter. Thus, in summary, it may be said that particle decay involves a process of not only destruction of matter particles but also creation of new matter.

Lifespan of Various Particles: Before passing on to the topic of annihilation, we will pause for a while and look a little more into the stability of matter particles in nature, and this information would also give us some insight into the concept of matter creation. By and large, all matter particles which are heavy are unstable. The most stable particles in nature are the lightest particles. The reader may look into the list of particles in Ch 3: Table 3.2 and see that only the 'first generation' particles such as up-quarks, down-quarks, and electrons are stable, and these are the lightest of the particles in nature. And we have seen that only these particles take part in the building of all atoms in the universe. The 'second' and 'third' generation' particles (Table 3.2) are massive and unstable, and they do not participate in matter formation but are identified only in high-energy collisions in nuclear labs. And even when they form, they rapidly break down into lighter particles and/or into energy.

In summary, we may say that our baryonic matter is exclusively formed from only three of the *least massive* of the particles of the Standard Model, *viz.* electrons, up-quarks, and down-quarks (forming protons and neutrons), and they are bound together by two of the *massless* force carriers of electromagnetic and strong interactions, *viz.* photons and gluons (i.e. all atoms in nature are formed by five fundamental particles—up-quarks, down-quarks, electrons, photons, and gluons). By this discussion, we may deduce that nature, in general, prefers lighter matter (or even free energy) to heavier matter.

Nature abhors heavy matter and prefers lighter matter or free energy.

In this context, we will consider the lifespan of various particles in nature. By far, the most stable particle in nature is perhaps an electron. They are known to be almost 'immortal'. Recent studies have estimated that electrons, when left alone, may survive for at least about 66,000 yotta-years (1 yotta-year is 10^{24} years)—that is about 5 quintillion times the age of the universe! However, electrons, when they are in the vicinity of their antiparticles—positrons—they annihilate instantaneously into pure energy. Similarly, it has been observed that a free proton is also very stable and is said to almost last forever,* and a proton consists of the lightest possible quarks in them (see Sec-E for explanation). However, a proton may undergo transmutation into a neutron under certain extreme conditions (such as described above).

A neutron, as long as it is a resident of a stable nucleus, is a stable particle. But then a free neutron travelling in space, in contrast to a free proton, is no longer stable. It decays in about ten or fifteen minutes into a proton, electron, and electron antineutrino (all of which are very stable). But are quarks stable? It must be noted that *free* quarks, by themselves, will never exist independently outside the protons or neutrons, and the reason for this becomes evident in the following sections.

It's a rule of thumb that atoms of almost all naturally occurring elements are stable. They may participate in various chemical reactions by exchanging electrons in their outer rinds, but their nuclei are stable, and they appear to live forever (for example, the element bismuth-209 has a half-life of about a billion times the age of the universe)! However, the heaviest of the atoms in nature are inherently unstable, and so they become radioactive elements which have a variable lifespan. Some decay in nanoseconds and some in minutes, hours, days, or years (Ch 3-E). The reader may notice that this is a recurring theme of nature—nature preferring the lightest particles or a state of free energy—and we will know why nature demonstrates this property in the following sections.

Now we will go into a direct method of disintegration of matter—called annihilation—and see how it is relevant to our present discussion of matter creation in nature.

Particle Annihilation: Annihilation happens when two matter particles, usually of opposite charge, collide into each other and

---

* Nevertheless, many grand unification theories (GUT theories, Ch 3-F) predict the spontaneous decay of proton into a positron and a neutral pion (i.e. a p-meson consisting of an up-quark and antidown-quark), but this is only a hypothetical suggestion, and none of the experiments have demonstrated this feature so far.

disintegrate into pure energy. The most fundamental example of annihilation is the *electron–positron annihilation*, and more importantly, e⁻/e⁺ annihilation is the most common pathway by which matter in nature is converted into energy. And thus, this is the most frequent type of annihilation we observe both in nature and in man-made nuclear reactions (e.g. nuclear fusion in the sun and gamma ray emission in PET scans). Matter–antimatter annihilation is also the speculated mode of gamma ray production after the big bang (Ch 5-G).

Annihilation is of two types. It may take place between particles which are either at rest or in motion. First consider this: when the annihilating electron and positron are at rest, their mass–energy (i.e. rest energy) is equal to mc^2 because their kinetic energy is zero, and so their momentum is zero.* Thus, the total energy liberated from disintegration of an electron (or a positron) is equal to the mass (m) that disappears multiplied by the square of the speed of light (c)—i.e. $E = mc^2$ (which is equal to 0.511 MeV in the case of e⁻/e⁺). This sort of annihilation is called *low-energy annihilation*, and it often generates two photons of gamma rays which travel exactly in opposite directions. And of course, the energy of each photon is equal to 0.511 MeV. Here we may say that the event has started with two particles with only mass but no kinetic energy and ended up with two photons with only kinetic energy (and no mass).

Annihilation strictly follows all conservation laws. However, after annihilation, most of the properties of the matter particle, such as mass, charge, spin, and momentum, disappear totally, but energy is always conserved. Whereas charge, spin, and momentum† may disappear, because of self-cancelling opposite signs, the mass of particles vanishes into pure energy. In other words, it is only the energy which is rigorously conserved after annihilation—the total mass of each particle being converted into pure energy.

---

* The energy–momentum equation $E^2 = (mc^2)^2 + (pc)^2$ shows that if momentum (p) is zero, then E becomes equal to mc^2.

† The reader may claim that each photon, though massless, has linear momentum (p), so momentum is also conserved. But this is not true because the *total* linear momentum of the event would be zero (photons move in exactly opposite directions, so their momentum cancels out as noted in Ch 7-C).

Mass, charge, and spin are not conserved; energy is conserved.

Now we will look into the fate of fast-moving electrons/positrons colliding into each other. This is an important situation because, more often than not, the colliding particles are not at rest but travel with tremendous velocities carrying great amounts of kinetic energy with them. This is called *high-energy annihilation*. In such a case of high-energy e⁻/e⁺ annihilation, the result is not the simple generation of two gamma rays, but the kinetic energy of the colliding particles is converted into a variety of heavier matter particles, such as muons and antimuons or heavier quarks and antiquarks. But then the generated heavy particles are highly unstable, and they exist only transiently before annihilating themselves into pure energy in the form of gamma ray photons. In the case of quarks, they form mesons (such as *pions* and *k-mesons*), and soon they also annihilate into gamma ray photons. In other words, it can be said that the extra kinetic energy of the high-energy collision is transiently incorporated into heavier matter particles before they annihilate into photons. Hence, we may state that high-energy annihilation of lighter particles (such as electrons) may result in the creation of some of the heavier matter particles in nature (such as muons)—or simply put, high-energy annihilation entails matter creation.

Matter is created in high-energy annihilations.

There are many other examples of annihilation. Virtually any particle can annihilate when it meets its antiparticle partner, be it a muon–antimuon pair, quark–antiquark pair, proton–antiproton pair, or atom–anti-atom pair or a 'man–anti-man' pair (so never shake hand with an anti-man from outer space, lest you should disintegrate instantaneously into thick light!). Experiments also have shown that as the mass of the colliding particles increases, the probability of new matter formation increases. For example, a proton (with three quarks) may annihilate with an antiproton (with three antiquarks) in a complicated way, resulting in a florid array of very heavy matter particles along with gamma rays. Here one of the quarks annihilate with an antiquark, resulting in the generation of a gluon which mediates between other quarks which can now undergo reorganization (a process called *hadronization*), and this results in the creation of mesons which, by themselves, are unstable and soon annihilate into stable particles, such as electrons and neutrinos apart from gamma rays. To sum up, we may say that in all annihilations, the final product is the creation of nature's lightest particles and free energy—because nature always prefers lighter matter or free energy.

Before concluding this topic on annihilation, we will examine an interesting feature. During the process of e^-/e^+ annihilation, there exists a transient phase of matter comprising of an electron and a positron spinning around each other before they fall into themselves and annihilate into two gamma ray photons. This phase of matter is called *positronium* (which is akin to a hydrogen atom except that the proton is replaced by a positron), and it has been observed that this transiently bound state exists with a mean lifespan of 0.125 nanoseconds. The mass of a positronium is about 0.511 + 0.511 = 1.022 MeV (minus a few MeV, of course, due to loss of binding energy, Ch 3-D and 6-D). Such positronium atoms are, in fact, transiently observed in labs when positrons (released in a nuclear reaction) slow down as they transit the surrounding matter and capture the ambient electrons.

Now we will go into a peculiar case of direct conversion of energy into matter—pair production.

Pair Production: We have seen all along in the above discussion that some kind of new matter is formed during the course of disintegration of matter in almost all nuclear reactions. However, there is a direct method in nature by which a light ray is converted into a pair of matter and antimatter particles. This is the case of *pair production* wherein a photon is converted into a pair of electron and positron (or much less commonly, into a pair of muon and antimuon or, as a matter of fact, any other such particle–antiparticle pairs). As the reader may see, this is just a reverse process of annihilation. But then for this phenomenon to happen, two important prerequisites have to be met. Firstly, this event always happens in the vicinity of an atomic nucleus, indicating that this event is driven by the powerful field force of the nucleus. In fact, the probability of pair production increases with the square of atomic number of the nearby atom, suggesting that the heavier the nucleus, the more chance there is of pair production. And secondly, the energy of the incoming photon must be *at least* equal to (or more than) the total mass–energy of the outgoing particles. For example, the incoming photon has to have an energy of at least 1.022 MeV or more in order to create a pair of e^-/e^+, and the energy of a photon in excess of this is converted into kinetic energy, which is manifested not only as rapid acceleration of the released particles (i.e. electrons and positrons) but also as a recoil of the nearby nucleus (thus conserving the momentum of the incoming photon).

In conclusion of this discussion on matter formation, we may say that creation of matter in nature is not generally favoured, and even when it happens, it involves very high energies. In other words, we may say that matter creation becomes possible only when the energy involved in an event exceeds a critical limit, and this generally happens in the vicinity of a powerful field force of a nucleus.

Matter forms in nature only when the input energy crosses a critical limit.

With this understanding, we will now look into a mechanism of formation of matter in nature.

SECTION D
The Fermionic Negentropic Model

So far, we have studied the nature of matter and the instances of formation of matter in nature. But then what is the exact mechanism by which matter is created? What is the connection between photons and fermions? We have seen in Ch 8 that force is the pivotal factor which causes curvature of spacetime, resulting in the generation of an EM wave, and in Ch 8-E, we have seen that increased magnitude of force applied on spacetime increases its curvature, thereby increasing the energy content of the resultant EM wave.

However, in Sec-C, we have seen that in any nuclear event, if the magnitude of input energy crosses a critical threshold, it creates matter particles. By putting all these features together, we may say that if we increase the force acting on spacetime beyond a critical limit, then it would initiate certain changes in the disposition of spacetime, resulting in the creation of matter. In the following sections, we will study the exact way by which this feat of matter creation is accomplished by the application of massive amounts of force on spacetime. Now we will categorically examine the fermionic negentropic model and see how this model accommodates all the described features of matter we have discussed in Sec-A.

Negentropic–Entropic Dissociation and Negentropic Knots: We have seen that gamma rays are the most energetic waves in nature, which means that an EM wave of gamma radiation has the maximum allowable curvature, and hence it has the maximum energy content. Thus, the negentropic phase of a gamma ray photon can be said to have the shortest possible curvature for an EM wave with negentropic–entropic equivalence (Fig 9.5a). But it may be presumed that when the force applied exceeds a *critical value*,* then the spacetime curvature of the negentropic phase increases exponentially, causing it to curl upon itself tightly with ever-diminishing sizes. This dense region of curled-up spacetime is called the *negentropic knot* (Fig 9.5b and c).

Force applied beyond a critical value forms a negentropic knot.

The reader may realize that the negentropic phase in a negentropic knot represents negentropic force, which is attractive in nature. Thus, this negentropic knot tends to curl up the spacetime further and further, causing it to contract and become smaller and smaller. But then we have seen that negentropy in nature is invariably accompanied by entropy, and thus, this negentropic knot

---

* Gamma rays have a wide range of wavelengths and energies. Gamma ray photons emitted from nuclear emissions are about 10,000 to 10,000,000 times more energetic than that of visible light, but the gamma ray photons from high-energy collisions (as in annihilation) are immensely more powerful (perhaps a million times more energetic). Even more energetic gamma rays are detected in the cosmic rays, but their source and mechanism of production are not clear. This shows that huge amounts of energies are needed to cross the critical value to create matter.

is also associated with an entropic phase. However, entropic force being repulsive in nature, the entropic response to the negentropic knot becomes expansive (Fig 9.5b). Thus, we will call this entropic phase of a negentropic knot as the *entropic expanse!*

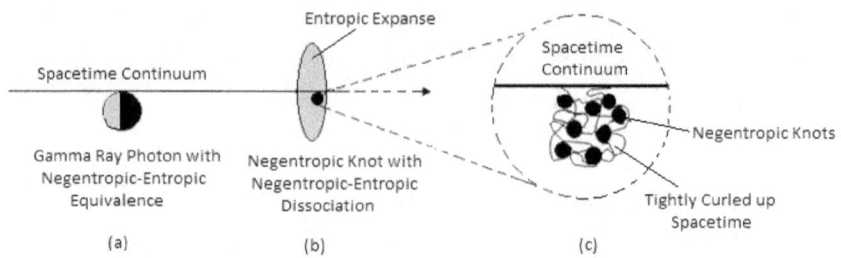

Fig 9.5: Negentropic–entropic dissociation
and negentropic knot formation.

This means that after the formation of a negentropic knot, there is a discrepancy between the negentropic phase and the entropic phase—one is contracting, and the other is expanding. In other words, we may say that in the case of negentropic knot formation, the negentropic–entropic equivalence (Ch 8.E) is lost completely. This discrepancy between the two phases in the spacetime curvature is called the *negentropic–entropic dissociation*. This is the basic premise in the mechanism of the formation of matter, and this is the quintessence of the fermionic negentropic model.

Negentropic knot results in negentropic–entropic dissociation.

A working model of matter formation is presented below which describes a complex interaction between these negentropic and entropic phases not only to give rise to various forms of matter particles ranging from quarks to leptons and others but also to form force particles such as gluons and W particles.

Negentropic–Entropic Dissociation and the Appearance of Space and Time: Because of the excessive force acting on the spacetime continuum, the spacetime curls up into a dense region of negentropic knot, and because of the cohesive nature of negentropic force, the knot tightens more and more, and the negentropic knot tends to attain smaller and smaller proportions. But then what would be the ultimate fate of such a negentropic knot? Does it disappear completely and turn into a miniature black hole? Or does it stop contracting at some point? If so, why? Consider this: a negentropic knot which has diminished into 'nothing' would cease to exist, and such an event is not conducive to the formation of matter, and hence, this consequence is not allowed in nature. We will see why such a collapse does not take place in nature under 'normal circumstances'* in Sec-E, but for now, we will proceed with the after-effects of negentropic–entropic dissociation.

We know that nature has an irrepressible tendency to respond to negentropy by exhibiting an equal and opposite entropy. In the case of the generation of light, as we have seen in the photonic negentropic model, the negentropic phase and the corresponding entropic phase are equal and opposite (Fig 9.5a), and because of this negentropic–entropic equivalence, the negentropic cycles of light were shown to propagate for an infinite period.

However, in the case of negentropic knot (which is under the influence of force beyond a critical limit), the spacetime gets warped to such an extent that the correspondence between the negentropic phase (i.e. negentropic knot) and the generated entropic phase is lost. Consequently, the negentropic knot and the entropic phase are no longer equal, and hence, they become dissociated (Fig 9.5b). This negentropic–entropic dissociation is the hallmark of the formation of matter because it brings about many fundamental changes in the nature of spacetime, which, in turn, leads to the development of fermions.

Negentropic–entropic dissociation is the hallmark of matter formation.

By far, the most important consequence of this negentropic–entropic dissociation is the segregation of the spacetime continuum into two separate entities—'space' and 'time'. Hitherto, we have seen that in the case of 3D light, space and time have a *constant*

---

* Under 'extraordinary circumstances', the collapse may proceed to result in a black hole. We will look into those extraordinary situations in Ch 12-C.

relationship denoted as 'space-and-time entity' (Ch 8-C). This means that even though space and time are segregated in a ray of light which makes them measurable (as discussed in Ch 8-C), both of them have a universally fixed value which is *unchanging*—which makes light travel by a fixed spatial parameter of 299,792,458 'metres' (i.e. space) in a fixed temporal parameter of 1 'second' (i.e. time). In other words, there could not exist any *change* in the space and time in 3D light (the proportionality of energy content (E) of a photon and the frequency of the EM wave (f) being constant by the equation $E = hf$, Ch 2-C and Sec-B). Thus, in the case of light, space exists with an absolutely fixed relationship with time—meaning that no change is possible in the disposition of space and time in light.

However, the first and the foremost step in the formation of matter is the occurrence of a change in the disposition of space and time. Thus, in the case of negentropic knot, space could change its spatial disposition in relation to time so that this spatially oriented negentropic knot appears to have an independent existence. In fact, now space has a relative existence with time in a negentropic knot. This dissociated negentropic knot with an exclusive spatial orientation appears to us as the smallest unit of matter! The negentropic knot, thus formed, has now occupied a discrete space in the hitherto monotonous 4D continuum.

Negentropic knot has an exclusive spatial orientation.

This dissociation happens because of the intrinsic dissimilarity between the two phases (Fig 9.5b). The negentropic phase, being cohesive in nature, makes the negentropic knot contract further and further. In contrast, the entropic phase, being repulsive in nature, tends to expand the spacetime curvature more and more. Thus, the contracting negentropic phase splits away from the expanding entropic phase.

Now consider this: in the case of a negentropic knot, the expansive entropic phase appears *after* a lapse of *time*—i.e. entropic expanse happens *after* the formation of a negentropic knot. In other words, we may say that the negentropic knot and entropic expanse now move *relative to each other*. This is the first indication of 'relative motion' in nature. In short, the relative movement of the space and time is not fixed any more. Hitherto, in the case of 3D light, the space and time travel in unison endlessly, but in the case of negentropic knot, they are not only segregated in their spatial orientation (i.e. 'space and matter' have started to exist from here). They are also segregated in their temporal orientation (i.e. 'time and motion' have started to exist from here)! The meaning of all this is that negentropic knots are now allowed to move in relation to each other—a feat which was not possible for negentropic phases in the case of light!

This initial time interval of this space and time dissociation may be reckoned as the smallest possible time in nature—i.e. *Planck time* (Ch 2-F)—and this is the smallest possible 'measurable time' in nature. In fact, it can be said that whenever a negentropic knot is created in nature, a *local time* is born. This is the fundamental concept upon which, of course, the whole concept of relativity is actually based – hence, it may be said that formation of negentropic knot heralds the origin of time.

Negentropic knot heralds the origin of 'local time'.

Thus, it may be concluded that each negentropic knot has its own 'timeline', and it has a 'history' of its own. In short, each negentropic knot has a specific 'place and time of birth'!

Formation of Quarks: The space, thus segregated from time, now attains an independent status as a 'unit of matter', which is nothing but a 'localized' knot of spacetime wrapping up a unit of force in the form of negentropic knot (Fig 9.5c). This is the first step in the formation of baryonic matter, and this smallest chunk of matter is represented as a *quark* in the Standard Model. The reader will soon understand why we should call negentropic knot as a quark (but not any other particle!). Thus, a quark can be considered as the first particle to form, and it represents the smallest possible unit of force (beyond the critical value) cloaked up in the spacetime continuum.

Quark represents the smallest unit of force cloaked up in spacetime.

The energy requirements to achieve this feat of negentropic knotting can be expected to be very large. After all, the equation *E*

$= mc^2$ showed us that an immensely large amount of force is needed to initiate matter formation. Thus, each quark harbours a lot of force wrapped up by spacetime.

The creation of quark may also be considered in another perspective. We may also say that the negentropic knot (or quark) represents a fermion, and the 1/2 spin is effected by repeated curling up of the spacetime in three-dimensional folds within the confines of a limited space (Fig 9.5c).

Negentropic knots have attained the status of fermions.

Thus, in conclusion, we may say that the formation of quark is the first step in the formation of matter in the hitherto monotonous 4D spacetime continuum—i.e. an 'eventless' spacetime has transformed into an 'eventful' matter particle with a 'local' space and time. As we will be discussing below, quarks initiate the building up of baryonic matter not only by triggering the formation of other matter particles, such as electrons and positrons, but also by forming force particles, such as gluons. Of course, we will see how the fermionic negentropic model explains the acquisition of all the properties of matter, such as mass, motion, spin, charge, and momentum in Sec-E and F.

Section E
Quark's Motion, Mass, and Work

So far, we have seen the mechanism of the formation of a negentropic knot (i.e. quark). Now we will see how a quark would acquire its peculiar properties related to its mass, movement, charge, etc. In fact, the following discussion would show us how a negentropic knot qualifies itself as being called a quark. But before going further, the reader is advised to recapitulate the essential properties of quarks and gluons from Ch 3-C so that the reader may get a good grasp of the new perspective presented below. We will start first by dealing with the fundamental problem of mass and motion of quarks.

Mass and Motion: The negentropic knot (i.e. quark) is now shown to interact with the spacetime in a complex manner, which would give the quark its motion and mass. First consider this: we have seen in Ch 3-G that the total amount of mass of any object is contributed almost exclusively by the nuclei of its constituent atoms, the contribution of electrons to mass of an object being nearly zero. And of course, the nucleus gets its mass from the collective mass of its resident nucleons (protons and neutrons). But then we have also seen that the rest mass of the quarks in a nucleon is merely about 1% of the mass of a proton/neutron (whereas the mass of a nucleon is 938 MeV/c^2, the total mass of the valence quarks in it is only about 5 to 15 MeV/c^2!). Thus, we may say that a quark has negligible intrinsic mass of its own.

Quarks have no intrinsic mass.

However, we went on to clarify in Ch 3-G that 99% of the mass of a nucleon is contributed by the kinetic energy of its resident quarks. Consider this: we have already seen in Ch 8-F that the negentropic phase in light is formed by the potential energy (i.e. force in the form of potential energy), causing a curvature of the spacetime. And in Sec-D, we have seen that a negentropic knot forms when a lot of force is applied to the spacetime. Speaking the other way, we may say that a negentropic knot is formed when a lot of potential energy is 'deposited' on to the spacetime. But then this stored potential energy in the negentropic knot (= quark) must give a lot of mass to the quark. What has happened to this inertial mass of the quark given by the stored potential energy?

We may explain this phenomenon of the absence of mass of a quark by a simple understanding. We have seen in Ch 7-D that potential energy is not favoured in nature and so is quite unstable. Consequently, it can be said that the potential energy is in a constant state of flux of getting converted into kinetic energy. This phenomenon makes a quark quite unstable, and thus it keeps moving at tremendous speeds—i.e. owing to the phenomenon of constant conversion of potential energy to kinetic energy, the quarks are driven into motion continuously at great speeds. In other words, we may state that motion is an innate property of quarks.

Motion is the intrinsic property of quarks.

But then any matter particle (or any particle with mass) in nature is known to travel only by non-uniform motion (because when it travels by uniform motion, it simply becomes a ray of light, Ch 7-F). Now consider this: for any object (or particle) to travel by non-uniform motion, it has to be driven by an agent called force, and we have already seen that a quark is an embodiment of potential energy (in the form of negentropic force) cloaked up in spacetime. And we have also seen that this potential force is converted into kinetic energy on a constant basis. Putting all this together, it means that a quark moves ahead by 'disintegrating' *itself* into an array of 'force particles'. We have already seen that these force carriers of quarks have a name—gluons! Thus, putting it this way, we may say that quarks moves ahead by converting themselves into gluons.

Thus, basing on the above discussion, we may pass this statement: it is the kinetic energy of the quarks which is represented as the field energy of gluons that we are calculating when we measure the mass of a nucleon. To be very precise, we must say that the conversion of potential energy into kinetic energy is measured as the mass of that matter particle. This means that the quarks are represented by potential energy (and so they are unstable) and gluons are represented by kinetic energy, and this, in effect, simply means that all the mass we measure is the field force of the gluons. Thus, finally, we may state that:

As the quarks move, they disintegrate and transform into gluons!

This is a radical statement—one which would surely require a thorough examination. But then the only way to support this statement is to offer some empirical evidence, and again, the only way to show evidence is to explain most of the unique properties, if not all, of quarks and gluons by using this new perspective. We will endeavour to do so in the following discussion.

The Quark–Gluon Cycle: To examine the correct relationship between quarks and gluons, we will have to study the peculiar quality of motion of quarks. What sort of motion does a quark demonstrate? Since any matter particle in nature is known to travel only by non-uniform motion, it may be construed that quarks also move only by non-uniform motion. This means to say that under the influence of force, quarks travel by curvilinear motion. Since this motion of quarks takes place at the most primary level, and since it is curvilinear in its course, we may call this motion a *primary matter wave* or, simply, a *quark wave*.

If we look at Fig 9.3, a photon, as it makes one full turn across the span of each electromagnetic wave, it takes a straight path without losing energy, and this results in its travel for indefinite periods and distances. In a similar way, each cycle of primary matter wave of a quark may also be represented in this fashion but with many significant differences. These differences in the motion of a quark, in fact, herald all other properties of matter, as we will see.

We have already seen that potential energy that is stored in the negentropic knot is in a constant state of flux of being converted into kinetic energy. This feature makes the quark an ephemeral entity without possessing any mass of its own—a quark holds its mass only transiently. In fact, this illustrates the feature that a quark does not exist independently in nature, and so a quark is represented in the diagrams as an *X-particle* (Fig 9.6a), indicating that it is only a virtual particle.

A quark holds mass only transiently; hence,
it has no independent existence.

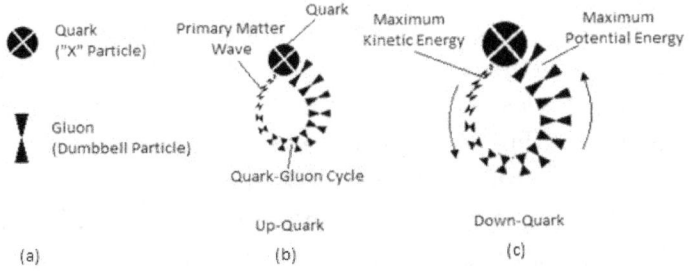

Fig 9.6: Quarks, gluons, and the quark–gluon cycle.

Now we will make an attempt to depict the movement of a quark in the form of a primary matter wave (Fig 9.6b). To do this, we will have to closely examine how exactly a quark is born, or in other words, we will analyze the actual kinetics that takes place during the formation of a negentropic knot. Consider this: a negentropic knot, at its initial stage of formation, has a great kinetic energy which is derived from the force of impact of the matter-generating event at source (Sec-C). However, as time proceeds, this kinetic energy is incorporated as potential energy in the spacetime. And as time proceeds further, the magnitude of potential energy increases (Fig 9.6b). This is like a stone thrown up into the sky on the earth, in which case the initial kinetic energy is maximum but is converted into potential energy gradually as it ascends up and attains its maximum as it reaches its maximum height, at which point the stone stops in its course (Ch 7-D). Similarly, the quarks start with a maximum kinetic energy initially, and as time proceeds, they undergo continuous *deceleration* (i.e. *negative acceleration*), and finally, they acquire the maximum potential energy. This property of deceleration, in fact, makes the quarks cohesive in nature, and therefore, they exhibit the property of attraction.

This sequence of events wherein the quark's kinetic energy is converted into potential energy may also be represented as a force particle, and this is nothing but gluon, its force carrier! Thus, in effect, we may say that this moving phase of a quark takes the form of a 'gluon'. However, because gluons decelerate, they may be predicted to behave more like spin-2 particles (rather than a spin-1 particle). For this reason, we will depict gluon as a *dumbbell particle* (Fig 9.6a) (the reader may look at Fig 9.1and see how a spin-2 particle should look like).

Now we may say that a gluon at the initiation of deceleration is small because, at this stage, the kinetic energy is more and the potential energy is less – hence, a gluon at the initial stage is depicted in its smallest size (Fig 9.6b and c). However, as the gluon decelerates, its potential energy increases, which increases its size progressively. And finally, the gluon achieves the maximum potential energy and reaches the maximum size and so becomes 'stationary' (Fig 9.6b)—as a stone thrown up stops in its course at one point. Having reached its maximum potential energy, this gluon may be considered to attain its 'resting' state, and now, at this resting state, it may again be called a quark! The quark, thus created, exists only for a transient period, with practically no intrinsic mass (see below) because this is only a 'passing' state, and it soon disintegrates by converting its potential energy into kinetic energy, once again generating gluons—and the process continues.

A gluon with the maximum potential energy becomes a quark.

However, it must be noted that, all along its course, the motion of gluons is not in a straight course but is curvilinear as depicted in Fig 9.6b. And because of its curvilinear course, over a period of time, this curved motion assumes the form of a cycle. This is the *quark–gluon cycle* (Fig 9.6c), and the resultant product is called the *quark–gluon complex*. Thus, in a quark–gluon cycle, a negentropic knot which starts its journey as kinetic energy continues in the cycle as gluons and ends up once again as quarks, only to start again in another quark–gluon cycle! In a nutshell, it may be said that a quark–gluon cycle represents the strong force!

Primary matter waves take the form of a quark–gluon cycle.
A quark–gluon cycle represents strong force.

Here, it is very important to notice the direction of the quark wave and that of the quark–gluon cycle. It is directed from right to left as shown in Fig 9.6b and c, or we may say that the quark–gluon cycle is moving in an *anticlockwise direction*. The significance of this feature will become apparent shortly.

What Gives Mass to a Particle? Now consider this: the quark–gluon cycle may really be represented as a universal cycle (Fig 7.5). The force that enters this cycle is, of course, the negentropic force, and the work done here is the displacement of a quark, and the entropy is represented by the expansive entropic phase (see Sec-F for details). However, the work done here in the quark–gluon cycle has a specialty: the displacement of quarks is always taking place *towards* each other because of the deceleration of gluons. This makes the quarks keep 'whirling' and contracting into smaller and smaller sizes until it reaches Planck sizes (the discussion below lets us understand why they do not attain smaller-than-Planck sizes). This cohesive property of the quark–gluon cycle not only makes quarks diminutive, but this feature also explains why protons shrink progressively in their measurements (as demonstrated by Randolf Pohl experimentally in 2010). Moreover, this property explains the diminutive nature of the nucleus (see Sec-H for details).

We will now deal with the mass of quarks. We have seen in Ch 7-E that resistance of an object to undergo non-uniform motion is called inertia. Thus, the deceleration (i.e. non-uniform motion) and displacement of a quark give inertia to the quark–gluon cycle, and this can be measured as inertial mass. This is the origin of mass of a proton/neutron, but since quark and gluons cannot be separated from the quark–gluon cycle, it is impossible to measure their mass separately. Hence, we may say that it is the deceleration which generates mass of a proton/neutron.

Why Particles with Different Masses? We have seen in the Standard Model (Ch 3-C) that there are several types of quarks, and each quark has a different mass (Table 3.2), the lightest being up-quark and the heaviest being top-quark. The following discussion explains this discrepancy.

Take a look at the quark wave. We may notice that as time elapses, the gluon travels for a greater distance, which increases the magnitude of potential energy. In other words, the more the 'length' of a quark wave, the more is the ultimate size of the gluon and, consequently, the more is the size of the resultant quark is (Fig 9.6c). In short, as the length of a quark wave increases, the quark's mass increases. Basing on this feature, we may say that the length of an up-quark matter wave is the least in nature, conferring it the least mass of all quarks (1.7 to 3.3 MeV/c²), and the down-quark matter wave's length is the next least (4.1 to 5.8 MeV/c²). And of course, we may say that the length of the top-quark matter wave is the lengthiest in nature, thus conferring it the most mass (172 GeV/c²).

As the length of a quark wave increases, the quark's mass increases.

In the above discussion, we have seen that the length of a quark wave depends on the magnitude of force, and this magnitude can be said to be dependent on the initial input of negentropic force on to the spacetime during the formation of a negentropic knot. This is in consonance with the finding we have seen in Sec-C that as the input energy increases in nuclear reactions, the chances of the formation of heavier matter particles increases. For example, in the case of 'high energy' annihilations, an array of exotic heavy particles such as muons are created. This is the reason for the existence of all heavy particles of the second and third generation (Table 3.1) of the Standard Model.

As energy input increases at source, heavier particles are created.

However, it is also seen that these heavy particles that are found in the nuclear reactions are only transient, and they soon break down into the lightest stable particles of nature. This process, of course, releases the stored potential energy in the heavy particles in the form of disseminating gamma rays. Thus, the input energy is first incorporated into the heaviest particles, and then they disintegrate and reassemble into smaller particles and free energy.

Formation of Protons and Neutrons: Now we will see how the lightest quarks arrange themselves to form stable nucleons (protons and neutrons). The unequal length of quark waves has an important bearing on the final architecture of protons/neutrons. Consider this first: if, for example, three up-quarks combine to form a baryon, we then notice that their quark waves will be of equal lengths, thus forming an equilateral triangular configuration (Fig 9.7a), and this configuration would soon contract inexorably and collapse upon itself to become a miniature black hole! However, if one of the quarks is of a different type, such as a down-quark, then the result is an isosceles triangular configuration (Fig 9.7b&c), which would not allow the baryon to collapse on to itself! We will see about this in Sec-G.

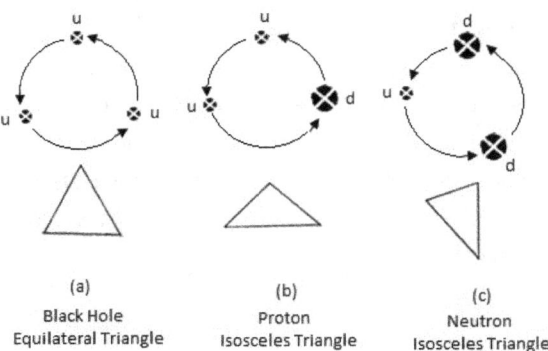

(a)
Black Hole
Equilateral Triangle

(b)
Proton
Isosceles Triangle

(c)
Neutron
Isosceles Triangle

Fig 9.7: Asymmetry of proton and neutron.

Having seen the relationship of quarks and gluons, we will now examine two intriguing nuclear phenomena for which our new fermionic negentropic model offers the correct explanation—asymptotic freedom and quark's nuclear confinement. The fermionic model also explains many other features of atom which are discussed in the later sections.

Asymptotic Freedom Explained: We have looked into the phenomenon of asymptotic freedom in Ch 3-C and learnt that as the distance between quarks *increases*, the attractive force between them becomes *stronger*, and *vice versa*. This phenomenon allows the quarks within a baryon to move freely and behave more like independent particles. We have also learnt that because the strong force becomes too great when quarks are pulled apart, it becomes ergonomically advantageous for the quarks to snap away from each other after separating for a critical distance, and the conserved energy is converted into matter particles (because of mass–energy conservation) in the form of a quark and an antiquark (because of charge conservation), and this quark and antiquark together form a meson. Of course, the meson annihilates, soon emitting a great flash of energy. This is one of the reasons why breaking up a proton in a nuclear reaction releases such tremendous amounts of energy (which is harnessed in our nuclear power plants).

The reader may now see that the very property of the primary matter wave (quark wave) explains asymptotic freedom in a very effective manner. In a quark wave, when the virtual gluon is in proximity (i.e. nearer to its origin), it has more kinetic energy, thus allowing its free movement (meaning a slackened strong force). But when the virtual gluon is at a distance (i.e. farther away to its origin), then it acquires more potential energy, thus reducing its movement (meaning a strengthened strong force). In other words, the property of deceleration of quarks results in the progressively increasing potential energy of the gluons as distance increases, which would make the force stronger.

Asymptotic freedom is an offshoot of quark-wave deceleration.

And of course, if the distance of separation is too large, then the size of the gluon in the quark wave becomes too redundant, and thus, it breaks up into smaller quarks and antiquarks, giving rise to mesons.

Short Range of Gluons Explained: We have also seen in Ch 3-C that gluons are short-range particles. They can travel only for ultrashort distances, about one femtometre (10^{-15} metres) at a stretch (unlike photons, which can travel for infinite distances to the end of the universe!). Thus, the strong force is really strong within the confines of the nucleus but can barely extend its influence beyond the boundaries of the nucleus. Why is this so?

We may explain the above phenomenon of the gluon's short-range spread by considering this. We have seen in Ch 8-D and E that a photon in a ray of light (3D light) spreads for infinite distances because of two reasons—its equivalent negentropic–entropic cycles and no-work-maximum-entropy feature. However, we have seen that negentropic knot formation fundamentally results in a negentropic–entropic dissociation, and it also performs work in the form of deceleration of quark–gluon cycle. In fact, their very property of deceleration and work makes the nucleons attractive and binds them together, and this property of work also results in the generation of entropic response, as we will be seeing in Sec-G. Thus, the gluons may actually be considered to travel in a *negative direction*, and this, in fact, makes the strong force an attractive force. And consequently, the travel distance of a gluon becomes curtailed and becomes limited to the confines of the nucleus! We will explore this property further in Ch 10-B and C.

Having examined the mass, motion, and other peculiar phenomena of quarks and gluons, we will now go into the discussion on charge formation of a matter particle.

Section F
The Origin of Charge

Charge is a fundamental property of matter. Charge is a unique concept to understand owing to the fact that, unlike mass, motion, and momentum, matter appears to possess no physical basis for showing up this property—charge is a quality of matter *per se*. We will now study the concept of charge in a new perspective as the reader may see below.

Charge has no physical basis.

We have stated in Ch 2-B that electromagnetic force is a 'chargeless' force owing to the fact that the fundamental unit of EM force, the photon, is a chargeless particle, and it is generally considered that, because of this feature, a light ray is unaffected by either electric or magnetic fields. However, we know that sometimes light behaves as though it possesses charge—as when the course of light is affected by charged fields (e.g. intense magnetism deflecting light as shown by Michael Faraday nearly two centuries ago)—or sometimes it is the other way round, as in the case of light affecting the course of charged particles (as, for example, in the case of a TV antenna where the received radio signals wiggle electrons of the antenna's metal). Moreover, we have seen in Ch 8-F and Sec-C that the generation of light usually involves the participation of charged matter particles or charged force fields. All these affairs of light allow us to suspect that photons surely contain some charge incorporated in them, but it appears that this charge is somehow masked away from our direct observation. We will now explore the phenomenon of charge at its basic level and also see how photons have charges incorporated in them.

Direction and Charge: Before going into the actual study of charge, we will review an important historical development which is of our interest here. In 1932, Carl Anderson noticed that some particles in the cloud chamber with all the characteristics of electrons have traced a path opposite the course of mainstream electrons, indicating that they were actually 'positive electrons' (the existence of such 'anti-electrons', however, was already predicted by Paul Dirac on theoretical grounds a few years earlier). These anti-electrons were subsequently named positrons. Nearly after a decade, Feynman improved the theoretical understanding of these positive electrons by ingeniously depicting them in his diagrams (Ch 3-E). Now consider this: in the Feynman diagram supplied below (Fig 9.8a), the time axis flows from left to right, and the vertical axis represents axis of the space. An electron travelling from the left, when impeded suddenly, emits a photon, alters its direction (because of conservation of momentum), and continues its journey towards the right as shown in the diagram. But then according to the theory of relativity, we may treat time dimension as any other space dimension so that we may swap their places freely. Thus, we are free to rotate the Feynman diagram by 90° anticlockwise (Fig 9.8b) and interpret the result all the same. Now by looking at the new diagram, we may say that an electron is *annihilating* a positron, generating a photon. Or we may interpret the Feynman diagram in this way: the electron is moving in a *forward* direction, whereas the positron is moving *backward* in time (Fig 9.8b)!

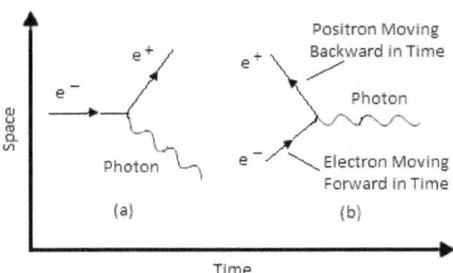

Fig 9.8: Feynman diagram showing charges and directions.

Thus, finally, it was concluded that an electron (or for that matter, any negatively charged particle) is a particle which is moving in a forward time direction and a positron (or any positively charged particle) is a particle which is moving in a backward time direction.

Negative charge moves forwards. Positive charge moves backwards.

But then what is the meaning of this 'forward' and 'backward' direction? We know that all the charged particles can exist independently in nature. For example, electrons behave like negatively charged particles even in isolation (even when its positive counterpart is not present in its vicinity), and positrons do so in the same way. In short, all electrically charged particles behave as though they are 'electric monopoles'. So what does *forward* mean to an electron and *backward* to a positron? Thus, it appears that the charged particles move in relation to some 'universal reference' (but not relative to one another). We will now go in search of such a universal reference!

The Universal Reference: This universal reference, as we may envisage, must be applicable to all particles uniformly. A 'local reference' will not work out as it is unlikely to bring consistency and uniformity to all the charged particles in all situations. In other words, this universal reference and the related charge must be applicable to all matter particles universally. Now we will see what this reference could be.

The reader may recall that in Ch 8-F, we have depicted a simplified diagram of the linear pathway of the natural sequence of events of the universe and showed that absolute uniform motion in nature takes a straight course and merges with the cosmic entropy. It is already noted that all material objects (with mass) travel by non-uniform motion, and this sort of non-uniform motion was shown to travel in the direction *opposite* the direction of cosmic entropy as depicted in the linear diagrams in Fig 8.5 and in Fig 9.9a. Now the reader may see that when we consider the uniform motion towards cosmic entropy as 'forward direction', we may consider the non-uniform motion as 'backward direction'. This backward direction, in fact, is curvilinear in its course, and it is pointing *away* from the direction of maximum entropy. The reader may now see that uniform motion travelling in a straight line is moving in the forward direction in contrast to the curved and backward direction of non-uniform motion.

Uniform motion is forward and straight; non-uniform motion is backward and curved.

We will now depict this feature of non-uniform motion and uniform motion in a cyclic fashion as in the case of the universal cycle (Fig 7.5), and the result is shown in Fig 9.9b. In this universal cycle, we have depicted work as an offshoot, and this offshoot of work may now be shown to take an opposite direction to the direction of entropy (Fig 9.9b).

Now consider this: the uniform motion in a straight line may actually be construed as causing maximum 'disorder', and on the other hand, the curved feature of non-uniform motion may be construed as causing 'order' to happen (see event 1 Ch 7-F and Ch 8-F). Look at Fig 9.9b once again. We will find that the entropic phase (uniform motion) takes a *clockwise* direction, whereas the work offshoot (non-uniform motion) takes an *anticlockwise* direction as shown in Fig 9.9c.

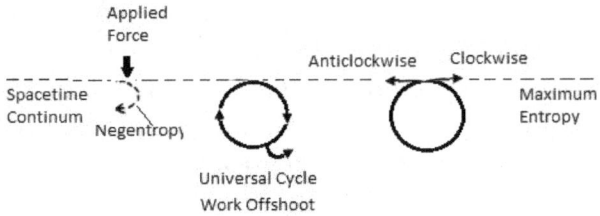

Fig 9.9: Negentropic and entropic forces in opposite directions.

We may summarize the above discussion and say that the work done by a quark takes a curved course travelling with non-uniform motion at subluminal speeds, moving away from cosmic entropy/inertia and finally causing an increase in order in nature by drawing quarks together. This can be depicted as an anticlockwise movement. All these features put together may be termed in a single phrase—'shift-to-left'—meaning that the shift of events in nature is to the 'left' of the linear pathway of natural events of the universe. In contrast, the resultant entropy takes the straightest possible course, travelling with uniform motion at the speed of light, moving towards cosmic entropy/inertia, and finally causing an increased disorder in nature. This can be depicted as a clockwise movement. All these features put together may be termed in a single phrase—'shift-to-right'—meaning that the shift of events in nature is to the 'right' of the linear pathway of natural events of the universe.

Thus, we may conclude and say that the cosmic inertia/entropy can be taken as a 'universal reference' which guides the particle to develop its charge.

Cosmic inertia/entropy is taken as universal reference.

The Meaning of Charge: The implication of the above conclusion is obvious – we can derive the formation of charges in nature by using the shift-to-right/left status of particles. It can be said that charges depict the direction of change that takes place in response to the applied force in order to accomplish the goal of travelling either towards or away from the cosmic entropy.

Charge is the direction towards or away from the cosmic entropy.

We will see how this rule explains charges of various particles. We will first take quarks and see how they develop fractional electric charges. The work done by the negentropic knot (= quark) can now be said to be taking the spacetime backwards in anticlockwise direction, increasing order or negentropy. In short, a quark has a tendency to shift-to-left. This state of a quark can be depicted as holding a 'positive' charge (Fig 9.10a). Because a quark always exists in an assemblage of other negentropic knots (in a proton/neutron/meson, etc.), it may be said to be 'tethered' to other quarks, and thus, it cannot take a full turn, and thus, it can be said to have a fractionated charge. This is the status of an up-quark.

A quark may also take the spacetime forwards in a clockwise direction, increasing disorder or entropy. In short, this quark has a tendency to shift-to-right. This state of a quark can be depicted as holding a 'negative' charge (Fig 9.10b), and it is also fractionated because of the reason stated above. This is the status of a down-quark.

This is the real meaning of the charge we observe in nature; it depicts the direction of change that takes place in response to the applied force to accomplish the goal of travelling either towards or away from the cosmic entropy. Positive charge of a matter particle indicates its present state to increase order and to cause negentropy. Negative charge of a matter particle indicates its present state to increase disorder and to cause maximum entropy.

Positive charge is negentropic; negative charge is entropic.

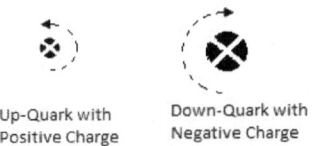

Up-Quark with   Down-Quark with
Positive Charge   Negative Charge

Fig 9.10: Depicting charges of quarks.

But then consider this: the electron is a negatively charged particle which hovers around the atomic nucleus in a bound state and sometimes existing in a free state (as in cathode rays). Now basing on the above statements, how do we interpret the negative charge of an electron? And also consider this: what of the positive charge of its antiparticle, the positron, and of course, what of all other charged particles in nature? How do we account for them? We will continue our discussion on charge in Sec-G wherein we will discuss about the expansive phase of a negentropic knot and its implications. But before going into this, we will examine an important question: is electromagnetic radiation a charged force? If so, where are the charges incorporated in a photon?

Photon as a Charged Particle: We have seen in Ch 8 that 4D spacetime is a no-space-no-time entity for which no mass, no motion, and no charge can be attributed. But the introduction of force into the 4D spacetime converts it into 3D light, which is represented by a photon, which, in turn, is composed of negentropic and entropic phases. By looking at Fig 9.11, we may say that the negentropic phase of a photon carries the force to the left (shift-to-left), and the entropic phase carries the force to the right (shift-to-right). Thus, in this simple way, we may conclude that the negentropic phase of a photon may be represented as positive; the entropic phase as negative.

Photon's negentropic phase is +ve; entropic phase is −ve.

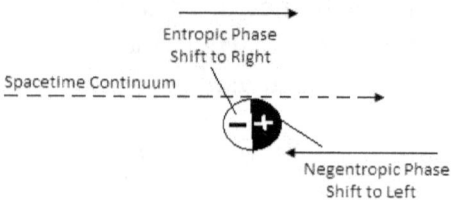

Fig 9.11: Representing charges in a photon.

However, for over nearly two centuries, we have known that energy of a light wave is represented by two field forces—the electric and magnetic field forces. We will now see what these field forces may really mean.

Magneto-Electricity: It is generally considered that electric and magnetic fields in an EM wave are two sides of the same phenomenon because of their interchangeability. We have examples of changing electric fields generating magnetism (e.g. electromagnets in a dockyard), and we have also examples of changing magnetic fields generating electricity (e.g. wind turbines). However, there has been some obfuscation in describing the real meaning of electric and magnetic field forces that constitute a photon. Now we will examine these issues in the light of our new understanding.

First consider this: we may depict the spin-1 state of a photon as its symmetric energy state (look again at Fig 9.3) because each spin state is represented by a field force represented as a negentropic phase and an equal and opposite field force represented as an entropic phase. Thus, we may say that a negentropic phase is a state wherein the energy imparted on spacetime is causing a change in the disposition of spacetime, which tends to travel in *non-uniform motion*, and entropic phase is a state wherein the energy is tending to travel in *uniform motion*—each following each in an interminable array of cycles which we call a ray of light.

However, these alternating phases of negentropy and entropy may also be represented as magnetic field and electric field, respectively (Fig 9.12), because of the following reason: the negentropic phase may be considered a representative of 3D motion (i.e. simulated uniform motion or non-uniform motion, Ch 8-C); hence, this 'moving' phase may be dubbed as magnetic field force. And the entropic phase is a representative of '4D motion' (where the concept of motion itself is not valid, Ch 8-C); hence, this 'stationary' or 'static' phase may be dubbed as an electric field force. Thus, in a nutshell, we may state that a negentropic phase can be depicted as a magnetic field, and the corresponding entropic phase can be depicted as an electric field.

A photon's magnetic field is negentropic; an electric field is entropic.

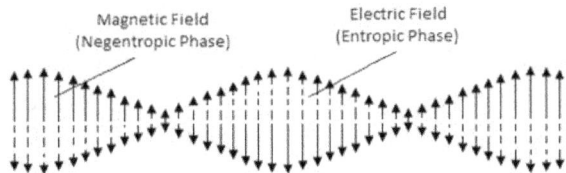

Fig 9.12: Magnetic and electric fields as negentropic and entropic phases.

For this reason, electromagnetism must be really termed as *magneto-electricity* because it's always the negentropic phase which happens first in nature, and the entropic phase is only reactionary to the negentropic phase. Hence, a photon may be depicted as a *magneto-electric unit*!

Having studied the creation of matter and development of mass, motion, and charge this way, we will now embark on to a discussion on the formation of electron and other features of an atom.

Section G
Electrons and Other Particles

We will now study the formation of electrons and other charged particles, and by the way of this discussion, we will be defining the very meaning of weak force in nuclear reactions. Moreover, in this discussion, the structure of the simplest atom in nature, the hydrogen atom, becomes evident.

We have seen in Ch 6 that disorder and entropy are the essence of nature and that all order and negentropy in nature must inevitably result in the generation of equal or more entropy. This entropic response to negentropy in the case of light, as we have seen in Ch 8, is equivalent, thus generating negentropic–entropic equivalence in the case of a photon. But in Sec D, we have seen that the entropic response to a negentropic knot is out of proportion, resulting in disparity between the two phases—negentropic phase being cohesive and diminutive and entropic phase being repulsive and expansive. We will now study this expansive phase.

One of the chief reasons for the development of an expansive entropic phase in the case of a negentropic knot is the magnitude of the negentropic force itself, which is large and beyond the critical limit. However, the magnitude of this entropic response is further augmented by the work done by the quarks in the form of their movement towards each other in the form of deceleration (thus increasing the local order, leading to the generation of an even greater entropic response). Because of these two reasons, the entropic reaction generated by a negentropic knot reaches exuberant proportions. Now consider this: the contracting negentropic knot tends to 'expel' the expansive entropic phase out of its locale into the surroundings because of the time difference between the formation of negentropic knot (= quark) and the entropic response owing to the complete segregation of space and time (Sec-D). This exuberant entropic phase, thus ejected out of a quark, is called an *entropic blowout* (Fig 9.13).

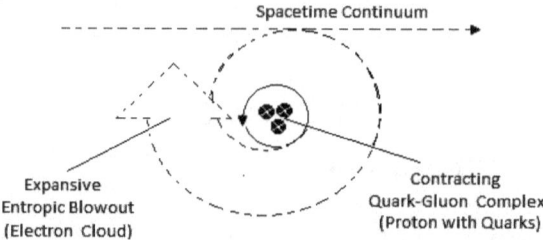

Fig 9.13: Entropic blowout and formation of an electron cloud.

We have already seen that if the spacetime is driven towards increasing disorder, it is travelling in the direction of cosmic inertia/ entropy—or in other words, it moves in a clockwise direction with a tendency to shit-to-right (Fig 9.13). This means that the entropic blowout takes on a negative charge. The reader may see that this entropic blowout assumes the form of an 'electron cloud' which spreads exuberantly around the nucleus.

Entropic blowout forms an 'electron cloud' which is negatively charged.

Before proceeding further, the phenomenon of entropic blowout must be understood in the right perspective. First, recapitulate the essential relationship between negentropic and entropic phases (as presented in Ch 8-E). A negentropic knot must be understood as an excessive 'folding' of spacetime resulting from the application of force beyond the critical limit, and the resultant entropic blowout must be understood as a spacetime 'unfolding' after the cessation of force upon it in an attempt to merge with the cosmic inertia.

There is another important feature of entropic blowout—they are merely spacetime aberrations without any real negentropic–entropic phases (as in the case of a ray of light which has regular negentropic–entropic cycles). In short, entropic blowouts are 'empty' spacetime aberrations in the vicinity of matter particles. These empty aberrations ultimately spread out and merge with cosmic inertia.

Entropic blowouts are empty spacetime aberrations.

With this fundamental understanding, we will proceed with events of nature which would lead to the formation of other matter particles.

Formation of a Hydrogen Atom: The reader may notice that, unwittingly, we have arrived at an important event of nature— the formation of the first element of nature, the hydrogen atom! Hydrogen atom may be now envisioned as a proton (i.e. nucleus) consisting of three negentropic knots bearing positive charge on it, and this negentropic phase gives out an exuberant entropic blowout, forming an expansive electron cloud outside the nucleus. We will look into the formation of complex atoms briefly in Sec-H below.

A hydrogen atom is a negentropic knot
surrounded by an entropic blowout.

The above discussion implies that an electron is formed as an entropic response to the negentropic knots residing in a proton, but then how does a 'free electron' exist (which we see abundantly in nature)?

Formation of Free Electrons: We have seen in Sec-D that the most important characteristic feature of any matter particle in nature is the appearance of a negentropic knot. And thus, the only way by which the electron cloud could attain fermionic status and become a free electron is by the way of formation of a negentropic knot of its own. However, the formation of a negentropic knot requires an input of a lot of negentropic force to warp the spacetime beyond the critical limit. Where does this energy come from?

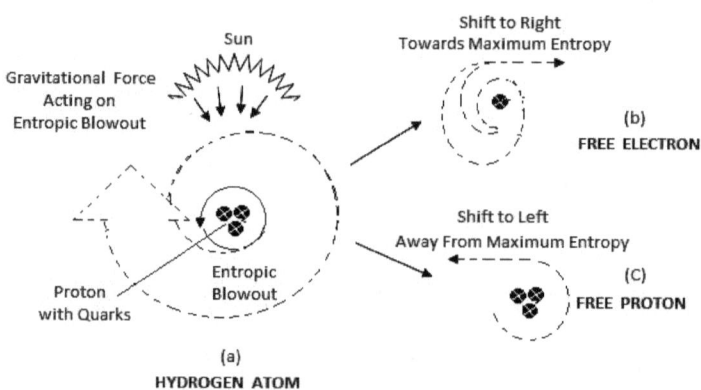

Fig 9.14: **Hydrogen atom breaking up into proton and electron.**

There are several ways by which free electrons form in nature. For example, they form inside a cathode-ray tube, during photoelectric effect on metals, in several nuclear reactions, inside stars during proton–proton fusion reactions, and so on. It may be noted that in all these processes, a great deal of force is utilized either in the form of powerful electromagnetic rays (as in photoelectric effect) or in the form of gravitational force (as in a star, Ch 4-D).

It may be suggested that the imparted force acts specifically upon the spacetime expanse of the electron blowout (i.e. electron cloud) of an atom, which curls its spacetime further into a negentropic knot (as we have seen in Sec-D). Thus, the imparted force is now incorporated into spacetime of the entropic blowout as a tiny negentropic knot within it. We will now take events in a star to show how this works out. In Fig 9.14, we have depicted gravitational force acting on the electron cloud inducing the formation of a negentropic knot, which soon splits up into a free proton and an electron. As soon as this feat is achieved, the electron cloud gets separated from the proton and becomes a free electron which zips fast into the surrounding space because of its great kinetic energy. This electron cloud along with its negentropic knot now has its own spatial orientation, and thus it has attained the status of a fermion—i.e. an electron which is capable of having an independent and stable existence.

A free electron has its own tiny negentropic knot.

Thus, in conclusion, it may be said that an electron bound to an atom exists in the state of an electron cloud with no definite spatial configuration, and thus, it carries no mass as long as it is orbiting a nucleus. Whereas a free electron that is ejected out of the atom exists as a matter particle with a definite spatial orientation, with a measurable mass and other characteristics of a matter particle.

Atom-bound electron exists as electron cloud;
free electron exists as matter particle.

And what would be the mass of a negentropic knot that is incorporated in an electron cloud? Of course, it can be assumed that its mass would be equal to that of an electron! Perhaps, the mass of an electron indicates the maximum possible mass for a negentropic knot to form in nature (whereas the least possible mass in nature is that of the quarks)!

Electron mass is the maximum possible mass for any negentropic knot.

But what is the fate of the bereft proton in the above process? A free proton, when it travels at low speeds, emits the entropic cloud around it, which once again becomes a hydrogen atom by itself! However, a very fast-moving proton remains as a proton along with its features of positive charge because the high velocity of a proton compensates for the entropic response itself (we often run our experiments with fast-moving protons so that we will not appreciate this feature of slow protons becoming hydrogen atoms!).

Having seen the formation of the most fundamental atom in nature, it leaves behind a herd of questions unanswered regarding the existence of other charged particles (e.g. positrons and muons), neutral particles (such as neutrinos), and weak force carriers. In order to understand these affairs, we have to further explore the entropic phase released by the negentropic knot. We know that all these exotic matter particles do not participate in the formation of an atom but are generated transiently during nuclear reactions as discussed in Sec-C. So we should take time to re-explore various nuclear reactions and study them again in the light of our fermionic negentropic model and see how this model stands out to answer all the questions.

Beta Decay Revisited: To do this, we will continue with our discussion on the next event that occurs in the star's core wherein, under the influence of huge gravitational pressures, protons are squeezed together, initiating a 'proton–proton fusion reaction' (here the reader may review details from Ch 4-D) which finally results in helium formation. When the proton–proton proximity reaches a critical distance of 10^{-15} metres, the protons are said to undergo 'quantum tunnelling' (Ch 2-D), and one of the two protons undergoes transmutation into a neutron, thereby forming a deuterium nucleus. We will study this process of beta-plus decay in light of the fermionic model and see how exactly this transmutation generates weak force and other matter particles.

Here we will study beta-plus decay, a non-spontaneous process requiring huge amount of energies, and see how exactly it becomes operative in the fermionic negentropic model. We have seen that a proton is composed of three negentropic knots (uud) with fractional electromagnetic charges on them, giving rise to an overall integer charge of +1. And in order to transmute a proton into a neutron, one of the up-quarks has to gain some mass (up-quark is lighter) and also change its electromagnetic charge to become a down-quark (Ch 3; Table 3.2).

But here is a paradox: in order to break down a proton, it needs a dissociative force (i.e. a force which causes repulsion, Ch 8-B). But then the huge gravitational force acting upon the proton is a negentropic force which tends to increase order by causing cohesion rather than repulsion! Then how does this splitting up of a proton into a positron and a neutrino become possible by the action of cohesive force, such as the gravitational force in the core of a star? How is this paradoxical event orchestrated by nature?

Nature seems to circumvent this problem by a simple contrivance: the gravitational force first initiates the formation of a very massive negentropic knot (with several times the mass of electron) which is incorporated into one of the up-quarks, and this huge negentropy, in turn, generates a powerful entropic response. In other words, by the effect of this gravitational force, the up-quark is converted into a huge negentropic knot with a fanning entropic phase (Fig 9.15). We have already known such a particle in beta-plus decay; it is the W^+ boson which is the force carrier of the weak force!

W^+ boson is a huge negentropic knot with a fanning entropic phase.

However, we know that the W^+ boson is highly unstable because its negentropic knot is many times more massive than the maximum allowed 'electron mass', and so it exists only very transiently during beta decay. Thus, a W+ boson breaks down into lighter and relatively stable particles. However, as we have seen, the total mass–energy and charge conservation laws involved in this event are strictly followed. Hence, the mass of this massive W^+ boson generated in the beta-plus is redistributed into an array of the following particles— three negentropic knots (quarks), udd, which reassemble to form a neutron; one negentropic knot with a shift to left generating a positive particle with one electron mass (which we call a positron); and one neutral neutrino (we will discuss the structure of a neutrino below) (Fig 9.15).

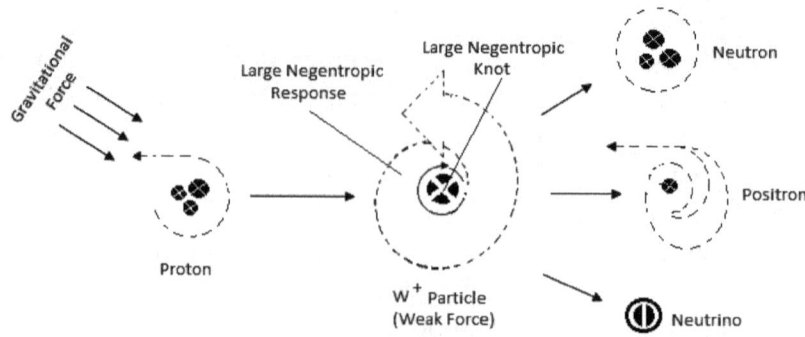

Fig 9.15: Beta-plus decay—formation of a W⁺ boson.

A note on beta-minus decay: the decay of a neutron into proton, electron, and antineutrino is a spontaneous process because of the inherent instability of the neutron. This is also mediated by a weak force particle, the W⁻ boson. Perhaps the spontaneous disintegration of one of the down-quarks converts its negentropic knot into a fanning entropic phase (which we call a W⁻ particle), and this complex assemblage soon splits up into lighter and stable particles following the laws of conservation—three negentropic knots (quarks), uud, which reassemble to form a proton; a negentropic knot with a shift to right generating a negative particle with one electron mass (which we call an electron); and a neutral neutrino with an opposite spin (i.e. an antineutrino).

Thus, it can be said that the weak force (mediated by W bosons) couples quarks with leptons. In other words, the W bosons act as 'intermediates' which transiently exist while the quarks transform into leptons.

W boson couples quarks and leptons and assists in their transformation.

Hitherto, we have seen the mechanism of the generation of almost all lighter particles in nature. It can be presumed that heavier matter particles are created when huge amounts of negentropic force is deposited on the spacetime. For example, in high-energy e⁻/e⁺ annihilation, the application of tremendous kinetic energy on spacetime induces the formation of many exotic heavy particles as seen in Sec-C. If the particles show an overall tendency to shift-to-left (away from cosmic entropy), then they behave as positive particles; on the other hand, if their tendency is to shift-to-right (towards cosmic entropy), then they behave as negative particles. We will now look into the structure of the most elusive particle in nature, the neutrino.

Neutrino—a Chimera Particle: Neutrinos are enigmatic neutral particles which are now thought to possess some non-zero mass travelling with a velocity almost near to the speed of light. They are commonly produced in nuclear reactions in the stars, and they are the most commonly occurring subatomic particles in nature spread throughout the universe (the reader may look into Ch 3-C for a detailed description). However, the nature of neutrinos remained elusive to us till now. We will look at neutrinos in a new perspective.

The fermionic model suggests that a neutrino is a particle with the smallest possible negentropic knot possible in nature, which is smaller than quark mass (whereas an electron has the largest mass allowed for a negentropic knot). Expectedly, this smallest negentropic knot may be nearer to the absolute length in nature—i.e. Planck length (Fig 9.15). And it may be presumed that because of its near-Planck length, the negentropic knot fails to generate an exuberant entropic response (as in the case of a quark), but it will only generate an equivalent entropic phase (as in the case of a photon). In short, a neutrino behaves just like a photon with an equivalent negentropic–entropic phase—albeit with some non-zero mass.

In simple words, a neutrino is the smallest matter particle possible in nature in which negentropic–entropic equivalence is maintained without causing dissociation—i.e. it's a photon with mass! This feature, in fact, confers a neutrino its duel properties of both light and matter, so it may be called a *chimera particle*.

A neutrino is a photon with mass; it's a chimera particle.

So far, we have discussed about the generation of quarks, electrons, and other matter particles in nature, and we have seen how gluons share a relationship with quarks and how W particles form and disintegrate. And finally, we have also seen how hydrogen atom forms. We will use this knowledge in Sec-H to see how more complex atoms are created in nature.

SECTION H
The Anatomy of The Atom

An atom is one of the most stable complex structures we find in nature. We find atoms in so many configurations ranging from lighter atoms to heavier atoms to radioactive atoms, but all the complex atoms share two general characteristics. The first is the exquisitely small size of their nuclei. The atomic nucleus is a diminutive structure occupying a volume of just 0.0000000000001% of an atom, and the space outside the nucleus is enormously expansive, which is about 100,000 times larger than the nucleus, which is occupied by electrons arranged in shells. The nucleus, though infinitesimally minute, is exceedingly dense with 99.97% of mass and matter concentrated in it. The second important feature of a complex atom is that of the ratio between the number of protons and neutrons—as the proton number increases in a nucleus, the number of neutrons in it also increases, indicating some relationship between them which confers stability to the nucleus. We will now examine these two issues in the light of our negentropic model.

The diminutive nature of the nucleus can readily be explained by the fact that the negentropic knots are extremely cohesive, which makes the nucleons collapse into each other inexorably (Sec-D). In fact, the nucleus would have plunged into a state of interminable collapse except for the asymmetric configuration of quarks (as discussed below). We have also seen in Sec-D that the mass of the quark–gluon complex represents all the mass of proton/neutron, which explains the concentrated mass in the nucleus. Moreover, in Sec-G, we have seen that electrons are nothing but entropic expanses which carry no mass at all within the nucleus, and this feature makes the nucleus diminutive and the electron cloud expansive.

The nucleus is ever contracting; an electron cloud is expansive.

Now we will look into the puzzling affair of neutrons making a nucleus stable. Before going into this, we will first see why free protons are stable and free neutrons are not so stable.

We have seen that the protons and neutrons are composed of quark waves which undergo constant deceleration, which makes them progressively diminutive. But why don't the protons/neutrons ultimately collapse upon themselves and disappear into nothing?

First we will see what will happen if a baryon (any particle consisting of three quarks) has all the three quarks of the same type (e.g. uuu or ddd).* If a baryon is composed of similar quarks with equal length of quark waves, then their quark–gluon cycles tend to create an *equilateral configuration* (Fig 9.7a), which would assume the form of a regular polygon, and this polygon would soon approximate into a circle. Because quarks move by deceleration, this circular movement of quarks whirls down into a progressively diminutive whirlpool which would collapse inexorably into its own centre and disappear into a state of 'nothing'. In other words, an absolutely symmetric equilateral triangle of quarks would soon turn into a miniature black hole (Fig 9.16)!

The isosceles configuration of a quark–gluon cycle prevents black hole formation.

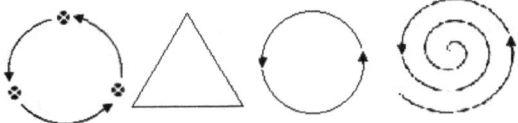

Fig 9.16: Equilateral triangle terminating into a black hole.

Now consider this: we have seen in Sec-E that protons and neutrons have three quarks (negentropic knots) with at least one of them belonging to a different type (uud in proton; udd in neutron). Thus, these quarks in a proton or a neutron, while participating in the quark–gluon cycle, form an *isosceles geometrical configuration* with unequal lengths of quark waves (Fig 9.7b and c), which will not allow them to collapse into black holes. In fact, even though many types of

---

*   Though some exotic baryons do form with such a configuration, they are very transient in nature.

baryons are known to scientists, almost all of them are composed of different quarks with different lengths of quark waves. Thus, we may say that baryons, in general, do not form regular polygons.

However, there is another reason why uud and udd are the most stable combinations, making protons and neutrons the commonest baryons in nature. The up-quark is the lightest of all quarks, and so it will participate in all our baryonic matter formation (because nature prefers quarks with the least mass). But then, since an up-quark alone cannot form a stable baryon (because as we have seen, they will turn into a state of interminable collapse), it needs at least one different quark, and this different quark must be the next lightest quark—that is the down-quark (Ch 3; Table-3.2). This is the most stable configuration with the least mass nature could achieve, and so we have up-quarks and down-quarks in all our baryonic matter in the universe! By the same token, three different quarks forming a 'scalene' triangular configuration may also be considered theoretically stable, but such a particle does not exist in nature because, once again, any combination other than uud is certainly heavier than a proton, so it becomes redundant and unstable. Thus, finally, we may state that:

A proton has the most stable combination of quarks with the least mass.

Of course, neutrons are the next stable baryon possible in nature; a free neutron could last for nearly fifteen minutes in nature. However, given time, it would tend to convert one of its down-quarks (by beta-minus decay) into an up-quark so that it could achieve the most stable combination of a proton.

Negentropic Burden and Neutrons: Now we will see the phenomenon of negentropic burden, which explains the stability of a complex nucleus. Before we venture into this intriguing issue, we will first look into the characteristics of the initial few elements, and this information throws light on the stability of the nucleus.

The simplest element in nature, hydrogen (also called *protium*), has an atomic nucleus which is just a single proton, and it is the most stable and abundant element in the universe; we have already seen its formation in Sec-G. The next possible element has a nucleus with two protons—a *diproton* (or also called *helium-2*), which is highly unstable (see explanation below). Thus, this diproton soon allows one of its protons to decay into a neutron, thus transforming itself into *deuterium*. Deuterium is a very stable nucleus and is abundant in nature; majority of deuterium is now thought to be generated at the big bang owing to the superlative temperatures and pressures existent at that time (which would mediate beta-plus decay of one of the protons in a diproton). We will proceed further. It is possible that two protons may combine with one neutron, forming *helium-3*. Helium-3 is a very stable nucleus, and it is also a naturally occurring element in the universe. However, it is also possible that one proton may combine with two neutrons, forming *tritium* (isotope of hydrogen), which is relatively unstable (half-life of about twelve years). And lastly, two protons may combine with two neutrons, generating *helium-4*, which is a highly stable nucleus, and it is, in fact, the second most common element in the universe! We know that more complex nuclei are formed in the universe as more energy is put in as it happens inside the cores of massive and supermassive stars as seen in Ch 4-B.

Now we will look into the stability affair of these nuclei by using the negentropic model. Consider a proton. It has two up-quarks which are positively charged, meaning that the proton tends to move *away* from the maximum entropy. This means to say that a proton would always tend to 'correct' this negentropy by the way of generating more entropy. In other words, a proton has a *'negentropic burden'*.

Up-quarks increase negentropic burden of the nucleus.

A proton corrects its negentropic burden of its nucleus (proton) in four ways. First, the down-quark compensates for the negentropic burden of its two up-quarks. Second, a proton attains stability because of its least total mass that is possible in nature as already discussed. Third, in the case of a fast-moving proton, its velocity itself compensates for the negentropic burden. Fourth, in the case of a hydrogen atom, the proton develops an expansive entropic blowout in the form of an electron cloud around it, which also counters its negentropic burden. All these four features add up to make a proton or a hydrogen atom highly stable and the most preferred particles in the universe!

In the case of a diproton, the situation is different. Diproton is highly unstable because the cumulative negentropic burden of the whole unit of diproton (two protons) becomes too much for nature to accommodate (see Ch 11-D for discussion on true nature of negentropy). And thus, a diproton readily allows the decay of one of its proton into a neutron to attain its stability.

A neutron inside a nucleus is stable because of the down-quarks; they tend to take the neutron towards the maximum entropy (however, a free neutron is unstable because of its mass). Thus, it may be concluded that as the number of protons increase in a nucleus, the negentropic burden increases exponentially owing to the nature of up-quarks (which tend to move away from maximum entropy), and such a nucleus demands more down-quarks in the form of neutrons to compensate for the increased negentropic burden!

Heavier nuclei need more down-quarks to compensate for the negentropic burden.

This simple explanation also holds good for the phenomenon of radioactivity we see in the case of very heavy nuclei or nuclei with abnormal proton/neutron ratio (Ch 4-E).

We will now ask a few fundamental questions: Why does nature not allow the phenomenon of interminable collapse of a matter particle into a black hole? Why should nature 'feel' an increase in negentropic burden, and why should it 'try' to compensate for it (for example by engineering decay of a large radioactive nucleus)? Or more fundamentally, why should nature prefer to merge with cosmic entropy? These fundamental questions are discussed in Ch 11-D and 12-G.

However, before attempting to answer these intricate questions, the reader may notice that we have discussed all the essential properties of matter in this chapter *except* for one major omission—the phenomenon of gravity! In Ch 10, we will look into the phenomenon of gravity.

☼ Carry-Over Points—Chapter 9 ☼

☼ A photon is a spin-1 particle with equivalent **negentropic—entropic phases.**

☼ **Negentropic—entropic dissociation** is the hallmark of matter formation. **Quarks**, in the form of **negentropic knots**, indicate the smallest unit of force cloaked up in spacetime. The unique properties of mass and motion of quarks are defined.

☼ Quarks are shown to disintegrate into gluons as they move, creating a **primary matter wave** (or **quark wave**). Quark waves, being curvilinear, generate **quark—gluon cycles** which move in anticlockwise direction. The length of the quark wave determines the mass of the quark.

☼ Normal matter is composed of **asymmetric baryons.** The isosceles triangular configuration of the quark—gluon cycle prevents black hole formation.

☼ Photons are defined as **magneto-electric units**, indicating their magnetic fields as negentropic phase and electric fields as entropic phase.

☼ **Charge** is defined—the direction towards or away from cosmic entropy is shown to represent negative and positive charges, respectively.

☼ **Hydrogen atom** is defined: an **electron cloud** around the nucleus is depicted as an **entropic blowout** in response to negentropy of quarks. **Free electrons** are formed when entropic blowouts acquire negentropic knots of their own.

☼ **W particles** are massive negentropic knots with fanning entropic phases.

☼ A **neutrino** is a **chimera particle** with dual properties. It has the smallest possible negentropic knot in nature with negentropic—entropic equivalence. Hence, it acquires photon-like properties though it has mass.

☼ An outline of the structure of a **complex atom** delineated. **Negentropic burden** explains the stability of the nucleus.

CHAPTER 10

Understanding Gravity
And the Play of Intermolecular Forces

Overview

The phenomenon of gravity has defied a proper explanation since its discovery by Sir Isaac Newton. Even though Einstein had showed us that gravity is not exactly a force but rather it is the result of spacetime curvature, the precise nature of gravity at the quantum level has remained an enigma for even today. In this chapter, we will undertake a quest for the origins of gravity first by putting forth an array of questions and statements which highlight the intricate relationship between light, matter, and gravity. A pointed discussion on the question of where exactly gravity originates in a group of molecules, as for example in a huge boulder of rock, reaps a wealth of theoretical insights for us to get to the roots of gravity as the reader will see in this chapter.

At first, we will conduct a discussion on intermolecular forces, which leads us to realize that they play a significant role in the origin of gravity. A diligent study of van der Waals forces actually allows us to conclude that they originate from within the atomic nuclei. Further discussion allows us to see how the fermionic negentropic model aptly explains the exact mechanism of generation of gravity in nature.

It is interesting to note that the above deep discussion on these issues of matter, gravity, and intermolecular forces fosters some unexpected dividends in the form of two additional bonanzas. One such reward is the comprehensive understanding of the mechanism of heat transfer in nature at the atomic and nuclear level, and the other is the accurate explanation of the phenomena of refraction of light in a medium as well as the phenomenon of bending of light in intense gravity. In the penultimate section, we will discuss the true meaning of mysterious gravitational waves. And finally, by the end of the chapter, not only does the exact meaning of quantum gravity become apparent to the reader, but the concept of gravitons is dispelled.

Now we will go amidst the molecules and look at their binding forces!

Section A
Initial Musings

In the photonic and fermionic negentropic models, we have seen that photons are generated from spacetime and force and that fermions are created by the action of excessive force on spacetime. And various fermions, in the form of negentropic knots, assemble to form atomic matter in the universe. Now consider this: apart from mass, motion, spin, charge, and other properties, this atomic matter has another fundamental property—the phenomenon of gravity. All matter attracts all other matter universally by gravity. In this chapter, we will undertake a study of gravity at its fundamental level. Before commencing our study, we will ask a few pertinent questions which will actually guide us in our investigation: What exactly is gravity? Where exactly is it originated in matter? Is gravity a fundamental force? If not, what sort force is gravity? And why is gravity always attractive? We will explore these issues here in this chapter categorically and attempt to answer the questions using the fermionic negentropic model. And the reader will see that as we wade through the chapter, we will also arrive at some very important ancillary conclusions which are even remotely connected to gravity!

We will first recapitulate the meaning of gravity as we have seen in general relativity in Ch 1-E. Gravity may be thought of as a distortion created upon the spacetime by an object with mass (Ch 1; Fig 1.6)—the more the mass, the more the distortion. Mass is a measure of the amount of matter in a substance. A fluff of cotton of a mass of 1 gram has the same amount of fundamental matter (in the form of subatomic particles) in it as that of a lead ball of 1 gram – and consequently, more matter distorts spacetime more acutely, thus causing the object to weigh more. But then here is a little confusing affair. We have also seen in Ch 8 and 9 that both light and matter are created from the distortion of spacetime by force, and here we are stating that matter generates gravity by distorting the spacetime once again. Thus, it means that light, matter, and gravity are all associated with force acting upon spacetime, and they seem to be somehow related – and we will now try to clearly delineate this relationship between them. And consequently, the following discussion would be the first step towards understanding gravity in fundamental terms!

Light, matter, and gravity are all spacetime
distortions and somehow related.

Having ascertained the relationship between matter and gravity, we will now explore the relationship between light and gravity. The following examples showcase this relationship. There are a few instances in nature wherein gravity is converted to light (i.e. electromagnetic energy). At first we will take a very common and accessible example of gravity being directly converted into EM energy. Take a boulder of rock falling from atop, hitting the ground, and generating heat and light (in the form of sparks of light). The gravitational potential energy of the boulder is first transformed into kinetic energy, which is ultimately converted not only into heat and light, but this boulder also does work in various other forms, such as breaking up the earth below, generating vibrations and sound. However, it must be noted that in this sort of conversion of energy, matter is conserved—matter is neither created nor destroyed (an issue which has a bearing in further discussion).

Now we will see one more example, and this is nothing but the instance of generation of electricity in our hydroelectric power stations. In the case of generation of hydroelectric power, the gravitational potential energy of water is first converted into kinetic energy, which is then utilized to produce electromagnetic energy (by the way of turbines) which lights up our homes, heats up our pizzas in ovens, cools our Coke in refrigerators, and does all the sundry works for us. Thus, in the case of hydroelectric power generation, we may say that it is the gravitational force of the earth (by the way of fall of water) that has transformed into EM energy. Here in this case too, matter is strictly conserved.

Earth's gravitational force may be converted into electromagnetic energy.

Now consider a third example: if we take the instance of the birth of a star, it is known that a protostar forms from the gravitational collapse of a nebula. It must be noted that this gravitational force was initially generated from the collapse of the small molecules of an interstellar nebula, and as the star grows up in size, this gravitational force builds up in strength and finally acts upon the molecules at the core of the star, resulting in increase in their kinetic energy which, in turn, initiates collisions and subsequent nuclear reactions (Ch 4-A).

Here are other examples. We have already seen a mechanism in Ch 9-G by which gravitational force is incorporated into a hydrogen atom to break it up into a free proton and a free electron (Fig 9.14), and we have also seen a mechanism by which gravitational force creates W^+ bosons (Fig 9.15). In other words, we may say that it is the gravitational energy that has fused hydrogen into helium which has ultimately led to the liberation of gamma rays—or to put in more simple words, here gravity is directly transformed into light!

In a star, gravity gets transformed into light.

Now we will look into the other ways by which gravity and light interact. We will give two examples here: *one*, a passing light ray bending under the influence of strong gravity, and two, more intriguingly, a light ray getting totally absorbed in the presence of

419

extreme gravity of a black hole. We will look into the bending of light in Sec-F, and the effects of black holes on light are studied in Ch 12.

Coming back to the original question of the origin of gravity, we will now put forth this array of questions: Where exactly has gravity originated in a group of molecules in the above instances? Where is the gravitational potential energy 'stored' in the molecules of, say, a large body of water overhead the waterfall, and how is this gravitational force transferred to the turbines and thence converted into electromagnetic energy? Or in the case of a boulder of falling stone, where is gravity located? Or consider this—where does gravity precisely begin to operate in the hydrogen molecules of a collapsing nebula of a protostar?

Quite obviously, a tremendous amount of force is somehow stored in the trillions and trillions of molecules in the body of water or a boulder of rock, which is ultimately manifested as gravity. But then we have already known of one scientific fact that these molecules are bound together by forces called *intermolecular forces*. What could be the relationship between these petty intermolecular forces and the mighty gravity we have seen in the above instances except that they both originate in a group of molecules? In order to study this possible association between these two seemingly dichotomous phenomena, we will have to first understand the various intermolecular forces in a proper perspective and then see if they could, in any way, initiate the generation of gravitational force.

Intermolecular Forces: Any material substance in nature—be it a massive planet, a heavy-duty steel ball, a body of water, or a cloud of gas—is held together by tiny intermolecular forces that operate between the constituent molecules.* If they are very strong, the substance forms a solid; if not strong enough, it becomes a liquid; and if weak, it forms a gas. The reader may note here that the molecules of a substance themselves are held by various strong *intra*molecular'

---

* A 'molecule' of a substance, in this context, is the smallest unit of that substance which retains most of the properties of the substance. For example, in case of water, its molecule is H_2O; in the case of a bar of iron, its molecule is a single atom of iron. Hence, in this discussion, we do not differentiate between atoms and molecules; rather, we use the term *molecules* for the smallest unit of any material object.

bonds, such as covalent or ionic bonds, and they determine the chemical properties of that substance. We will not study these chemical bonds here. In this chapter, we are interested only in the *inter*molecular' bonds that bind the molecules of a substance which are relatively weak in their strength. And these weak intermolecular bonds, in fact, determine the various physical characteristics of that substance, such as its density, melting point, boiling point, viscosity, surface tension, ductility, and malleability.

The mechanism of operation of intermolecular forces is fairly well-studied at present. The intermolecular interactions are chiefly attractive in nature resulting from the electrostatic attraction between positive and negative charges on the molecules of a substance. The strength of these coulombic attractions is obviously dependent on the strength of polarity of charges present over the molecules. If the molecules are more polar, the intermolecular forces are strong; if the polarity is weak, the forces are less strong. Thus, these intermolecular bonds determine the three states of matter—*solids* have the strongest intermolecular bonds, *liquids* have weaker bonds, and *gases* have the least powerful bonds. But then how do molecules show up their polarity?

Some of the molecules are highly charged owing to the electron deficiency or electron excess in their outer shells. These molecules are called *ionic*, and they represent strong intermolecular interactions. Our common salt (NaCl) is a typical example wherein the molecules have strong *ion–ion interactions* between them, and so our table salt remains solid at room temperatures. Some molecules in certain substances are bound by some still more powerful bonds (such as *covalent bonds* and *metallic bonds*), and these substances form even stronger solids—metal, rock, wood, etc.

However, molecules in most substances are electron sufficient, and so they remain neutral and non-ionic in nature. But then even these neutral molecules show some polarity because of the differential charge distribution as explained below. These molecules are called *dipolar*, and they are weaker than molecules with ionic bonds. A dipolar molecule has a charged pole with a slight electron excess (see below), making it *partially negative* ($\delta-$) at that pole, whereas the other pole is comparatively electron deficient, making it *partially positive* (δ+). The leading example for such a polar molecule is a water molecule. Water has *dipole–dipole interactions* between its molecules, and these are weaker bonds. Thus, water remains fluid at room temperatures. And when we dissolve salt (NaCl) in water (H_2O), the salt solution forms *ion–dipole interactions*, which are somewhat stronger than dipole–dipole interactions of water but are weaker than ion–ion interactions of salt. This is the reason pure water has a lower boiling point (see below) than salt water!

Lastly, in certain substances (see below for examples), the constituent molecules remain *non-polar*, and these molecules do not show any permanent polarity on them. However, these molecules are also bound together by some loose intermolecular bonds. These very weak bonds were first described by Johannes van der Waals in 1873 and hence are called the *van der Waals forces*. These weak forces are of our interest in this chapter, and we will make a brief study of them below.

Van der Waals Forces: Van der Waals forces are the fundamental forces of all molecules in the sense that *all* molecules attract *all* molecules by the way of van der Walls interactions, which means to say that all molecules, including ionic and polar molecules, demonstrate these forces apart from their respective charged interactions. Simply put, van der Waals forces are the ubiquitous forces exhibited by all matter.

Van der Waals forces are fundamental to all molecules.

But then in the case of ions and polar molecules, van der Waals forces are effectively masked away by their own powerful charged forces, whereas in the case of non-polar molecules, van der Waals forces become overtly manifest. Thus, in the case of ions and polar molecules, the physical properties, such as boiling point and melting point, are chiefly determined by the charged forces, whereas in the case of non-polar substances, van der Waals are the chief determinants. Many of our very familiar substances are non-polar in nature, such as sugars, oils, wax, paints, proteins, plastics—the list is endless. And in all these cases, the weak van der Waals play a role in their physical properties!

How do these non-polar molecules attract each other? What is the mechanism of these weak van der Waals? It has been theorized that the electron distribution in any molecule at any given time tends to be slightly shifted to one portion of the molecule for transient periods, thus making this side slightly more electronegative (i.e. partially negative, depicted as $\delta-$), and the other side of the molecule becomes temporarily electropositive (i.e. partially positive, $\delta+$). For a simple understanding, we will take the hydrogen atom. It may be assumed that the swirling electron of a hydrogen atom, at any given moment, is located on to one side, making that side transiently electronegative and at the same time rendering the opposite side slightly electropositive. This electronegativity in the case of hydrogen may be a very tiny force, but in the case of large molecules such as organic molecules (e.g. hydrocarbons and sugars), the magnitude of the collective electrostatic charge is also more owing to the presence of a large number of electrons in outer shells, which may shift to one side at a given moment. This transient lopsided placement of electrons induces polarity to some significant extent, thus making the molecule dipolar (hence they are called *induced dipoles*). And these charge-induced molecules, in turn, induce polarization in the adjacent non-polar molecules by displacing (i.e. repelling) electrons in them. It must be noted that these induced dipoles are not only transient but are also partial charges, and they are far too weaker than the other kinds of forces described above. These transient dipoles are also fluctuating because of the inherent instability in the position of electrons, and this further weakens the van der Walls forces. However, these forces are the chief reason for the solid or liquid states of non-polar substances, and this is also the reason why increased temperature renders these molecules to dissociate (because when heated, the outer electrons swirl faster and change positions faster).

Because the shift of electrons takes place more readily as their number increases, it may be said that *polarizability* of a molecule, in general, increases with the size of the molecule. We may consider the boiling points of noble gases to illustrate this point very clearly. For example, helium atom has only two electrons, and so it is very difficult to be polarized. Thus, it tends to exist in a gaseous state, and it has the lowest boiling point (4 K) among ideal gases. As the

423

atomic number of a noble gas ascends, the polarizability increases, thus increasing the boiling point—neon's boiling point is 27 K, argon 87 K, krypton 121 K, and so on (see below).

Now consider this: the amount of heat that is required by each substance to achieve dissociation of their molecules naturally depends on the strength of van der Waals—the stronger the intermolecular force of a substance, the more is the energy needed to break them, and consequently, the higher is its melting/boiling point. For example, if we cool a gas sufficiently, the van der Waals forces dominate over the intermolecular space, and the molecules become more cohesive—turning the gas into liquid (below its boiling point) or even into solid (below its melting point)! It is worth repeating here that, in general, as the size of the molecule increases, the strength of van der Waals increases, and so it needs more energy to break them, thus raising their boiling and melting points. For example, radon (which has the largest molecule in the class of noble gas) has the highest boiling point for a noble gas at 211 K (–62 °C) because the molecules are bound by stronger van der Waals. In the case of hydrogen gas, the smallest atom in nature, the van der Waals is so weak that it needs to be cooled down to about 21 K (–252 °C) to turn it into a liquid state, but it needs a further drop of temperature to about 14 K (–259°C) to keep it in a solid state.

The reader may see a paradox here: the intermolecular forces of helium are weaker still, so it has to be cooled to 4 K (–269 °C) to bring to its boiling point and nearly to 1 K (–272 °C) to get down to its melting point. This means that the boiling/melting point of helium is the lowest of all elements in nature—i.e. it is lesser than that of hydrogen. But then obviously, hydrogen being the smallest atom, we may expect it to have the lowest strength of van der Waals with lower boiling point. This obviously bespeaks of the fact that there are other factors which determine the physical properties of various substances in nature. As we have mentioned earlier, one of the most important determinants is the polarizability of a molecule—the more the polarizability, the more powerful are the van der Waals force between molecules. Now we can see that the helium atom is more symmetrical with two electrons swirling around the nucleus, and so it is less easily polarizable, whereas the hydrogen atom is a naturally lopsided atom with one electron on one side, so it is more readily polarizable. Hence, van der Waals forces of helium are the least powerful in nature, and thus it has the lowest possible boiling point of all elements.

However, as already pointed out, the reader must realize that the physical properties of any substance is a complex issue and is dependent on many other factors,* but we will not go into the details here, and the interested reader may look into various other sources to get further information on these properties.

With this theoretical background, we will now proceed to understand not only the true nature of van der Waals forces in reference to the negentropic model, but we will see, step by step, how these tiny forces initiate ripples of gravity at the fundamental level.

SECTION B
Van Der Waals Forces and The Nucleus

The first step in this discussion is to see how we may visualize van der Waals forces according to the negentropic model, and this analysis would sequentially lead us into the exact location where ripples of gravity take their origin in matter.

Van Der Waals as Negentropic Force: At first we will pass a flat statement that van der Waals forces represent negentropic force, and then we will see why! Consider this: it may be said theoretically that at the near absolute-zero temperatures, all substances in nature exist as solids, and their molecules are placed at their closest possible distance. Thus, it may be stated that at near-zero temperatures, the van der Waals are maximally powerful (because all the molecules are tightly bound together), and it may also be said that at this state, the EM energy between the molecules is practically absent. As thermal energy is supplied, the molecules drift apart from each other by increasing their intermolecular distance. Each substance has its own point where they dissociate themselves from solid to liquid state and from liquid to gaseous state. Thus, as the EM energy in

---

* The size, shape, and other properties such as the state of 'intramolecular' bonds also determine the boiling point, but we will not go into the details here.

the form of heat is supplied, the van der Waals slackens in strength, and the intermolecular distance grows. Now the reader may pause and recapitulate one fundamental feature here: We have seen in Ch 8-B that electromagnetic energy is described as an entropic (or repulsive) force that increases disorder in nature, and we can see here that, true to its definition, the EM force behaves like an entropic force by repelling the molecules and increasing the space between them. Moreover, in Ch 9-G, we have seen that the repulsive entropic phase manifests itself as an entropic blowout which spreads out and surrounds the matter particle. And we may say that this blowout is actually responsible for the molecular separation. In Sec-D, we will study the role of entropic blowouts in the mechanism of heat transfer, but for now, we will continue with the attractive van der Waals forces.

Thus, by the above analysis, we may say that van der Waals is a negentropic force because it is fundamentally an attractive force.*

Van der Waals is a negentropic force.

In summary, we may state that a substance, when heated, expands by increasing the intermolecular distance, which is achieved by an increase in electromagnetic energy content and by a concomitant decrease in the strength of van der Waals. Here in this section, we will continue with our discussion on the origin of van der Waals forces, and in Sec-D, we will deal with electromagnetic energy.

We will now examine van der Waals in a slightly different perspective, which actually shows that they take their origin from the nucleus. Consider this: in Sec-A, we have claimed that the tiny van der Waals forces are ubiquitous in nature, meaning that these attractive forces constitute the fundamental property of matter, and we have stated that *all* molecules attract *all* molecules by the way of van der Walls interactions.

---

* The reader may argue that at some places, the van der Waals may be repulsive in nature. For example, the van der Waals may become repulsive when 'like poles' of the molecules come in contact with 'like poles', but the fact remains that their overall effect is attractive.

All molecules attract all molecules by the way of van der Waals.

But then the reader may note that we have also stated in Ch 8-B that *all* matter attracts *all* matter by the way of *gravitational* force, indicating that gravity is also a fundamental property of matter. In this context, it may be noted that these two forces—gravity and van der Waals—are somehow related. However, it may be said that though the relationship between these two forces is apparent, it has largely been left unexplored in the scientific literature,[*] but now we will undertake a study of this with the assistance of the fermionic model and see how the tiny van der Waals fits into the big picture of gravity!

Gravity and van der Waals are somehow related.

We have already discussed the standard explanation that is put forth for the origin of van der Waals in Sec-A. The origin of this attractive intermolecular force is stated to be electrostatic in nature. The explanation that was given there is that the electrons temporarily shift to one side so that this side becomes electronegative, and this would render the other side of the molecule relatively electropositive. Below we will present an alternative explanation for the van der Waals.

We will first look at the origin of van der Waals from a different angle. Consider a huge boulder of rock such as a large asteroid in space. It has trillions of molecules in it which are all held together by the van der Waals forces (which are considered as electrostatic forces as discussed above). But then we know that this asteroid also has a certain magnitude of gravitational force (however weak that may be) which operates for over a certain practical distance (in fact, theoretically speaking, gravity acts on infinite distances). Now we will try to answer the following questions: What is the nature of this asteroid's gravity? Where does this gravity originate in this boulder of matter? How is this gravitational force related to the electrostatic forces (i.e. if we consider van der Waals as electrostatic)?

---

[*]   To the best of the author's knowledge, the relationship between van der Waals forces and gravitational force has only one prominent reference in the literature, and that is the study of Lei Zhang of Shandong University, China.

If we consider that these electrostatic charges cause gravity, then how can we theorize that electrostatic forces 'spread out' of the asteroid in the form of gravitational force to *attract* another object at a distance (say a smaller piece of rock)? The reader may see that it is highly improbable for us to presume that these electrostatic charges (whatever their strength may be at the molecular level) 'jump' and propagate out of the asteroid to give the asteroid its gravity (in fact, we have seen in Ch 8-B that electromagnetic charges are generally repulsive in nature, not attractive).

It is worth noting here that there has already been an alternative explanation for van der Waals forces wherein nuclei are considered to generate these forces. Fritz London had pointed out that at any given time, in a molecule, the centre of negative charge of electrons may not coincide with the centre of positive charge of the nucleus, thus making the molecule polar. The reader may see that this explanation views the positive charge as an independent entity—not simply as a 'relative electropositive' side of the induced electronegative charge, as we have seen above. So now we may state that the partial positive charge on a non-polar molecule may not be merely an electron-deficient electrostatic charge, but it is being 'actively' generated by the nucleus as an independent attractive force, albeit very weak in strength.

We will now consider another possible explanation in accordance with the negentropic model. Consider the following argument first: all matter in nature is known to attract all matter. In fact, more matter (i.e. more mass) causes more attraction. Now consider the fact that more than 99% of mass of an atom is concentrated in the nucleus, and consequently, we may expect that nuclei are the source of attraction in a given substance, the contribution of electrons being expectedly negligible. Because matter attracts matter, it can be said that nuclei attract nuclei, and thus, we may say that nucleus is the source of van der Waals. Of course, van der Waals being a very weak force, we have seen that the other intermolecular forces described above, such as covalent bonds, ionic bonds, and dipole interactions, would easily override these tiniest of the tiny forces operating between the atoms (or nuclei). But even so, it may be expected that the cumulative effect of millions and millions of nuclei in an object may play a significant role in the long range.

The nucleus is the source of van der Waals forces.

But then here is a caveat: how does a nucleus, which is known to generate strong force, generate the weakest of the forces such as van der Waals? In other words, how does strong force (the most powerful of the four forces) transform itself into van der Waals? We will explore this issue in a categorical manner, and in fact, this is the starting point for our theoretical investigation regarding the origin and nature of van der Waals and gravity.

How does the nucleus with strong force generate weak gravitational force?

Now consider this: the reader may remind himself/herself of one theoretical phenomenon that may become relevant to our understanding of the generation of van der Waals in the nucleus. In Ch 3-D, we have seen that a set of forces operates within the nucleus, which goes by the name of *residual strong force*. Residual strong force is a progressively weak force which is suggested to result from certain 'leakage' of strong force from the nucleons (protons/neutrons). This leaked strong force is thought to spread out of the nucleons for over some small distances, and this helps the nucleons to become 'sticky'. This stickiness, in turn, is thought to bind nucleons together in a nucleus. However, we will see in the upcoming discussion that this 'leakage hypothesis' is theoretically redundant, which need not be invoked at all to explain the stickiness of nucleons. We will now look into a hypothesis which explains the nuclear binding force basing on the fermionic negentropic model, and by the way, this also explains the exact location of van der Waals/gravitational forces inside the nucleus as the reader will see in the following discussion.

SECTION C
The Secondary Matter Waves

So far, we have seen that van der Waals forces represent negentropic forces and that they are generated in the nucleus of an atom/molecule. We will now explore a mechanism by which van der Waals forces take roots of their origin from inside the nucleus. And this discussion would automatically pave a way for understanding the phenomenon of generation of gravity in a bunch of molecules in an object. In short, below is a mechanism by which the nuclei generate van der Waals, which would ultimately transform into gravitational force.

To start the discussion, we will take hydrogen atom as our example. We know that for liquid hydrogen to become gas, it needs heat. In Sec-B, we have seen that electromagnetic energy (in the form of heat) causes a substance to expand, the chief player involved in this process being the entropic blowout which causes the molecules to drift apart, resulting in expansion. And we have also seen that at near-zero temperatures, hydrogen atoms are held closely together, and in such a state, the entropic blowouts of hydrogen atoms are virtually absent. And thus, we may state that at near-zero temperatures, van der Waals forces take over and bind hydrogen atoms together. This means to say that at near-zero temperatures, the intermolecular forces are chiefly represented by forces from within the nucleus. We will study the exact location of the origin of these weak negentropic attractive forces emanating from the nucleus.

We will continue with hydrogen as an example for our simple understanding. The nucleus of a hydrogen atom is a proton which consists of quarks moving in eternal quark–gluon cycles. These cycles, as we have seen in Ch 9-D and E, are never-ending cycles of deceleration, making the proton ever contracting towards its centre. These quark–gluon cycles are basically represented as primary matter waves or quark waves (Fig 9.7, Ch 9). This means to say that the strong force (represented by the way of primary matter waves in a quark–gluon cycle) always acts in the *negative direction*—i.e. pulling matter particles towards it. Thus, it may be said that this feature of the quark–gluon cycle makes the protons attractive, and it may be envisaged that these primary matter waves spread out in the negative direction (i.e. into the space surrounding the nucleus), and this phenomenon makes them progressively diminutive in their strength as they spread out. In other words, the contracting quark–gluon cycles can generate attractive negentropic cycles not only inside the nucleon but may also spread out of the nucleon in the negative direction. The quarks in a quark–gluon cycle may be compared to the blades of an exhaust fan which, when they rotate, suck in the air towards it from all around (Fig 10.1).

The reader may notice that by proposing these decelerating waves, we may actually obviate the necessity of 'leakage hypothesis' of residual strong force to make the nucleons (protons/neutrons) sticky!

All in all, this means that the basic negentropic nature of strong force will continue to show progressively diminutive attractive effects even outside its parent nucleon (theoretically speaking, for infinite distances). However, because the ripples of negentropic force are spreading in the negative direction, they become progressively less attractive (less sticky) as they pass away from the parent nucleon for infinite distances (Fig 10.1).

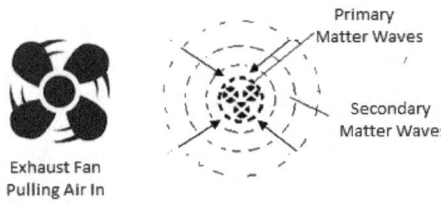

Fig 10.1: Secondary matter waves.

These spreading quark waves, as they travel out of the atoms and molecules, are actually represented as ripples in the empty space surrounding the atom/molecule. Because these ripples in the intermolecular space are secondary to the primary matter waves, we may call them *secondary matter waves* (Fig 10.1), and they represent the van der Waals. And thus, van der Waals forces are the secondary matter waves which travel in a negative direction to make the molecules attractive and sticky. We will look into the exact nature of these secondary waves below.

Secondary matter waves represent van der Waals forces.

Two Sets of Intermolecular Forces: From the above discussion, it becomes apparent that there are two sets of forces plying in the intermolecular space—one, the blowout waves, and the other, the secondary matter waves. We will now see how they interact with each other to determine the physical state of an object (to exist in solid, liquid, gaseous, or plasma state). We may say that in a state of near-zero temperatures, we have only the negentropic secondary matter waves around the molecules, making the substance solid. However, as heat is supplied, the entropic blowouts expand, and the blowout waves start to spread out into the surrounding spacetime. And we know that these blowout waves travel in a direction opposite to the direction of the secondary matter waves. The reader may realize that the entropic blowout waves being the derivatives of the entropic phase (Ch 9-G), they always tend to spread towards the spacetime continuum (thus attempting to merge with cosmic inertia), and the entropic blowouts always tend to cause more molecular disorder by increasing the intermolecular distance. In contrast, secondary matter waves, being negentropic in nature, always tend to spread in the direction away from the cosmic inertia, and thus, the secondary matter waves always tend to cause more molecular order by decreasing the intermolecular distance.

In simple words, we may say that entropic blowout waves tend to cause disorder by driving the molecules apart, and the negentropic van der Waals waves travel in the opposite direction, causing order by causing molecules to come near (Fig 10.2).

> *Blowout waves bring about disorder; secondary*
> *matter waves bring in order.*

Thus, the spacetime arena around a molecule in a substance is a complex conglomeration of waves travelling in both directions, but the net effect on a group of molecules obviously depends on the strength of each force at a given state of a given substance. If the net effect of secondary waves dominates, the substance exists in a solid state, and if the net effect of blowout waves dominates, the substance transits progressively from solid to liquid to gas and plasma states as discussed above.

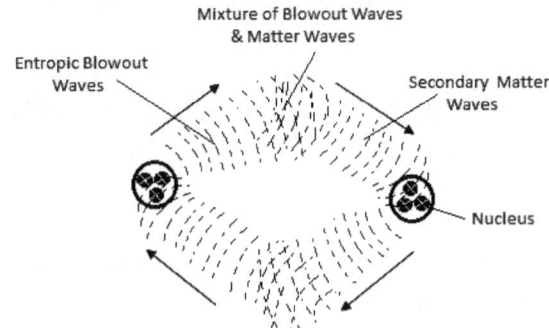

Fig 10.2: Entropic waves and secondary matter waves.

Secondary Waves and Blowout Waves as Spacetime Aberrations:
But then what exactly is the nature of these waves of entropy and negentropy? We have already discussed about the fundamental nature of the blowout waves in Ch 9-G. The blowout waves are actually envisioned as 'empty' aberrations in the surrounding spacetime of a molecule as the entropic blowout merges with cosmic inertia, and these blowout waves are devoid of negentropic and entropic phases as we see in the case of light. In the same way, secondary matter waves may also be considered as distortions of spacetime in the surroundings of a molecule which travel in a negative direction—i.e. they propagate towards the generating source and hence cause attraction of matter particles (Fig 10.2).

These empty spacetime waves signify the fact that these intermolecular aberrations have no negentropic phases in them so that there are no reactionary entropic phases, and hence, these blowout waves or secondary matter waves may actually be considered as 'after-effects' of the actual entropic and negentropic events that happen in a negentropic knot (in Ch 11, we will understand more about the actual nature of these events). However, the reader may realize that these spacetime ripples exist merely as virtual waves, which render them 'invisible' to our observation! Thus, these blowout waves and secondary waves put together are called *spacetime aberrations*, which are really virtual waves without actually being discernible to our senses or detectable to our daintiest experiments.

Blowout waves and secondary waves are virtual spacetime aberrations.

To sum up, we may say that these virtual 'forces' do not really carry any real force with them, but they show their effect only when they encounter any matter particles in their travel. These issues are further discussed in Sec-F.

Van der Waals as Scalar Quantity: The inference of the above discussion is obvious: if the size of the nucleus is more, more matter exists in the nucleus, resulting in more quark–gluon cycles, and consequently, more primary and secondary matter waves are generated. Thus, the more the size of the nucleus, the more is the strength of the attractive intermolecular forces (i.e. van der Waals forces). And consequently, it takes more amount of heat (i.e. EM force) to disrupt them, which means that such a substance has a higher boiling point (as seen in Sec-A in the case of noble gases where radon has the highest nuclear mass and hence has the highest boiling point). In the same manner, we may also state that as the number of molecules increases in a given substance, the number of atomic nuclei increases, thus increasing the total number of quark–gluon cycles, all of which finally increase the strength and extent of secondary matter waves. In short, we may say that the more the mass of an object, the more is the strength of the secondary waves and the more heat is needed to disrupt them.

The more the mass of an object, the more is
the strength of secondary waves.

And lastly, to get a complete picture of these secondary matter waves, it is very essential for the reader to remind himself/herself that all the molecules in a given substance are bathed in a sea of spacetime continuum in all directions in a 3D fashion, which means that the effects of entropic and negentropic forces on the spacetime are manifested in every possible direction around the molecule (see footnote in Ch 1-E). Thus, we may say that van der Waals represents a scalar unit.

With all this intricate knowledge of intermolecular forces in the background, we are now ready to examine the real nature of gravity. However, before going into the actual discussion of gravity, we will first touch upon one important phenomenon of nature that has bothered us in Ch 7-C, and this phenomenon is the question of how a photon could, without a magnitude of its own, mediate energy transfer in nature. This discussion of heat transfer actually becomes relevant here in connection with intermolecular forces as the reader shall see as he/she reads on.

Section D
Mechanism of Heat Transfer

We will first recapitulate the problem of heat transfer here briefly (the reader may get more details by going through the paradox of heat transfer from Ch 7-C). First consider this: we have seen in Ch 1-C that according to the special theory of relativity, all matter, when heated up, gain mass, which means that electromagnetic energy, when supplied to a sample of matter, gets incorporated into it and increases its mass. But then even though we know that EM energy carries energy from place to place to heat up the matter on which it impinges (this being the essential property of EM radiation), we could not account properly for the incorporated mass in the heated matter because all our experiments have conclusively shown us that photons are essentially massless. And so the question remains of wherefrom the heated matter gains mass! Or in other words, where is this mass lurking inside a photon?

In addition to the above paradox, the reader may also consider the following enigma. We have seen in Ch 8-F (Fig 8.5) that force exists only as long as an object travels at subluminal speeds, and once it attains the speed of light, force disappears, and the object is transformed into a ray of light. However, in Ch 7-C, we have seen that force is nothing but the rate of change of momentum ($F = p/t$), and this should mean that when force is zero, momentum is zero. And thus, a light ray should have no momentum! But then if there is no momentum of its own, how does a photon transmit energy to matter and increase its mass?

Before proceeding with an explanation, we will also consider these two intriguing phenomena that we have encountered in the preceding chapters. In Ch 9-C, we have seen that in the process of pair production, particles with mass are directly created from a ray of light in the form of subatomic matter particles. But then by what mechanism did a ray of light transformed into a particle with mass? Thus, this phenomenon of pair production also showcases the feature of mass appearing from 'nowhere'! In the same way, consider this: the sun generates huge amounts of light by disintegrating mass by the way of nuclear reactions, and the sun loses about 4 million tons of mass each second (Ch 4-D). But when light does not carry any mass of its own in its photons, then where is the mass of the sun disappearing into?

And also consider this: when the phenomenon of change itself is not a property of light (as discussed in Ch 7-C and Ch 9-D), how does electromagnetic radiation bring about a physical change in matter, such as molecular separation, upon heating? Thus, in summary, when a photon has no force, no momentum, and no mass wherefrom the heated matter gains mass, and how does a ray of light transmit energy to matter? Or putting it this way, we may ask simply, how does a beam of light perform its work in the universe?

Though we have offered a mathematical derivation in Ch 7-C (see the Compton effect) from which a photon could acquire its momentum, there is no clear theoretical or conceptual understanding on how a photon could acquire momentum and transmit energy (and mass) to matter. However, we will now collate all the information we have from the preceding two chapters and arrive at the exact mechanism by which electromagnetic radiation heats up a bar of metal and increases its mass.

Before commencing our discussion, we will recapitulate the relationship between electromagnetic (entropic) force and van der Waals (negentropic) force: as a substance is gradually heated, the entropic force (i.e. EM energy) dominates over the intermolecular space, and the negentropic force (i.e. the van der Waals forces) diminishes in strength, resulting in molecular dissociation.

In a heated object, entropy dominates, and negentropy decreases.

We will start by first understanding the various changes that occur in a substance when electromagnetic energy is supplied—i.e. when heated. To understand these changes at the fundamental level, we will examine the arrangement of molecules in a substance in its solid state at near-absolute-zero temperatures. For the sake of simplicity, we will take a block of hydrogen atoms at near-zero temperatures as our substance under examination. First consider this: it is already seen in Ch 9-G that a hydrogen atom is depicted as a structure containing three valence quarks bunched up together in an eternal quark–gluon cycle (constituting its nucleus, i.e. proton) with an entropic blowout representing its electron cloud. As this entropic blowout spreads out, it may be envisioned to merge with the surrounding spacetime continuum (Fig 10.3a). In fact, it must be realized that the essential purpose of an entropic blowout is to merge with cosmic entropy.

Now consider this: this spreading entropic blowout causes an aberration in the empty intermolecular space surrounding the atom, which may be represented as a spacetime aberration which is actually a spreading blowout wave. This blowout wave may be envisaged to spread out into the surrounding spacetime for infinite distances, of course dwindling out in its strength as it spreads out in distance so that its effect becomes barely perceptible after a certain distance (depending on the total mass of the negentropic knots in a nucleus which determines the magnitude of the resultant entropic blowout because entropic blowout is nothing but a response to negentropic burden). However, as already noted in Sec-C, a blowout wave must not be considered as a negentropic–entropic wave with a defined negentropic and entropic phases, such as in a light wave. It is only an 'empty' disturbance in the surrounding spacetime of an atom as the entropic blowout merges with cosmic inertia.

Blowout wave is an 'empty wave' without negentropic–entropic phases.

However, in the case of our hydrogen atom at near-zero temperatures, the entropic blowout may be expected to be closely situated around the nucleus without much of a spread, and this situation may be envisioned as a 'tightly wound watch spring' (Fig 10.3a). And consequently, the surrounding blowout wave may also be expected to be of the barest minimum magnitude possible. This bunch of hydrogen atoms with little or no blowout waves surrounding them would pack densely closest to each other, resulting in its solid state. Thus, we may conclude that at near-zero temperatures, the hydrogen atoms may be imagined to bunch up close to each other with minimum or no intervening spacetime continuum in between because of the negligible span of blowout waves (Fig 10.3a and b).

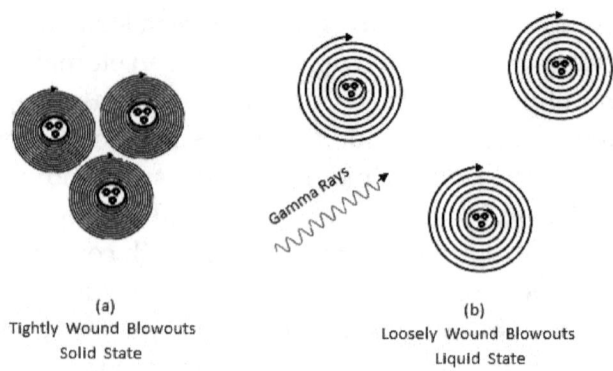

(a)
Tightly Wound Blowouts
Solid State

(b)
Loosely Wound Blowouts
Liquid State

Fig 10.3: Hydrogen in solid state and in liquid state.

However, when energy, in the form of photons, is supplied to a substance in solid state (in our example a block of hydrogen atoms), the following changes occur: the negentropic phase of a photon (if it has sufficient magnitude of negentropic force as in the case of any powerful ionizing radiation, such as gamma rays or X-rays, see Table 2.1) gets incorporated into the nucleus of hydrogen (perhaps in the form of a few 'non-valence' quarks). This incorporated negentropic phase of the incumbent photon curls up the spacetime inside the nucleus (i.e. proton), thus creating new non-valence negentropic knots (or quarks), and this increases the mass of the hydrogen atoms (without disturbing the atomic number of the nucleus). Thus, it is the incorporated (or 'absorbed') negentropic phase of the photon which contributes to the mass of a heated object. And needless to say, this increased magnitude of negentropic knot increases its negentropic burden, which, in turn, induces a greater entropic expanse, resulting in a more expansive entropic blowout. And this further distorts the surrounding spacetime in the form of blowout waves.

Supplied energy is incorporated into the nucleus as non-valence quarks.

This expansive entropic blowout and its blowout wave may now be envisioned as a watch spring which has wound out and expanded into the spacetime (Fig 10.3c and d). In effect, this increased distortion of the surrounding spacetime is reflected as increased intermolecular distance. This whole sequence of increased mass and consequent increase in entropy may be likened to a burning matchstick— the burning head representing the nucleus of an atom, while the surrounding flame representing the electron cloud (the cloud of flame increases as the mass of the matchstick head increases). This expanded entropic blowout and its wave represent the liquid state of hydrogen. Thus, in the liquid state, the intermolecular span of spacetime is larger than in solids.

As we keep injecting energy into the system, more negentropic knots join the nucleus, and the resultant blowout waves spread out in size, thus increasing the spacetime between the molecules and thereby increasing the intermolecular distance (Fig 10.4a). And of course, this represents the gaseous state. And finally, it may construed that as the temperature of the substance keeps increasing to some superlative degrees, the spacetime continuum straightens further and further, and the negentropic knots start getting separated, forming free quarks—i.e. in such a state, matter attains plasma state (Fig 10.4b). And if the temperature crosses beyond the plasma state, then the entropic knots will unfold back and disappear. In fact, in this infinitesimally superlative state, the spacetime takes the shape of regular, uniform cycles of a negentropic–entropic wave—i.e. the matter gets transformed into light (Fig 10.4c)! This is the way all matter—be it a bar of metal or a pot of water or a cloud of air molecules—gets heated up and expands when heated. And this is the way matter annihilates into rays of light!

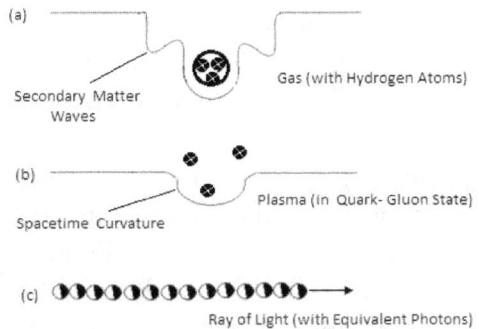

(a)

Secondary Matter Waves

Gas (with Hydrogen Atoms)

(b)

Spacetime Curvature

Plasma (in Quark- Gluon State)

(c)

Ray of Light (with Equivalent Photons)

Fig 10.4: Hydrogen in gas and plasma states and turning into a light ray.

It may be noted that as the frequency of the incident ray increases, the magnitude of negentropic force of the impinging photon increases, which enhances the chance of photons getting incorporated into the nucleus as negentropic knots. This is the reason why a gamma ray is more energetic and ionizing, and so it can heat up a substance more readily. And for the same reason, a ray of radio wave is less powerful and least ionizing, and this is also the reason why a radio wave is more penetrating and passes through objects more readily than gamma rays (Ch 2-B).

The reader may note that the above explanation of heat transfer also efficiently explains the phenomenon of photoelectric effect. We will discuss this explanation in *Question No. 7* in the *Epilogue* at the end of the book.

In this section, we have utilized the negentropic model to explain the mechanism of heat transfer and annihilation of matter into light. In Sec-E, we will continue with the mechanism of generation of gravity in a group of molecules.

Section E
The Birth of Gravity

Now we will examine the original question posed at the beginning: where exactly does gravity originate in a group of molecules? To begin with, we take an example of a situation where gravity plays a chief role. We have already known that a star is born by the gravitational collapse of the interstellar gas cloud called nebula. We will now closely examine the initial events of a collapsing nebula in the process of the formation of a protostar, which highlights the generation of gravity at the fundamental level in a group of molecules.

A nebula consists of gas molecules, the composition of which varies from place to place. In the interstellar location, it consists chiefly of molecular hydrogen as H_2 gas, helium, and other trace elements, such as carbon, silicon, oxygen, and some other rare elements derived from stellar 'dust'. On the other hand, in the *intergalactic* voids, the nebulae are composed chiefly of hydrogen in atomic form or even in ionic form (Ch 4-B). A star is usually born by the collapse of an interstellar nebula, and it is generally thought that the 'dust' particles within it trigger the gravitational collapse among gas molecules. We will concentrate on the very initial stages of this gravitational collapse and see where exactly gravity originates.

We will first hypothetically suppose that there is a vast group of molecules consisting of only hydrogen atoms suspended in the empty void somewhere in the cosmos. It is a monotonous spread of hydrogen atoms with a certain intermolecular distance, and it may be envisaged that this intermolecular distance is maintained in equilibrium by a delicate balance of the intervening blowout waves and secondary matter waves. Provided there is no input of any electromagnetic energy into this *atomic gas cloud*, we may consider that it remains stable for indefinite periods of time. However, as time passes, the ambient temperature may decrease gradually (perhaps because of the expanding space), which allows the entropic blowouts to slacken in their strength. This in turn allows the secondary matter waves to overpower blowout waves, and this drives the atoms in the cloud together to form diatomic hydrogen molecules (H_2). This process of atomic binding releases energy out into the ambience, which increases the temperature of the *molecular cloud*, and this leaves the molecular cloud once again in a stable state of equilibrium. Here the reader may note that the intermolecular forces (both blowout waves and secondary waves) in such a molecular gas cloud may be presumed to spread out equally in all directions around the molecules without any particular direction.

Intermolecular forces in a gas cloud in equilibrium have no direction.

Origin of Gravity: This stable state of molecular cloud may continue to exist until the monotony of hydrogen molecules is somehow disturbed. Now imagine that there exists a different element somewhere in the molecular cloud (in the interstellar space), such as a molecule of silica or carbon or copper (common ingredients of interstellar dust). This 'foreign' particle would certainly cause a disturbance in the eternal equilibrium because of the discrepancy in the nuclear sizes of the molecules in the molecular cloud. In other words, this large foreign molecule has a larger nucleus with more number of negentropic knots, and so it has a greater strength of secondary matter waves in its vicinity. This disturbs the monotony of the molecular cloud, initiating a gradual collapse of the entire molecular cloud from all directions *towards* this foreign molecule (Fig 10.5).

We will analyze the nature of this collapse. Hitherto, the intermolecular forces surrounding the molecules in the molecular cloud which suspends the molecules in a state of equilibrium have no particular direction; these intermolecular forces behaved like scalar units. However, as the molecular cloud collapses, the intermolecular forces travel towards a particular foreign molecule, and this sets in a direction to them. In other words, we may say that the intermolecular forces now behave like vector units. And several of these intermolecular forces travel towards the foreign molecule, thus creating a *core* at the centre of the molecular cloud.

Putting it otherwise, we may say that the molecular cloud is now undergoing a *gravitational* collapse. Or in simple terms, we may say that *gravity* is driving all the molecules of the cloud towards its centre. The reader may notice that we are now referring to the same van der Waals as 'gravitational force' when these forces are directed towards the centre of the nebular cloud! Thus, we may conclude that the hitherto scalar units of several intermolecular forces have transformed into vector units of gravitational force.

Van der Waals is scalar; gravity is vector.

We will call these miniature gravitational waves which exist *within* the substance of a given object as *proto-gravitational waves* (Fig 10.5) or simply *proto-waves*. And all these proto-waves travel towards the centre of the object in question. Of course, here we may point out that this sort of molecular attraction happens in any object in nature, be it a massive star or a huge planet or a boulder of asteroid or a small piece of rock. All objects have their centre of gravity at which these proto-waves concentrate their strength! Obviously, the strength of these proto-waves is dependent on the mass of that object—the more the mass, the more is the strength of secondary matter waves, and the more is the strength of the proto-waves.

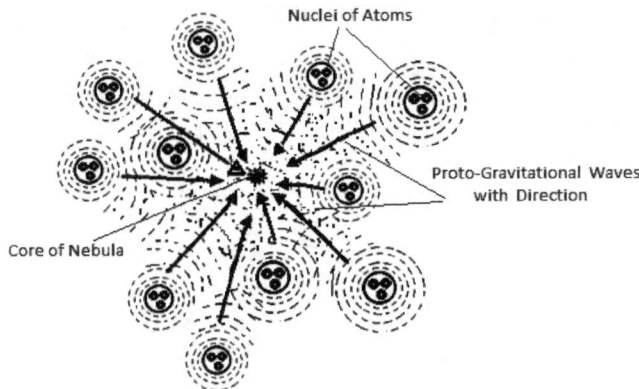

Fig 10.5: Origin of gravity in a nebular gas cloud.

We will now draw a complete picture of gravity. All these primary waves, secondary waves, and proto-gravitational waves are all negative waves, and thus, they spread out of their locales of origin in a diminutive manner. And this means that the proto-waves spread out of the object and manifest themselves as the attractive gravitational force. In simple words, we may say that proto-waves are attractive forces within an object, and gravitational force is the attractive force manifested on the outside of the object (Fig 10.6).

The inside proto-waves are manifested as outside gravitational force.

Of course, we must remind ourselves of the fact that all these are empty spacetime aberrations, signifying that they are virtual waves which have no negentropic–entropic phases in them. Thus, the proto-waves and gravitational waves are virtual waves.

Now imagine the surface of a large object (like a planet) containing trillions and trillions of molecules. The generated proto-waves from inside the object may be envisioned to spread out from the surface of that object as its gravity, and this gravity propagates into the surrounding space for infinite distances (Fig 10.6). These waves which emerge from the surface of the object not only are negative and negentropic waves (i.e. attractive) but they are also progressively diminutive. When these negentropic waves encounter a smaller object in its surrounding space (say, an asteroid in the vicinity of a planet), then these negentropic waves become warped.

This spacetime warpage surrounding the smaller object (asteroid) alters its pathway by dragging it towards the centre of the larger object (planet). This gives us the impression of a gravitational pull (Fig 10.6). This is the real meaning of gravity.

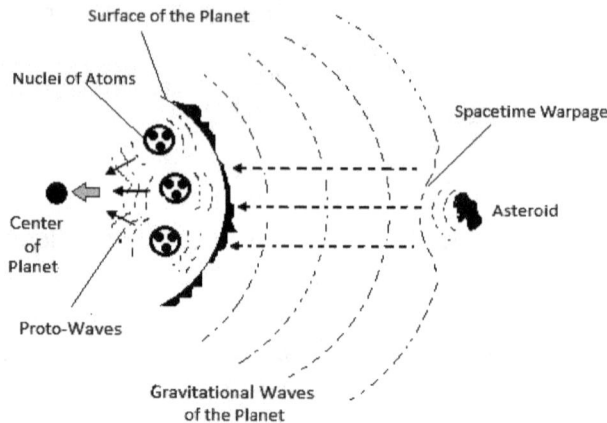

Fig 10.6: Primary matter waves à proto-waves à gravitational waves.

Thus, in conclusion, it may be said that gravitational force represents the collective force of several intermolecular forces projected in the general direction of the centre of an object (to be precise, it is in the general direction of the centre of a group of constituent molecules of an object). In other words, gravitational waves are the representatives of secondary matter waves extending in a general direction towards the centre of the object concerned. And of course, the proto-waves being negative waves, the effects of gravity are always attractive, and they are noticed to be spreading out infinitely into the surrounding spacetime of the object— their strength increasing as the mass of the object increases and diminishing in their strength as the distance increases. And needless to say, these diminutive negentropic waves originating in the matter of any object dutifully obey Newton's law of gravity, bearing the formula, $F = G\,(m_1 - m_2)/r^2$!

We may sum up the negentropic waves and say that there are four types of negative waves (Fig 10.6): (1) primary matter waves (quark waves) which are generated within the nucleus, (2) secondary matter waves which are generated in the vicinity of a molecule,

(3) proto-gravitational waves which are collective secondary negentropic waves that occur within the object which travel towards the centre of that object, and (4) gravitational waves which spread out of the object for infinite distances with diminishing strength, attracting all the objects in the vicinity.

Quark waves à Secondary matter waves à Proto-waves à Gravitational waves

We will now dig further into the manifestations of gravity, and the following discussion would highlight several ambiguous issues in new light.

Gravitational Waves, Gravitons, and Quantum Gravity: Having studied gravity in this detailed way, we will now try to understand the phenomena of gravitational waves and gravitons in clear terms. Ever since general relativity was proposed, the phenomenon of gravitational waves has come into the limelight. And even though their existence was established by LIGO in 2015 (Ch 1-F), there is no clear theoretical understanding of the relationship between the phenomena of gravity and gravitational waves. In the following discussion, we will explore gravitational waves and see how they are related to gravity.

We will first recapitulate the concept of gravitational waves from Ch 1-F: gravitational waves are ripples in the fabric of spacetime created as an object accelerates in space, and they are thought to travel by the speed of light. These spacetime aberrations are extremely weak disturbances, and thus they are hard to detect. And because of this extremely weak strength, they become detectable only when they are created by some immensely violent cosmic events in the universe, such as collision of some supermassive black holes or huge supernova explosions. Elusive though they are, gravitational waves are finally detected by LIGO from deep space in 2015. Now we will analyze these waves in the light of our new understanding of gravity and see how gravitational force may be equated to gravitational waves.

First consider this: gravitational waves are generated when any object (be it a minuscule matter particle or a large planet or star) travels in space in non-uniform motion. Speaking the other way, we may say that when an object travels by uniform motion, it cannot generate gravitational waves simply because it then assumes the status of a ray of light, which has negentropic–entropic equivalence and hence can no longer generate any ripples in spacetime. But then we may say that all matter particles create negative waves in their vicinity as soon as they are born simply because we have seen that all negentropic knots (quarks), as soon as they are born, generate primary matter waves (quark waves), and these quark waves transform into secondary matter waves which propagate into the surrounding space (in a negative direction). In short, this clearly shows that all objects in nature generate secondary matter waves (or gravitational waves) as they travel in space. Of course, smaller matter particles generate negative waves of negligible strength, which may carry no practical significance in the events of nature!

All matter generates gravitational waves as they travel in space.

Thus, in accordance with the fermionic negentropic model, gravitational waves may be envisioned as the 'after-effects' of matter formation. They are merely ripples in spacetime created first as primary matter waves in the negentropic knots, which then give rise to secondary matter waves which, in turn, give rise to proto-waves and gravitational waves. As we have seen that gravitational waves are spacetime aberrations which travel in a negative direction, this causes two bodies to attract each other. This must prompt us to think that these gravitational waves correctly depict gravity, and thus, we may say that gravitational waves and gravitational force are merely two terms of the same phenomenon.

Gravitational force and gravitational waves are one and the same.

Finally, having concluded that gravitational force and gravitational waves are two sides of the same coin, we may state that gravitational force (as we colloquially call gravity) is a measurable force which we measure and feel between two bodies (such as the earth and an apple); gravitational waves are exceedingly faint ripples of spacetime that are too weak for us to detect (and thus they became detectable only when emitted by powerful sources, such as celestial collisions, as seen above). In short, when nearer, we call it gravity; when afar, we call them gravitational waves!

Gravitational force is a stronger short-range manifestation.
Gravitational waves are weaker long-range manifestations.

Before concluding our discussion on gravity, there are two important related issues that need to be threshed out—the speed of travel of gravitational waves and the existence of gravitons. The equations of general relativity predicted that the gravitational waves travel in space at the velocity of the speed of light. We have seen that secondary matter waves are empty spacetime aberrations without any negentropic–entropic phases in them, and we will see in Ch 11 that it is the innate property of spacetime itself to set the maximum limit to the speed of light. So we are justified in assuming that these empty spacetime aberrations also travel at the maximum speed allowed by nature, i.e. the speed of light.

Gravitational waves travel at the speed of light.

Now consider this: because these virtual waves have no negentropic phases, they do not have entropic phases either. Thus, a unit of gravitational wave would never assume the status of an ideal universal cycle (i.e. light), and consequently, it is meaningless to suppose that it would attain the status of a boson (Ch 9-B; Fig 9.3). And as a result, they may never assume the status of a force particle! Hence, the concept of gravitons is superfluous and redundant, and we may state further that because of this reason, the phenomenon of gravity may never be incorporated into the realm of quantum theory, and it is a futile attempt to do so!

Gravity has virtual waves; gravitons do not exist.

A Note on Quantum Gravity: Ever since the inception of quantum theory, and ever since the need to merge quantum theory with general relativity arose, scientists have been attempting to explain the meaning of the concept 'quantum gravity' because quantum mechanics necessitated the existence of gravity in a packet of force, i.e. graviton! However, we have already seen that gravitational force does not exist in packets (or quanta), and so it is pointless to investigate for the existence of gravitons. By the same token, it may be said that it's also a futile attempt to consider the phenomenon of quantum gravity because of the very fact that gravity itself originates at the quantum level and spreads out of the matter in the form of spacetime aberrations, as seen in this chapter. But if the reader insists, then he/she may reckon the primary matter waves themselves as the representatives of quantum gravity, but then it is pointless to attempt to quantize empty spacetime aberrations (which are virtual waves)! We will discuss the significance of these minuscule negative forces at the time of the big bang in Ch 12.

Primary matter waves may be considered to represent quantum gravity.

Having understood gravity in this way, we will now look into the various theoretical implications of gravitational force in nature. It is interesting to note that we have some unexpected dividends out of the above discussion on matter and gravity, as we will see in the following section.

Section F
Implications of Secondary Matter Waves

In this section, we will embark on the study of the effects of secondary matter waves (or gravitational waves) in various situations of nature. First consider this: gravity is known to affect the events of nature in many ways. Apart from acting on matter by the way of causing attraction, gravity is also known to act on a ray of light to cause bending of its course. Not only this, but gravity also interacts with light peculiarly in certain situations, such as in the case of a black hole wherein gravity causes an irretrievable collapse of light into its depths. We will now study the exact mechanism by which gravitational waves affect light by using the fermionic negentropic model and see how the secondary matter waves discussed above play a pivotal role in these situations.

To understand the interactions between gravity, light, and matter at the fundamental level, we have to once again look into the intermolecular forces and see how they become relevant in our discussion. In this context, we will take up an interesting issue of refraction of a light ray in a medium, and this would serve us as a starting point for further discussion.

Refraction and Spacetime Aberrations: We will take up the common phenomena of refraction of light for our scrutiny. Consider this: when a light ray falls on matter, any of the three things may happen—it may get absorbed, reflected, or refracted. Each of these events is dependent on two factors—the energy state of the molecules in the substance at a given time and the energy of the impinging light ray.

Here we will only highlight the theoretical relationship between absorption, reflection, and refraction of light. When atoms have low-energy electrons in their outer shells, the impinging light excites these electrons to a higher level, and thus, the energy is absorbed (we have already studied the absorption of light by matter in Sec-D and understood the mechanism by which electromagnetic energy causes expansion of matter). When the atoms have high-energy electrons in their outer shells, the photons cannot get absorbed (because the electrons have no higher place to shift), and the photons are either transmitted as in the case of refraction or the photons may be reflected back. The example of graphite and diamond illustrates the point very well. Atoms of carbon in *graphite* have many electrons hovering freely in their conduction bands (Ch 2-E), which can absorb energy readily, and so they become easily excitable by photons of almost all wavelengths. Thus, all visible light impinged on graphite is absorbed, and so it appears black (in fact, this free electron movement makes graphite a good conductor of heat and electricity). On the other hand, atoms of carbon in a *diamond* have their electrons chiefly in the valence bands (Ch 2-E) and need a lot of energy to be lifted into the conduction band, and thus, they are not so easily excitable. For this reason, light impinged on diamonds is either reflected or refracted (depending on the angle of incidence), giving them their resplendent lustre! Anyway, if high-energy light (such as ultraviolet light) is shone on to a diamond, it gets absorbed to a large extent. Hence, should we look at a diamond in UV light, it would appear jet black (much to the consternation of the prospective buyer!).

We will not study the phenomenon of reflection any more here. Rather, we will concentrate on the mechanism of refraction in this section, which actually becomes relevant for our further discussion. And now we will understand refraction using our fermionic negentropic model and the secondary matter waves.

We have seen in Sec-C that the intermolecular space is an arena for both blowout waves and secondary waves as they spread in opposite directions as spacetime ripples (Fig 10.4). The net effect of these spacetime ripples, as already observed, is dependent on the relative strength of each of these waves—at near-zero temperatures, the secondary waves dominate, giving solid state of matter; and at the other extreme, the entropic blowout waves dominate, resulting in plasma state. For simplicity's sake, we will collectively name these ripples of blowout and secondary waves as spacetime aberrations.

Now imagine a glass pane with vacuum on both sides. When high-energy EM rays (e.g. UV rays or X-rays) impinge over the glass molecules, the negentropic phases of these high-energy photons are incorporated into the nuclei of the silicon atoms, and this absorption of light by the nuclei augments the blowout waves, thus increasing the spacetime aberrations in the intermolecular space, as already discussed in Sec-D. However, when a light ray of greater wavelength (lesser energy) impinges upon its surface, the incident light ray gets bent—i.e. it undergoes refraction. Now we will study the mechanism of refraction.

The course of light becomes bent when it encounters the molecules of glass because the space around the molecules itself is bent. In other words, the molecules of glass are surrounded by spacetime aberrations (Fig 10.7) caused by the complex interaction between blowout waves and secondary matter waves. The path of the incident light ray in this situation tends to take the path of least resistance in the ambience of glass molecules, but then the path of least resistance for the light ray to take is nothing but the geodesic (Ch 1-E) of the spacetime aberrations. In fact, this geodesic is the straightest possible path for the light ray to take, and thus, the light ray follows this course. And the light ray passes along the medium of glass from one geodesic of a glass molecule to the other—on and on—until it reaches the other surface of the glass (Fig 10.7).

Light ray passing from one geodesic to the other causes refraction.

Having discussed refraction this way, we will see what happens to the speed of light in the case of refraction. Of course, the speed of light remains the same all along its passage in vacuum and inside the glass, but it has to take a longer course in a curved path (i.e. along the geodesics of spacetime aberrations), so it appears to slow down. And this explains the phenomenon of light losing its speed while travelling in a medium. Needless to say that as the light passes out of glass, it once again encounters vacuum, and it travels straight and so regains its absolute velocity!

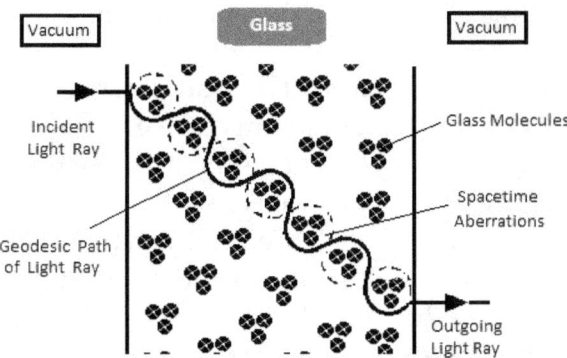

Fig 10.7: Refraction of light.

This is the reason why a light ray is refracted when it passes from one medium to another medium. The angle of this bending of light, of course, is dependent on the degree of curvature of the spacetime aberration (Fig 10.7). And obviously, the degree of curvature of spacetime aberrations is dependent on the relative strength of blowout waves and secondary waves as discussed above, which, in turn, is dependent on so many factors, such as the size of the atomic nuclei, the chemistry of the medium's molecules, the temperature of the medium, and its density.

This explains a common phenomenon of increased temperature of the medium decreasing the refractive index (which implies that the light within a medium travels at greater speeds as it gets heated). This happens because when heat is supplied, the blowout waves of the molecules of the medium take over the spacetime aberrations, which, in turn, cause their expansion. Now the light ray can take a relatively straighter path through the stretched out geodesics of spacetime aberrations!

Of course, the other most important determinant of refraction is the wavelength of incumbent light itself. If the wavelength of the impinged light ray is short (as in blue light), the light tends to take the geodesic of the smaller inner waves (i.e. waves surrounding the immediate vicinity of the glass molecules), and so it is deflected more. Thus, blue light has a narrower angle of refraction. In the case of red light (with longer wavelengths), it tends to take the geodesic of larger outer waves, and thus, it is deflected with a wider angle. Hence, red light has a wider angle of refraction. This is the reason why a prism refracts light in the dazzling colours of VIBGYOR.

Gravity and Bending of Light: The reader may now see that the phenomenon of bending of light in massive gravitational fields can be explained by invoking the same analogy as we have seen in the case of refraction. Fig 10.6 shows us that all objects generate secondary matter waves, which are finally manifested outside the object as gravitational waves. In other words, gravitational waves are spacetime aberrations that spread out from the surface of all objects— the greater the mass, the greater is the strength of aberrations, which means that more matter generates more gravitational force. And thus, it may be presumed that a very large object (such as the sun) generates much stronger spacetime aberrations in the immediate vicinity of that object. As a light ray passes in the vicinity of a large astronomical object with great mass, the light takes the course along the geodesic of the gravitational wave (because here the spacetime itself is curved). Of course, the nearer the ray of light passes to the object, the more is its bending (because of greater spacetime curvature near the surface).

Gravitational Red Shift: What is gravitational red shift? It means simply that a photon, as it passes along its course in strong gravity, increases its wavelength, which means that a light ray shifts to the red side of the spectrum in the vicinity of gravity. Gravitational red shift not only signifies a decreased frequency of light, but it may also be interpreted as a slowing of time. This phenomenon is more a reality than myth. As we have already seen in general relativity (Ch 1-F), this slowing of time by the effect of gravity has an important bearing on the precision of our GPS functioning.

The reader may see that gravitational red shift may be explained invoking the spacetime aberrations. If the gravity is sufficiently strong, then the spacetime curvature of the gravitational waves will be so acute that it may not only bend the light but drag upon the negentropic phase of the photons, causing a widening of its wavelength (and decrease of frequency) and leading to a red shift.

Equipped with this knowledge of the structure of electromagnetic wave and photon, structure of matter, and the nature of gravity, we are now ready to examine the universe at a large scale and understand its beginnings, its progression, and perhaps its fate in a new perspective. We will take up these issues in Ch 12, but before going into these cosmic affairs, we will take upon an important concern which may be lingering long in the minds of our readers— the very fundamental meaning of spacetime and force! And this is the subject of our discussion in Ch 11.

☼ *Carry-Over Points—Chapter 10* ☼

☼ *Van der Waals forces* are chiefly attractive. *All molecules attract all molecules by the way of van der Waals forces. Van der Waals forces and gravity are related.*

☼ *Primary matter waves* (*quark waves*) *spread out beyond the atoms into the intermolecular space* (*in a negative direction, making them attractive*) *to become* **secondary matter waves**. *Van der Waals forces represent secondary matter waves.*

☼ *The* **intermolecular space** *is a complex arena of both* **blowout waves** *(which are entropic) and* **secondary matter waves** *(which are negentropic)—when blowout waves dominate the intermolecular space, the substance expands; when secondary matter waves dominate, the substance contracts. Both the blowout and secondary waves are empty spacetime aberrations, and so they are* **virtual waves with no negentropic—entropic phases.**

☼ *The exact mechanism of* **heat transfer** *between objects is clearly described. At near-absolute-zero temperatures, van der Waals dominate the intermolecular space—as the temperature of the object rises, entropic blowout waves take over, thereby increasing the intermolecular distance and causing the object's expansion. Heat energy supplied to the object is incorporated into the atomic nuclei in the form of* **non-valence quarks,** *which* **increases the mass** *of that object.*

☼ **Gravity** *is an after-effect of primary matter waves. Secondary matter waves of molecules of an object extend outside the surface of an object as* **proto-gravitational waves,** *which extend out as* **gravitational waves.** *Gravitational waves and gravity are one and the same.* **Gravitons** *do not exist because gravitational waves are virtual waves representing empty spacetime aberrations.*

☼ **Quantum gravity** *is represented by primary matter waves.*

☼ **Refraction** *and* **gravitational bending of light** *are explained in clear terms.*

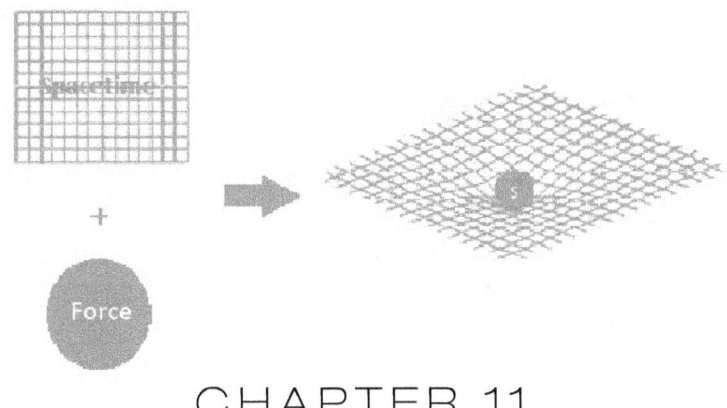

CHAPTER 11

The Two Primordial Entities
And the Fundamentals of Fundamentals

Overview

All along the preceding chapters, we have seen that spacetime and force played a pivotal role in the generation of light, matter, and gravity. But then we have no clue as to what exactly constitutes these two entities. In this chapter, we will study these fundamental entities and try to understand the real meaning of spacetime and force.

At first we will start with the characteristic features of spacetime and force, and this discussion would lead us to certain universal truths of nature. In the subsequent sections, we will see how the unique features of spacetime and force explain the most intriguing phenomena of the quantum world, such as wave–particle duality and matter waves. Surprisingly, we will see that these quantum explanations also throw some light on one of the most diabolical phenomenon of science, the observer's effect. In the next section, we will see how the cardinal principles of quantum theory, such as Heisenberg's uncertainty principle, sum-over-pathways, and quantum entanglement, are explained by using the negentropic model. Later on, the problem of vacuum energy in empty space is taken up and studied in a new light.

In the later section, the real meaning of energy, work, motion, mass, inertia, and entropy is examined, and this analysis coincidentally explains the universal phenomenon of thermal radiation emitted by all matter at all times! In the final section, we will see what sets the fundamental constants in nature and what determines the various sizes of atoms and subatomic matter particles as we see in the Standard Model.

Now starts our journey to explore the seemingly unexplorable spacetime and force!

SECTION A
Defining Force and Spacetime

Having explored light, matter, and gravity in detail in the preceding chapters, we may conclude that they are all merely different manifestations of force acting upon spacetime. Force, when acting equivalently on spacetime, results in light; when excessive force acts, it results in matter. And the after-effects of matter on spacetime result in gravitational waves. However, though we have been harping on these two entities all the time in the previous chapters, we have absolutely no idea as to the real meaning of force and spacetime. Here in this chapter, we will try to decipher the fundamental nature of force and spacetime themselves as diligently as possible, and the reader may see that this understanding will lead us to comprehend some of the most difficult concepts in theoretical physics at their fundamental level.

Spacetime and Force—The Two Primordial Entities: We will now critically examine the phenomena of force and spacetime and arrive at some important generalizations. If we analyze the discussion in the preceding three chapters, we may say that all the phenomena of generation of light, creation of matter, and the causation of all events of the universe, like gravity, motion, work, charge, entropy, inertia, and all others, can be simply traced back to only two most fundamental entities—spacetime (which may also be referred to as spacetime continuum, Ch 1-E) and force. In other words, every object in the universe or every event in nature may be understood as force acting upon spacetime, which causes it to fold or unfold in various ways. In the case of light, the spacetime folds with mathematical regularity, resulting in negentropic–entropic equivalence; in the case of matter, the folds are in irregular units, separating space from time; in gravity, the spacetime folds in empty aberrations that are virtual, which tend to spread for infinite distances. The tendency of this curved spacetime to revert back to its absolutely straight state is manifested as entropy/inertia. All in all, we may say that force and spacetime are the most basic constituents of nature. And hence, we may call spacetime and force as the *primordial entities* of the universe.

Spacetime and force are the primordial entities.

Thus, it may be said that the entire universe and all the continuous array of events that happen in it from the time of the big bang to the present day are caused merely by a continuous array of interactions between these two primordial entities operating in the background. In other words, all the events of the universe involving light, matter, motion, gravity, energy, work, entropy, etc. are orchestrated by these two primordial entities. But then what exactly is this primordial spacetime, and what exactly is primordial force? What is their fundamental nature? How can we define them in scientific terms?

We have already seen in Ch 8 that 4D setting is a representative of spacetime, and force converts 4D into a 3D setting. Thus, in order to understand spacetime and force individually, we will first recapitulate the essential differences between 4D and 3D settings. In Ch 8-D and C, we have stated that the 'relativistic' state of the 3D universe takes its origin from an 'absolute' 4D continuum by the introduction of an agent called force. And this 3D state of the universe is continuously striving to return to the original 4D status by eliminating the force imparted upon it but without ever achieving it. We went further on to describe the 4D spacetime as an infinitely spreading 'universal scaffold' which transforms, under the influence of force, into a localized '3D space-and-time grid', and this space-and-time grid always strives to eliminate force and merge again with 4D continuum.

Now consider for the time being that there is no force at all acting on the spacetime. The spacetime in this state is absolutely straight and flat without any folds, and so it is quite characterless. In such a case, there is no way by which we may be able to know the existence of spacetime, and there would be no experiment to detect this characterless spacetime. In other words, the nature of spacetime is known only when its course is 'folded' (or 'altered' in any way) by force. Thus, we may say that imagining spacetime without force becomes 'unintuitive' to the human mind. However, by the same token, consider this argument: how can we know the existence of force? It may be said that only an alteration in the disposition of spacetime bespeaks of the existence of force. Thus, we may conclude that imagining force without spacetime is 'intangible' to the human senses as well. Hence, we may state that both of these primordial entities exist only in relation to each other and cease to exist once they are separated. In fact, it is worth noting here that we have no way of separating these two primordial entities, and so we have no way of understanding them as independent entities. The reader may recollect John Wheeler's aphorism from Ch 1-E regarding the relationship between mass, spacetime, and curvature: 'Mass grips space by telling it how to curve, the space grips mass by telling it how to move.' We may modify this and say, 'Force grips spacetime by telling it how to curve; spacetime grips force by telling it how to move.' In short, spacetime and force have no independent existence.

Spacetime and force have no independent existence.

Thus, force and spacetime are abstract entities, and they are independently non-existent. Or we may say that the presence of force becomes tangible *to us* only when it *warps* (or 'contorts') the spacetime, and conversely, the presence of spacetime becomes apparent *to us* only when it *wraps* (or 'ensheathes') the local force (Fig 11.1). In short, spacetime 'wraps' force and makes its presence felt; force 'warps' spacetime and makes its presence felt.

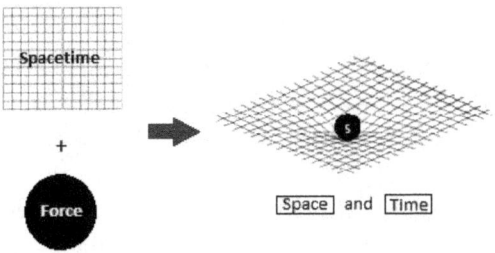

Fig 11.1: Spacetime and force forming measurable 'space' and 'time'.

Here is an important note: this 'to us' in the above statements represents the human standpoint, and these primordial entities have no independent existence *only* when considered from the subjective standpoint of a human being. This is because human beings can only decipher the events of nature by using their limited 3D sensory faculties, and so the 'true' and objective nature of primordial spacetime and primordial force would remain obscure to us forever. However, it may be said that, should a '4D being' exist somewhere in the universe, if ever, then that exotic alien might be able to sense the world exclusively in spacetime continuum alone!

Spacetime and Force—The Two Opposing Entities: Now consider this feature which is of paramount importance to us: even though we have seen that spacetime and force are not independently known to us, we have one important general feature of spacetime and force which radically differentiate them from each other. In fact, we have discussed about this feature all along in the preceding chapters. And this is the phenomenon of their contrasting disposition—spacetime tends to expand and merge with infinity; force tends to contract and become localized. In other words, spacetime has an essential entropic disposition, whereas force has a definitive negentropic disposition.

We have already seen that in the case of light (EM radiation), the negentropic phase represents non-uniform motion, takes a curved course, and causes an increased order. All these features are the features of force. In contrast, the entropic phase represents a release of spacetime from force, which tends to take the spacetime once again to uniform motion in a straight course, causing it to merge with cosmic inertia (Ch 7-E). This shows that force has a tendency to cause order by causing localization, and spacetime has a tendency to cause disorder by causing generalization. In short, force is localizing and contracting, and spacetime is generalizing and spreading.

Force is localizing and contracting; spacetime is generalizing and expanding.

In the case of matter formation also, we have seen that the negentropic phase of a negentropic knot becomes contractile, whereas its entropic phase becomes expansive in the form of a blowout. We may analyze the negentropic and entropic phases in the following way: the spacetime, which is under the influence of force, tends to contract; and once the influence of force dwindles, it enters into the entropic phase, and the spacetime now tends to expand. This shows that these two entities are opposing each other. Thus, spacetime may be said to possess an innate property of increasing disorder, tending towards entropy, and becoming expansive (its ultimate objective is to join cosmic entropy). It just behaves like an infinite sheet of trampoline (Fig 11.1) which, once weight is taken off, recoils out to become flat (Ch 1-E).

Now imagine this property of spacetime and extrapolate it on to the vast dimensions of the universe. Because of its feature of infinite spread, spacetime becomes all-pervading and universal. Thus, the fundamental property of spacetime is to spread across the universe limitlessly. In fact, this infinitely spreading feature of the spacetime continuum contributes the essential property of cosmic inertia/entropy (Ch 7-E).

Spacetime is infinite and all-pervading.

Thus, all entropic responses (be it the entropic phase of a photon or the entropic blowout of a quark or spacetime aberrations in the case of gravity) are the features of the primordial entity of spacetime. This, of course, is the reason why all entropic reactions tend to have a direction towards cosmic entropy.

On the other hand, force has an innate property of increasing order, tending towards negentropy, and becoming contractile. It takes the spacetime curvature in a direction opposite to the direction of cosmic entropy and tends to increase order (Ch 8-F; Fig 8.5). Thus, force 'acts locally' by folding spacetime, whereas spacetime 'reacts generally' by unfolding. The more force acts locally, the more is the folding, and the more is the general reaction of spacetime by unfolding. Thus, the fundamental property of force is to act locally and finitely.

Force is finite and localizing.

We may also put it this way—force is a finite primordial entity which tends to be exacting and 'gritty'; spacetime is an infinite primordial entity which tends to be ill-defined and 'vague'. Thus, spacetime is depicted as disappearing lines in Fig 11.1, and force is depicted as a localized sphere. Spacetime is inherently expansive— it tends to spread out and out. Force is inherently cohesive or diminutive—it tends to contract further in and in.

In summary, we must understand that each and every event of the universe, such as motion, work, and entropy, can be understood as the interplay between these two opposing entities—one opposing the action of the other. Nevertheless, these two components are immiscible like water and oil and have a tendency to exist independent of each other. They interact with each other transiently, but they tend to roll back into their independent forms sooner than later. Thus, the most preferred state of spacetime and force in nature is to exist separately in the universe. This is the state of 4D spacelessness and timelessness (Ch 8-C); this is the 'cosmic principle' of nature (see also Sec-D).

The most preferred state is for spacetime and force to exist separately.

However, it must be noted that this complete segregation is only mythical and is impossible in nature. Even if such a separation is possible, it is not tangible to our human minds, as we have already discussed. But then at any rate, that's certainly not the universe we want! But if such is the great tendency of spacetime and force to exist independently, then what has caused them to interact so that a tangible universe was created, or in other words, what has caused our universe to appear? We will get into these affairs in Ch 12 where this discussion becomes relevant. For now, we will continue with the study of interaction between these two primordial entities.

Light as a 'Next-Preferred State' of Nature: Having passed a statement that the most preferred state of spacetime and force is to exist independently, and having asserted that this is only a mythical state, we will now examine what is the next best state for them to exist! Here we will make an interesting point to consider: spacetime and force having 'failed' to achieve their independent states, nature prefers (or chooses) the next best state of nature, and that is the 'state of equivalence' wherein the contribution of spacetime and force is equal. We have already observed this state of equivalence in Ch 8, and this is nothing but the state of a ray of light! In simple words, it may be stated that this state of perfect synchrony of spacetime and force exists only in the case of light, and this is the most preferred state in nature (of course, next only to the mythical independent state!). This is the reason why nature has an inexorable tendency to exist in a state of free energy, or simply put, this is the reason for free energy being the most preferred state of nature (Ch 9-C).

Failing independent state, the next-preferred state is to exist as light.

Now consider this: whenever such a perfect synchrony of force and spacetime (in the form of equivalent negentropic–entropic phases of light) is disrupted in nature, the innate nature of spacetime and force fights to win back this perfect synchrony. We have seen that during the formation of matter, the perfect synchrony of force and spacetime is lost by the way of negentropic–entropic dissociation, and thus, nature would attempt to restore the synchrony at all times by disintegrating matter into light! This is the reason why nature abhors matter and prefers to exist in a state of free energy. In other words, nature has an inexorable predilection for free energy rather than matter. Even if matter is formed in nature under the influence of excessive force, it has a tendency to disintegrate into equivalent units of energy (as dictated by $E = mc^2$).

Nature prefers free energy and abhors matter
because nature prefers perfect synchrony.

But then here is a crucial question: if nature prefers such a perfect synchrony, then why does nature allow the formation of matter in the first place? Putting it the other way, why should matter occur in the universe at all? The universe could have simply existed in a monotonous state of light throughout its expanse! The universe might have conveniently existed as an infinite sheet of light, with equivalent proportions of spacetime and force, without any lumps or bumps of matter! We will explore this intriguing question in Ch 12-G in connection with the formation of the universe.

So far, we have examined the characteristic features of spacetime and force as primordial entities. Now we will proceed to re-examine the various interesting and ambiguous phenomena we have discussed in the quantum theory (Ch 2) in the light of our new understanding of the nature of spacetime and force and try to explain these older concepts in a more meaningful way. The first and the foremost phenomena we would examine is none other than the stronghold of quantum theory, the wave–particle duality!

Section B
Wave–Particle Duality Revisited

Before embarking on the study of various quantum phenomena in the perspective of our negentropic model, we will first summarize the achievements of the negentropic model in explaining the behaviour of light and matter in nature.

The photonic negentropic model has already shown us why light travels at a fixed and finite speed, spreading in an infinite range, and why energy content of a light wave is proportionally related to its frequency (Ch 8-E). This model also has explicitly shown us the exact mechanism by which a ray of light is generated at its source (Ch 8-F). If the reader ponders over the ideas presented in Ch 8, he/she may find that the characteristic features of the primordial entities of force and spacetime have played an indispensable role in all the explanations offered in that chapter.

The fermionic negentropic model also has made elaborate propositions by elucidating the precise structure of a photon; by describing the exact mechanism of creation of a matter particle; by delineating the formation of gluons, quarks, electrons, and other complex particles; by outlining the anatomy of a simple atom (Ch 9); and by explaining the very meaning of gravity (Ch 10). Moreover, the negentropic model has given us a profound understanding on the fundamental phenomena, such as space, time, mass, motion, charge, work, strong force, and weak force. The negentropic model also has detailed the exact mechanism of interaction between matter and energy and has shown us how a slab of metal gets heated up and expands by the input of energy (Ch 10-D). The reader may see that all these are very important achievements of the negentropic model in the field of theoretical physics. And once again, we may reiterate that all these concepts are dependent primarily on the concept of 'expanding spacetime' and 'contracting force', and thus, we may say that the primordial entities of spacetime and force occupy the centre stage of all the events that take place in the universe.

In the following sections, we will see how these newer concepts would also profoundly affect our understanding of some of the quantum phenomena we have discussed in Ch 2. We will now take them one by one and examine them categorically in the light of our new understanding of spacetime and force. Before proceeding with the discussion, the reader may recollect the statements presented in Ch 2-B, C, and D regarding wave–particle duality, matter waves, uncertainty principle, and various other features of quantum mechanics.

Explanation for the Wave–Particle Duality: As mentioned in Ch 2-B, a light ray behaves both like an array of particles or waves at a given time, and in fact, we have seen that this ambiguity has insidiously led to the development of the quantum theory. But then why should light demonstrate this peculiar property of travelling in waves but behave like particles at the same time? So far, we have no logical explanation. We have simply studied the wave–particle duality without a proper conceptual explanation.

Moreover, consider this: in the case of matter waves (Ch 2-C), we have seen that not only light but also all objects in nature (from the smallest subatomic particles to the largest cosmic structures) move by waves. And in addition, we have seen that according to the quantum theory, an electron which hovers around the nucleus of an atom may be considered either as a particle or as a wave packet of energy (Ch 3-A). An electron travelling in space also behaves both like a particle and a wave – it behaves like a particle when observed, thus generating no interference pattern in double-slit experiments, and behaves like a smeared-out wave function when not observed, thus generating interference pattern dutifully (look into the 'observer's effect', Ch 2-C). All these quantum phenomena have been adequately documented by experiments, but we have no rational explanation for all these phenomena. Putting it squarely, nature has taken us for a wily subjugation into accepting these intriguing properties of light and matter as indomitable principles without a proper explanation. The modern-day scientists have simply come to terms with these intriguing properties of nature. And here lies more trouble—because wave–particle duality and matter waves are essential keys upon which many of the later quantum phenomena are based, all these later developments have also become perplexingly enigmatic. In simple words, we may say that we lack a proper conceptual understanding of the very grounds upon which the edifice of quantum theory is based!

However, the reader may see that all these ambiguous concepts described above may have some very valid explanations in the negentropic model. In fact, by this time, the reader, being connected with the negentropic model already, might have some sort of inkling of the following explanations presented hereunder!

At the outset, we will pass a statement and follow it up with an explanation. According to the photonic negentropic model, a light ray may be considered to possess the representatives of both matter (representing particles) and energy (representing waves), both existing in exactly equivalent proportions. In fact, it may be said that the basic feature of wave–particle duality stems from this dual property of light possessing equivalent proportions of matter and energy. To understand this statement, consider this: the negentropic phase of a light ray has many properties which may qualify it to be considered as the representative of matter, and consequently, the negentropic phase behaves like a particle; and the entropic phase has many properties which may qualify it to be considered as the representative of energy, and consequently, the entropic phase behaves like a wave. The following discussion makes the point clear.

The negentropic phase manifests itself as a matter particle because it is actually a packet of primordial force wrapped up in spacetime, and this packet possesses the characteristic properties of matter, such as non-uniform motion and curvature, which, in turn, manifest as mass and momentum. On the other hand, the entropic phase represents the primordial entity of spacetime because, as seen in Ch 8-D, the entropic response to the negentropic phase is actually a phenomenon of 'getaway' from the negentropic force, and it tends to merge with the cosmic entropy, which is the essential property of spacetime itself. This entropic phase, in fact, manifests as energy with a tendency to unfold and travel in a wave form to finally merge with cosmic entropy.

In consequence, we may suppose that if we conduct experiments that are designed to detect the entropic phase in a ray of light, then these tests would depict light as a wave; the leading example of such an experiment is the double-slit experiment! In other words, this experiment is a 'spacetime detector'.

Entropic phase of light allows us to study light as a wave.

However, if we conduct experiments that are designed to detect the negentropic phase, then these tests would depict light as a particle, and the leading example here is the experiment with photoelectric effect! In other words, this experiment is a 'force detector'.

Negentropic phase allows us to study light as a particle.

The above discussion clearly shows that a light ray may 'legitimately' act as both a wave and a particle—a wave which behaves as an infinite and spreading entity carrying energy and a particle which behaves as a localizing and contracting entity carrying mass. Or putting this the other way, it may be said that primordial spacetime represents energy, and primordial force represents mass.

It all depends on which way we consider it to be and in which way we design our experiments.

In fact, because of this dual property, light gets its unique property of showing up a dual behaviour of *finite speed* and *infinite extent* at the same time. The primordial force in the negentropic phase makes it travel by a finite speed (which is the very quality of primordial force), and the reaction of spacetime in the form of the entropic phase allows it to spread as an infinite extent (which is the very quality of spacetime). Thus, in other words, the entropic phase allows a light ray to behave more like 4D light, and the negentropic phase allows light to behave more like 3D light. And this property of the negentropic phase gives it the thermodynamic capacity to transfer mass and energy to matter, as we have discussed in Ch 10-B. This conceptual understanding of light has some profound implications as we will see in subsequent sections. To sum up, we may say that in a ray of light, primordial spacetime allows it to behave like a wave with energy, and primordial force allows it to behave like a particle with mass.

In a light ray, spacetime behaves like a wave; force behaves like a particle.

Having explained the cardinal feature of the quantum theory, the wave–particle duality, we will now take up the phenomenon of matter waves for our examination.

Explanation for Matter Waves: In Ch 2-C, we have seen that all matter particles demonstrate an intrinsic property of travelling in waves; we will now recapitulate the essentials. It has been suggested, basing on their behaviour in double-slit experiments, that electrons may be described as 'smeared out' clouds travelling in space which could pass through both the slits *simultaneously*. Moreover, we have seen that Max Born had proposed that the exact location of the electron in an atom must be considered more as a probability than certainty and described the matter waves as wave packets. Subsequently, these wave packets were conceptually thought of as wave functions by Erwin Schrödinger, the wave functions being a combination of waves of different frequencies with a high probability of cresting together at the centre of the wave packet where the energy is likely to be concentrated (rather than at the periphery). Hence, the probability of finding a moving particle is greater at the centre of a wave packet than at the periphery (Fig 2.3).

We will now invoke the fermionic negentropic model (with its negentropic knots operating in the background of all matter particles) and look at these ambiguous probability statements afresh and arrive at a conceptual explanation. We will take the example of an electron and see how it exhibits the phenomenon of matter waves. We have seen in Ch 9-G that a free electron travelling in space exists as a negentropic knot surrounded by an entropic blowout (Fig 9.14b). Now consider this: we may reckon an electron as a concentrated form of primordial force in the form of mass represented by a negentropic knot. Or we may also reckon an electron as a repeatedly curled up tangle of spacetime in the form of an entropic blowout. But then is an electron (or any matter particle, for that matter) a unit of concentrated mass composed of specks of concentrated primordial force which tends to contract and move constantly away from cosmic inertia, or is it really a tangle of spacetime composed of waves which tend to drift constantly towards the cosmic inertia (Fig 9.5c)? In short, is an electron a speck of force warping the spacetime or a sheet of spacetime wrapping up a speck of force?

Is a matter particle a speck of force or a tangle of spacetime?

According to the negentropic model, it isn't a proper question. The correct way of putting it is to say that it is both a speck of force and a sheet of spacetime. Both of them exist at the same time in a state wherein they work in unison to create a matter particle. We, the human beings, with our limited 3D faculties, have no capability to separately appreciate either force or spacetime, as already discussed in Sec-A. So it is not correct to look at a matter particle as either a point of force or a tangle of spacetime; it is both. We may once again recall the aphorism presented in Sec-A: force grips spacetime by telling it how to curve; spacetime grips force by telling it how to move.* Thus, a matter particle must be visualized as both spacetime and force working together to create a negentropic knot.

A matter particle is a conglomeration of spacetime and force.

To understand matter waves, we will go a little further and examine the phenomenon of motion of a matter particle. Motion is the fundamental property of anything with mass (Ch 9-A). We have already seen in Ch 8-C that force segregates spacetime into 'measurable space' and 'measurable time' (the reader may refresh his/her understanding of this phenomenon from Ch 8-C). We may now say that the essential purpose of spacetime is to 'move' the primordial force imparted on to it so that it can 'eliminate' it and allow itself to straighten and to merge with the cosmic inertia/entropy. Thus, this property of spacetime actually imparts the inevitable property of 'motion' to any particle with mass in nature (see Sec-D for further discussion). This motion is wavy owing to the repeated folding and unfolding of spacetime, and so, we may state that the motion of any mass particle in space at the fundamental level is a wave-like motion and the path of its motion becomes curved! In the case of light also, the spacetime fights to eliminate force in a wave-like motion, but the path of light takes a straight course because of equivalence.

There is another intriguing affair which has baffled us most in all quantum physics, and that is the phenomenon of observer's effect (Ch 2-C). We are now ready to look into this phenomenon.

The Observer's Effect: The reader may recapitulate from Ch 2-C and D that it has been experimentally shown that an electron behaves as a wave in the double-slit experiment, which would spread out and cause interference—only when it is *not* observed. When observed, the wave functions collapse into a single trajectory which is now

---

* Here we may make an additional note: the reader may imagine force to represent some kind of tinier of the tiniest physical body, depicted as dark dots in Fig 9.5c, which is really not a correct idea! Force may only be depicted as an 'unseen effect' of spacetime—much like spacetime being depicted only as an 'unseen effect' of force. However, it may also be realized that such an imagination is beyond human capability!

observed as the path taken by the electron. All along, we have reckoned this observer's effect as a diabolical phenomenon of nature, which is completely beyond human comprehension. In fact, we have considered this as some sort of mystic phenomenon in the realm of modern physics, which has utterly defied all our scientific wit and logic.

Before we proceed to analyze the observer's effect, we will note down a striking feature—when *light* is used in a double-slit experiment, the interference pattern appears invariably with both the cases—with observation or without observation. In other words, this means that observer's effect does not play a role in the double-slit experiment in the case of light! Or putting this in another way, we may say that observer's effect appears only when we experiment with matter particles. We will now explore why observer's effect does not take effect in the case of light.

Observer's effect is seen only with matter particles but not light.

We have already explained the wave–particle duality of light by considering that a light ray is represented by a negentropic phase which behaves like a matter particle and an entropic phase which behaves like an energy wave, but both of them are equivalent in their magnitude. And when we conduct a double-slit experiment using these equivalent cycles of light, we will get no observer's effect. We have seen that a negentropic phase has to first lose its negentropic–entropic equivalence to become a matter particle (Ch 9-D), and when we conduct a double-slit experiment using matter particles, they show up the observer's effect. This shows that as long as the negentropic–entropic equivalence is maintained, the observer's effect is null and void; and as soon as negentropic–entropic dissociation happens in the form of negentropic knots, the observer's effect starts to appear. Thus, it may be concluded that the causative factor for observer's effect is the negentropic–entropic dissociation.

Observer's effect happens when there is
negentropic–entropic dissociation.

Now consider this explanation: while we are *observing* a double-slit experiment, we are actually conducting *two* sets of experiments—one is the double-slit experiment itself, and the other is the observational experiment using a gadget (e.g. a Geiger–Müller counter or our eye–brain sensory complex). We will discuss the characteristic features of these two sets of experiments separately.

We have already seen above that it all depends on which type of experiment we conduct that determines the outcome of our observation. Now consider this: the first set (e.g. double-slit experiment) is designed to detect waves which are really the representatives of primordial spacetime. In other words, this experiment is a 'spacetime detector'. Thus, in this first set, the electrons show interference pattern because they are seen as waves in consonance with the essential feature of spacetime. In the second set, our observational experiment (e.g. a Geiger–Müller counter) is designed to detect particles which are really the representatives of primordial force—in other words, this experiment is a 'force detector'. Thus, in this second set, the electrons do not show an interference pattern because they are seen as particles in consonance with the essential feature of force.

Here the reader may note that many of the observational experiments (if not all) we conduct are force detectors; examples are our eyes sensing an image, our ears detecting sound, a Geiger–Müller counter detecting a charged particle, etc. This is simply because of the fact that we are all made up of 3D stuff, which can detect only the exacting features of finite force but not the vague features of infinitely spreading waves! In short, our gadgets (and ourselves) can detect the manifestations of force more readily and directly because we are the embodiments of force. And we can make the observation of waves only in an indirect fashion (as in the case of a double-slit experiment).

We can detect force directly because we are the embodiments of force.

In the case of a double-slit experiment with *light*, it does not matter which way we conduct our experiment because both spacetime and force are in equal magnitudes, and consequently, both sets of experiments yield the same results. Thus, light continues to show interference pattern with or without making an observation. However, particles, being embodiments of negentropic–entropic dissociation (owing to negentropic knots), represent force in greater magnitudes. And when a gadget designed to observe force is employed, then it makes an observation of the force, thus collapsing all the infinite trajectories of the spacetime so that the trajectory of the said matter particle under observation becomes tolerably certain (the reader may get a better insight into this phenomenon from an explanation of sum-over-paths presented in Sec-D).

Having explained the wave–particle duality of both light and matter particles in this manner, we will now go into the other intriguing quantum phenomena and look at them in a renewed perspective.

SECTION C
Exploring The Quantum Phenomena

Now we will go into a few of the myriad quantum phenomena and see how they may be understood using the concept of primordial entities of spacetime and force. First we will examine the most prominent of the phenomena, such as the Heisenberg's uncertainty principle, sum-over-histories, and quantum entanglement, and later we will examine the other issues.

Heisenberg's Uncertainty Principle: As we have seen in Ch 2-D, Heisenberg's uncertainty principle describes the theoretical inability to determine both the position of a particle and its momentum in space with any precision at the same time. In other words, it means that if we could determine the precise location of a particle in space at a given time, we would not be able to determine its velocity with precision at the same time. Thus, a matter particle in space tends to oscillate constantly because once a stationary position is obtained in space (which indicates a zero velocity), its absolute position could be precisely determined, and this is not allowed in accordance with Heisenberg's uncertainty principle. And as a consequence of this principle, it may be stated that no matter particle in nature occupies a stationary position, and it is in a state of constant motion! For example, theoretically speaking, if we place an electron in a box and squeeze the size of the box to a smaller and smaller size to define its position precisely, we would expect the electron to wriggle with faster and faster velocities. With this understanding of the uncertainty principle, we will now investigate into its real meaning.

Consider this peculiar situation: this indomitable principle of non-certainty becomes 'absolutely certain' in only one situation in nature—that is, in the case of a ray of light. That is, if we consider the negentropic phase of an EM wave as a representative of matter with properties such as non-uniform motion, curvature, and increased order, and if we consider the entropic phase as a representative of energy with uniform motion tending to merge with the spacetime continuum, then we may say that the absolute state of these two conjugate pairs (Ch 2-D) can be precisely ascertained in a ray of light. In fact, their association in the case of light is an absolutely fixed entity as defined by the fixed speed of light. In other words, we may say that in the case of light, the relationship between the position determined by the negentropic phase and the velocity set by the entropic phase may be precisely ascertained with absolutely certainty! Thus, we may say that both these phases—negentropic and entropic—exist in strict unison throughout the course of light, which may be interpreted to mean that the particle's position and velocity are fixed and absolute.

Heisenberg's uncertainty is null and void in a ray of light.

Now consider this: we have already seen that negentropic–entropic equivalence is disrupted in the case of matter formation. And we may say that, along with this negentropic–entropic dissociation, the absolute certainty of the wave–particle association is also lost. Hence, it may be said that at every instance of the birth of a matter particle, the phenomenon of Heisenberg's uncertainty makes its appearance! This is the reason the Heisenberg's uncertainty principle emerges as the essential property of all matter particles in nature. But then why does this uncertainty creep up into existence in the case of matter? We will now see how this uncertainty may be conceptually explained in accordance with the fermionic negentropic model.

We will begin our explanation with the concept we have already seen in Sec-B that a matter particle can neither be depicted as a speck of force nor can it be depicted as a tangle of spacetime; it is rather represented as a collective manifestation of both spacetime and force which work together to create a negentropic knot. We have also seen in the above section that spacetime tends to behave like a wave, and force tends to behave like a particle. Now we may say that the spacetime component of a particle (being represented in a particle such as an electron by an expansive entropic blowout) depicts the wave nature which, in turn, depicts the precise velocity with which the particle can move in space (see Sec-D for a discussion on the relationship between spacetime and motion). In contrast, the force component of the particle (being represented by the negentropic knot) depicts the particulate nature of matter particle, which, in turn, depicts the precise position of the particle in space. This shows that a particle, being a conglomerate of both spacetime and force, the quantum state of the particle in space may be described *only* as a *combination* of both position and velocity, and they can never be determined independently! In simple words, force determines position; spacetime determines velocity. And because a matter particle is a conglomeration of both, its position and velocity are functions of spacetime and force. This way the basic tenet of the quantum theory—the uncertainty principle—stands, explained by using the fermionic negentropic model.

> *The quantum state of the particle is a combination*
> *of both spacetime and force.*

So far, we have explained the wave–particle duality, matter waves, and Heisenberg's uncertainty principle by employing the negentropic model. Now we will go into the other mystical affairs which have shrouded the quantum theory and try to develop a proper conceptual understanding of these ethereal phenomena.

Sum-over-Histories and Quantum Entanglement: Before proceeding further, the reader may revise these phenomena from Ch 2-D wherein we have discussed that, in order to explain the queer results of the double-slit experiment, Feynman had conceptually revised the wave–particle duality and matter waves and theorized that a particle (e.g. an electron), as it travels in space from destination to destination, actually traverses all the possible paths in transit simultaneously (Ch 2: Fig 2.4). This is known as sum-over-histories. But then more importantly, as we have seen in Ch 2-D and Sec-B, it has also been stated that when an observer makes an observation of the trajectory of an electron, these infinite trajectories collapse into a single predictable pathway which is nothing but the pathway as predicted by the good old Newton's laws of motion. And once again, when the observer is not making an observation of the event, the electron takes infinite trajectories! Therefore, it has been stated that unobserved, a moving particle in space may exist in all the possible states simultaneously, but when observed, it takes only a single state.

It has also been experimentally shown that particles which have interacted at some time in the past tend to retain their association and remain entangled in pairs for indefinite periods of time, and this association has been shown to extend itself for infinite distances. This is the phenomenon of quantum entanglement! This means that in the case of two entangled particles, if one particle makes a spin, say, in the 'up' direction, the entangled particle would spin in a 'down' direction; or if one particle rotates in a clockwise direction, its partner would rotate in an anticlockwise direction; and so on.

These two phenomena, without doubt, indicate weirder of the weirdest propositions in the history of science! How can we theoretically justify such outrageous hypotheses as infinite and indefinite communication between the particles in space and the particles' interminable quantum association? What underlying mechanism could possibly support these surreal suppositions?

If we carefully consider, the common theme of sum-over-histories and quantum entanglement is that both of these phenomena defy the upper limit of the speed of light. In the case of sum-over-histories, the particles go over infinite places hypothetically before they catch up with the old Newtonian trajectory in time, and for this to happen, we must suppose that some sort of queer faster-than-light phenomenon has to happen. And also, in the case of quantum entanglement, for the entangled particles to pair up and communicate instantaneously over infinite distances, it needs a faster-than-light phenomenon to happen. In short, both these phenomena defy the big order of special relativity—the upper limit of the speed of light! So now, in order to have a toehold of explanation, we have to first examine this unique feature of the speed of light and see what really has set its upper limit.

To begin with, consider a state in which no force acts on spacetime. This state is nothing but spacetime existing as an independent entity (which, in fact, is the most preferred state of spacetime, as we have seen in Sec-A!). To understand this state (of course, from the human standpoint), consider the following thought experiment: imagine a tiniest object such as an electron (i.e. a negentropic knot consisting of a simple conglomeration of spacetime and force) travelling in space. The electron travels in space in non-uniform motion at subluminal velocities (because as we have seen in Ch 8, only light with negentropic–entropic equivalence could travel by uniform motion at the speed of light). But then as the electron attains the speed of light, it may be envisaged that its negentropic knot unfolds itself, and the spacetime and force spread over each other in regular chains, at which point the electron transforms itself into a ray of light (Fig 11.2).

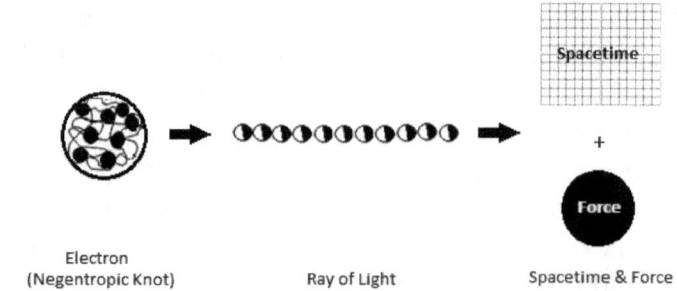

Fig 11.2: Matter particle disintegrating into a ray of light.

And now hypothetically consider what happens if force ceases completely to act on spacetime. Because of the natural tendency of spacetime to get rid of the superimposed force, the electron, which has hitherto transformed into a ray of light, may proceed further to shed force once for all and enter into a state of maximum entropy. In other words, from a state of fixed speed of light, the light ray now jumps to an infinite and indefinable state wherein it gets separated into spacetime and force completely (Fig 11.2). This state really cannot be called superluminal speed because there is no state in nature that may be called superluminal; it is just an infinite or indefinable or an indescribable 'no state' entity!

There is no 'superluminal velocity'; it is only an infinite 'no state' entity.

This infinite velocity is not only an indefinable entity but also an immeasurable entity because light's speed cannot surpass this, and so no numerical value can be attributed to this infinite state! And it is infinite just because of the intrinsic property of spacetime; the 'plastic property' of spacetime stretches only to the extent of the speed of light, beyond which it is meaningless even to make any conjecture (see Sec-E for a full analysis of this plastic nature of spacetime). Thus, it is a futile attempt on our part to search for an object/particle travelling at faster-than-light velocities because no gadget could pick up this no-state entity! Thus, we may state that the absolute value of the speed of light is set by the very property of spacetime. In short, it may be said that speed of light is a function of spacetime!

The innate nature of spacetime sets the speed of light.

Whereas primordial spacetime sets in speed of light, primordial force appears to determine many properties of matter, such as work, entropy, and inertia, of which we will discuss in Sec-D and E.

Before attempting to explain sum-over-histories and quantum entanglement with the help of the above understanding, we will examine the relationship between zero and infinity—a relationship which comes in handy in our discussion.

The Problem of Zero and Infinity: The values of zero and infinity, though very ancient these concepts are, have very flimsy definitions, which render some sort of slippery idea in the minds of the readers. This is because both of them may appear to mean diagonally different ideas, yet they mysteriously appear to convey the same idea! However, we will now interpret them in a new perspective which would make their association (or their distinction) clear.

First take the example of the mass of a photon: we know that the mass of a photon either can be taken either as zero (and say that a photon is a massless particle) or may be mathematically shown to gain infinite mass and momentum (because at the speed of light, it needs infinite amounts of force to propel it, Ch 1-C)! Now we will try to consider the problem with a metaphysical mind. In this connection, it is worth recounting the philosophical musings of Sir Eddington highlighting the relationship between the two antithetical parameters of zero and infinity: in the context of a debate on the events at the time of the big bang, Sir Eddington had remarked that the universe may be considered to last for eternity, and such a universe was equivalent to nothing, adding that 'to my mind *undifferentiated sameness* (signifying infinity) and *nothingness* (signifying zero) cannot be distinguished philosophically'.

Now consider this: it seems to indicate that though zero means 'nothing', it may actually signify everything because there could really be 'no nothing' in nature. And on the other hand, though infinity means 'absolutely everything', it may actually signify nothing because the concept of everything at any one given time and place is equally absurd. With this imbroglio at hand, how do we scientifically interpret the relationship between zero and infinity?

In order to have a clear grasp of this dilemma, we may look at zero and infinity in the perspective of the two primordial entities, spacetime and force. Now consider this: when we measure the mass of a photon from the 4D point of view, it may be said to be zero because we are viewing it from the standpoint of infinity, and so we may be able to 'measure' the value of a photon and place it at zero. In contrast, when we measure its mass from the 3D point of view, then it may be said to be infinite because we are viewing it from the standpoint of zero, and so its mass becomes 'immeasurable' to us! In other words, immeasurability of spacetime reckons the photon's mass as a measureable entity (i.e. zero); measurability of force reckons it as an immeasurable entity (i.e. infinity).

Immeasurability of spacetime reckons mass
as a measureable entity (= zero).
Measurability of force reckons it as an immeasurable entity (= infinity).

In other words, we may say that in a 4D setting, the concept of mass is in reference to force and so is zero, and in a 3D setting, the concept of mass is in reference to spacetime and so is infinite. We may extrapolate this statement and say that infinity is the function of primordial spacetime and zero is the function of primordial force. This is the reason why we, the 3D creatures (but 'sufficiently sapient creatures' who are erudite enough to know of a background 4D existence!) are always in a dilemma as to the real meaning of these two seemingly antithetic parameters! And because of this educated confusion, we would always find it hard to realize that zero and infinity are really one and the same.

Zero and infinity are merely the two sides of the same coin.

With this theoretical insight, we are now able to interpret the real meaning of sum-over-histories and quantum entanglement. First, consider this: because a matter particle is already shown to have formed by both spacetime and force (Sec-B), it may now be said that any moving particle in space may be considered to either exist in a state of zero (represented by force) or in a state of infinity (represented by spacetime). Thus, theoretically, it is possible that an electron travelling in space (as shown in Fig 2.4) may disappear (i.e. by becoming zero) at any given time and may go on to exist at any number of places (i.e. by becoming infinite) before it can rematerialize into an electron once again at the point of destination. As a matter of fact, any moving object in nature (be it a small particle or a large planet) may also be considered to either exist at one place (when observed with our 3D minds and gadgets) or may be considered to exist at all places (when not observed) at the same time. It may be noticed that the property of appearance or reappearance of a matter particle at any place may be considered to be either the property of the localizing primordial force or that of the all-pervading primordial spacetime because both force and spacetime put together make it possible to 'materialize' a matter particle from 'nowhere'!

By the same token, it may be said that the phenomenon of spooky quantum entanglement happens because of the machination of spacetime over force. Though spacetime has an infinite distribution, it may 'remember' the events of force being applied on to it (to create either a photon or negentropic knot) for some variable periods of time before merging with the infinite continuum. This means that any two particles that were created at one time may be presumed to share similar spacetime 'signatures' of force (whatever that may mean!), and so they become entangled for a certain period of time for infinite distances. In this context, we may say that force is the pen with which a signature is depicted on the sheet of spacetime! By extension, it may be said that all the events of force that occur in nature imprint their 'signatures' on the spacetime for some time before they completely vanish into the continuum. In short, we may say that the queer quantum phenomena of sum-over-paths and entanglement may occur because of the localizing primordial force acting on the all-pervading primordial spacetime.

In summary, we may say that all the weird quantum phenomena hitherto studied are the result of the duality of the matter particle, matter particle being the conglomeration of spacetime and force.

All the quantum phenomena happen owing to
the duality of the matter particle.

Now we will look into an equally mysterious aspect of quantum theory, the phenomenon of vacuum energy, and see how the fermionic negentropic model explains it.

Empty Space and the Problem of Virtual Particles: In Ch 3-I, we have theoretically concluded that nature does not allow the existence of an absolutely empty space, but it is actually teeming with constant energy fluctuations. In fact, we have seen that empty space exists with a minimum energy at all times, which is called the vacuum energy (the reader may review the explanation given in Ch 3-I). Also, we have seen that the empty space consists of virtual particles (Fig 3.7) which may move rapidly with faster-than-light velocities and may present themselves as antiparticles which annihilate with particles to give back the energy that was transiently stolen from the empty space (called the zero-point energy). It was seen that these virtual particles may transiently violate the law of conservation of energy by transforming into heavier particles (which, of course, annihilate immediately to pay back the energy deficit). The reader may note that even though the existence of virtual particles and vacuum energy is proved beyond doubt experimentally (as presented in Ch 3-I), all these phenomena are extremely confusing and counterintuitive to the human mind, which require a goodish bit of imagination on our part. In other words, we may say that there is no simple and valid theoretical explanation which would logically explain the occurrence of vacuum energy in nature. Hereunder, we will see a simple and direct explanation which is based on the features of spacetime and force hitherto studied.

In Ch 10-C, we have seen that the intermolecular space is streaming continuously with two sets of spacetime aberrations which act in opposite directions (Fig 10.4). One set are the blowout waves that emanate from the entropic blowouts of the atom which flow away from the atom, and the other set are the secondary matter waves (or proto-gravitational waves) which are attractive and so flow towards the atom. We have seen that both of these waves are virtual waves in the sense that they are merely empty aberrations of spacetime without negentropic–entropic phases.

Now consider this: this scenario of complex virtual waves in the intermolecular space is further complicated by the ubiquitous electromagnetic waves (i.e. light rays) that trespass the empty space throughout the universe (Ch 8-A). This means that these intermolecular spacetime aberrations distort the passing light ray and turn it over into some myriad sort of waves surging upwards and downwards in the quantum space ('quantum foam', Ch 3-I). Or the photons may turn into some exotic and heavy virtual negentropic knots and entropic blowouts which pop out of the quantum space in pairs represented as transient particles and antiparticles which soon annihilate each other to give back the zero-point energy. This energy is again redistributed in the form of regular cycles of negentropic–entropic phases. In other words, the light ray reforms in the empty space and continues its onwards journey as if nothing has happened!

Spacetime aberrations interact with photons, generating virtual particles.

With all these explanations and clarifications, we will go into Sec-D wherein we will examine a few more implications of the concept of primordial entities.

SECTION D
The True Meaning of Energy/Work, Mass/Motion

In the above sections, we have discussed about the role played by the two primordial entities in many quantum phenomena related to the behaviour of matter particles and waves and explained quantum issues, such as wave–particle duality, Heisenberg's uncertainty principle, superposition, and entanglement. In this section, we will delve into the more practical issues of nature, such as energy, work, mass, motion, inertia, entropy, and other attributes of matter and see how our knowledge of these primordial entities lead us into a critical understanding of these issues. Moreover, this critical analysis would lead us to understand the evolution of fundamental physical constants in the universe in a better way, as the reader may see in Sec-E.

To start with, we will briefly recapitulate the relationship between force, motion, energy, work, and entropy from Ch 6-A and Ch 7-C and D wherein we have seen that *force* is the agent which causes *motion* (or displacement, or change) in nature. And this phenomenon of motion is measured as *energy*, and a unit of this energy is understood by us as the *work* done in the universe. We have also seen that any work done in nature is accompanied by *entropy*.

Perhaps the most important definition we may derive from the above discussion on primordial spacetime and force is the fundamental meaning of energy and work! Energy may be defined as the amount of spacetime warpage in a particular locale of the universe, and work is the measure of uncurling of this spacetime curl—the more the curling of spacetime at a given locale, the more is the energy (a radio wave has less curl and so has less energy, a gamma ray has more curl and so has a greater energy, a unit of matter has the greatest curl in its negentropic knots and so has much more energy stored in it!). And of course, the more the curl, the more is the effort of the innate property of spacetime to uncurl, which, in consequence, shows up as more work done in that given locale!

Energy is defined as the amount of spacetime curl.
Work is the measure of the uncurling spacetime.

We have depicted all the above events of nature in the form of a universal cycle (with uniform motion as its fundamental basis) wherein force enters the universal cycle, inducing non-uniform motion (Fig 7.3), and it exits as work done so that, finally, the cycle continues as uniform motion in the form of entropy (which is nothing but uniform motion represented as electromagnetic energy).

Force causes motion, and measured as energy and work.
Universal cycle continues as entropy.

Now we will examine the phenomena of motion/mass and see how we may understand them in more fundamental terms. This discussion also paves the way to understand the relationship between mass, inertia, work, and entropy, thus giving us a holistic view of the affairs of matter and energy in nature! First we will analyze the exact nature of motion.

Meaning of Motion: To begin the discussion, we will ask a fundamental question: how does a matter particle move in space? Or in other words, what causes a matter particle to move?[*] We will invoke negentropic model to understand the very basic concept of motion. We have seen in Ch 9 that a matter particle is really a negentropic knot suspended amidst a sea of surrounding spacetime. We have also realized that this negentropic knot always experiences some negentropic burden (Ch 9-H) because it is formed by the application of excessive force upon spacetime, and the innate property of spacetime (Sec-A) is to correct its curvature and to straighten itself within the shortest period of time. We will now

---

[*] This is relative motion. We have already seen in general relativity that absolute motion is a meaningless concept. After all, a particle cannot move in relation to itself! Moreover, if you imagine a solitary particle in the universe, it will have no dimension, and consequently, it has neither motion nor mass!

see how this innate property of spacetime becomes manifested as motion and other changes in nature.

We will now study the response of the negentropic knot to the negentropic burden. There are three ways by which the negentropic knot brings about change in the surrounding spacetime: (1) it takes the form of entropic blowouts, which are empty aberrations (travelling away from the negentropic knot); (2) it generates rays of light with equivalent negentropic–entropic phases; and (3) it generates primary/secondary matter waves which are also empty spacetime aberrations (travelling towards the negentropic knot). Each of these phenomena brings about a few important changes in nature which, in fact, illustrate the basic physical and thermodynamic properties of matter. In this section, we will look into the empty entropic blowouts and the equivalent negentropic–entropic waves (i.e. a light ray) and see why these things happen and what sort of changes they bring about to the negentropic knot. We will not go into primary/ secondary waves as they are already dealt with in Ch 10.

First we will discuss the entropic blowouts: the exuberant entropic response in the form of a blowout has an important consequence, and this is the change in the disposition of the negentropic knot itself. This means that the entropic response, represented by spacetime aberrations, causes a physical displacement of the negentropic knot in space (in relation to the other matter particles*) so that a change in its orientation takes place. In short, entropic blowouts cause motion of the matter particle, and this is a very important attribute of matter in nature.

Entropic response initiates motion.

Here the reader may notice that the birth of a negentropic knot splits spacetime into space and time, as we have discussed in Ch 9-D. Thus, the negentropic knot may be considered to have not only attained a spatial orientation but also achieved a temporal orientation. In other words, the negentropic knot has not only moved

---

*   Here also we indicate only relative motion. After all, a negentropic knot cannot move in reference to the primordial spacetime itself because of its infinite extent!

in space but also effectively created time in nature! Thus, we may say that the very presence of a negentropic knot in nature bespeaks of the existence of time, and if time stops, the very concept of existence of a negentropic knot (i.e. matter particle) becomes meaningless. This relationship between matter and time is the mainstay in our understanding. As long as matter exists, time exists, and as long as matter and time exist, motion continues to take place (because the concept of time is nothing but matter in motion!). In simple words, the very existence of an object makes it move constantly in nature, thus making the Heraclitus doctrine of 'nothing is permanent but change' invincible! So finally, we may state that the indomitable principle of constancy of motion may be traced back as the fundamental result of the innate property of spacetime!

Now we discuss the generation of equivalent negentropic–entropic waves—i.e. the generation of electromagnetic waves. Apart from eliciting an entropic response from spacetime, the negentropic knot exhibits another important feature—the negentropic knot is unstable owing to its negentropic burden, and thus, the spacetime attempts to free itself of this redundancy of excessive primordial force in some way. The ideal state of nature is for both the primordial entities of spacetime and force to separate themselves completely, but this state being impossible in nature, the nature settles with the next-preferred state (Sec-A), which is the state of equivalent negentropic–entropic phases. And this is nothing but the generation of a ray of light! This implies that a negentropic knot always tends to 'break' up gradually and disintegrate itself into smaller negentropic units which join some of the blowout waves and convert themselves into rays of light. Putting it in another way, this means that all matter always have an inexorable tendency to convert into rays of light, as we have already observed in Ch 9-D! And this phenomenon of disintegration of matter is manifested in the form of another invariable phenomenon of matter in nature—that is, the phenomenon of emission of thermal radiation by all matter at all times!

Thus, in summary, a matter particle, as soon as it is born in nature, not only strives to move in space but also constantly loses its mass into the surroundings in the form of rays of light. Therefore, all objects in nature above the absolute-zero temperature have an invariable property of emission of infrared rays (Ch 2-C)!

Matter always tends to emit thermal radiation at all times.

Though this explains the thermal radiation, one intriguing question remains: why is it that nature prefers infrared light over other ranges in the spectrum? The simple explanation for this feature is that infrared light falls in the Goldilocks zone of the entire spectrum of electromagnetic radiation. We will elaborate to explain this feature (Ch 2-B; Table 2.1). First consider why gamma rays are not the candidates for thermal emission: though the short and powerful waves of gamma rays take away negentropic burden of a negentropic knot in significantly large 'chunks' of negentropic force in a short time, it needs a lot of acute bending of spacetime in order to generate the gamma rays, and this very sharp curvature is generally discouraged by the innate nature of spacetime. Now consider why thermal radiation is not radio waves: though generating radio waves needs less bending of the spacetime curvature, it carries significantly lesser amounts of force from the negentropic knot, so radio waves are also not preferred by nature. The ideal waves which carry much of the force from negentropic knots by bending spacetime only in a modest way are the infrared waves! Thus, infrared light is the most preferred band of thermal radiation by all matter in nature.

Infrared light is the preferred band of thermal radiation.

Having discussed about motion in this fundamental way, we will now embark on studying the other phenomenon related to matter, such as mass, inertia, work, and entropy.

Meaning of Mass: Motion and mass have an inextricable relationship—without motion, the concept of mass is meaningless, and without mass, the concept of motion is meaningless! In all measurements of mass, metres and seconds creep in, indicating the involvement of motion in mass. And when mass disappears, the object tends to travel in uniform motion at the speed of light, which is timeless and so motionless (the reader may once again go through the discussion presented in Ch 8-C). We will now see how the phenomenon of mass is also intricately related to other fundamental phenomena of nature, such as energy, work, inertia, and entropy. In fact, by the following discussion, we will be arriving at the conclusion that all these phenomena are the manifestations of a single grand cosmic principle!

In Ch 7-E, we have seen that the inevitability of entropy and the invariability of inertia are one and the same, and they are the consequence of the same cosmic principle. But then what is that cosmic principle which guides both inertia and entropy to behave like they do? Therein, we have stated that both inertia and entropy happen in nature in accordance with the natural principle that all events in nature attempt to join with the natural state of uniform motion at the speed of light!

Now having gone through all the preceding discussion, we will be able to answer the above question in a slightly different way. When a negentropic knot was born, space and time were split. It may now be said that the cosmic principle is an attempt of the primordial spacetime to restore its original state of spacelessness and timelessness (Sec-A). In other words, the attempt of nature to attain total segregation of primordial spacetime and force is the most indomitable cosmic principle that guides all events of nature—failing this total separation, nature settles down with an equivalent spacetime–force cycles, which is a ray of light!

The cosmic principle is the attempt of nature
to segregate spacetime and force.

We will now examine the relationship between mass and inertia. Do all objects have the same inertia? Of course, no! The more the matter in an object, the more is its tendency to resist acceleration and hence the more is its inertia, and consequently, the more is the mass of that object. Therefore, we observe that a heavy metal block offers more resistance, which makes it harder for us move than a lighter wooden box. However, a piece of wood of 1 kilogram and a slab of metal of 1 kilogram contain the same amount of fundamental matter in them—i.e. the same amount of matter in the form of subatomic particles (protons, neutrons, and electrons put together), though the atomic and molecular arrangement differ very much from wood to metal. Thus, both of these objects exert the same inertia, and this naturally confers the same mass to them. All objects with mass (or matter) demonstrate this property of inertia—a tendency to achieve uniform motion—failing which they move in non-uniform motion. Once an object (be it a small matter particle or a large planet) attains uniform motion at the speed of light, it has achieved cosmic inertia, and so it becomes a ray of light. Thus, a ray of light has infinite inertia, or we may say that it has zero or no inertia at all!

Now we will see how the above cosmic principle guides the phenomena of entropy and inertia. The entropic reaction to negentropic action on spacetime is nothing but the manifestation of the cosmic principle, and the magnitude of effort put out by the spacetime is understood by us as *inertia*, and a measure of this effort is calibrated by us as *mass* of the object. Therefore, the more the negentropic burden (Ch 9-H), the more is the entropic response, and so the more is the inertia and mass of that object. Simply put, more the matter, more the mass! In other words, we may say that mass of a matter particle is the function of its entropic response.

Mass is a function of entropic response.

In a thermodynamic setting, we do not measure the inertia of the event (which takes place at a particular locale); rather, we will measure the total entropic response of the event, and that is measured as the entropy of that system. Thus, both inertia and entropy measure the total entropic response of the event under observation, and so, they are one and the same.

We will now move on to the next section wherein we will discuss other implications of interaction between primordial spacetime and force.

SECTION E
Other Implications of Primordial Entities

Having arrived at the end of our discussion on the primordial entities, we will now try to comprehend spacetime and force in more clear terms. We have seen all along in this chapter that force and spacetime have no independent existence, and it is only their interaction that makes up all the events of our universe. We have also noticed that spacetime is a limitless entity (in the sense that it is a generalized entity) and force is a limited entity (in the sense that it is a localized entity). When we consider the primordial entities in these terms, we may arrive at an important conclusion that force denotes a *conserved unit,* whereas spacetime denotes a *non-conserved phenomenon.* This means to say that force represents a unit which can neither be created nor destroyed, and spacetime represents a phenomenon (or state) where it is utterly meaningless to suppose either its creation or destruction.

In nature, primordial force can neither be created nor destroyed.

When we consider primordial entities in this way, it also derives another important conclusion—in any given event in nature, it is the magnitude of force which chiefly determines the outcome because force is the limiting factor (being finite) but not spacetime (being infinite)! This should mean that all the values we record in nature— be that the mass of a fundamental particle or the velocity of an object or the thermodynamics of a system or the speed of light or the Planck constant, and all other happenings of the universe—are chiefly determined by the force but not spacetime.

However, there appears a theoretical hitch when we consider spacetime to be the non-limiting (infinite) factor, which is as follows: the speed of light, as we have seen in Sec-C, is determined by the spacetime reaction (i.e. entropic reaction) to the applied negentropic force (Ch 8-E). But then if the spacetime is a non-conserved, infinite phenomenon, then the speed of light (the fundamental constant upon which, in turn, dictates all other physical constants of nature) must be infinite! In other words, when the reaction of spacetime to a given magnitude of force becomes infinite, then the speed of light also must become infinite, and consequently, along with the speed of light, all our other measurements, such as mass and motion, would also become infinite! But then we know that this is not true—speed of light is fixed at 299,792,458 metres per second, and it is absolute!

When spacetime reaction is infinite, speed of light should also be infinite.

How can we overcome this theoretical impasse? The simplest way to explain the fixed speed of light is to assume that the reaction of primordial spacetime to a given magnitude of force is not infinite. Rather, the spacetime has a 'fixed plasticity' with which it can respond to a given magnitude of force. This fixed plasticity is, in fact, the innate nature of the all-pervading and omnipresent primordial spacetime.

Thus, finally, as it turns out, the primordial force may be said to be an agent which can initiate a 'graded action' on spacetime so that the spacetime itself may respond with a 'graded reaction'. Putting it in another way, we may say that both spacetime and force are the determinants of all events that have happened in the universe. We may imagine this by imagining a trampoline with a steel ball—the trampoline representing spacetime and the steel ball representing force (Fig 11.1). We cannot say that the mass of the steel ball alone decides that curvature of the trampoline; the curvature is also determined by the innate plasticity of the trampoline. In fact, Einstein had incorporated this principle in his general relativity equation—$R_{\mu\nu} - \frac{1}{2} R g_{\mu\nu} = 8\pi G.T_{\pi\nu}/C^4$ (Ch 1-E)—wherein the left side of the equation depicts the extent of curvature of spacetime that is allowed by certain magnitude of force applied, which is depicted on the right side of the equation. The reader may see that this graded action–reaction duo not only determines the speed of light but also sets in all other physical constants in nature.

This graded action–reaction has another very important implication—apart from setting the value of the speed of light, this relationship dictates the maximum value of mass a matter particle may have in nature. We have seen in Ch 9-E that in the case of the formation of a negentropic knot (or quark), the more the length of the quark wave, the more is the mass of that quark (Fig 9.7). This means to say that after reaching a certain critical maximum value of primordial force, the spacetime curvature 'shuts off', and this will not allow any more force into the negentropic knot. In other words, it may be said that the maximum amount of primordial force that can be accommodated by the primordial spacetime is determined by the innate nature of spacetime (i.e. its plasticity). And thus, this innate plasticity of spacetime dictates the maximum mass a matter particle in nature may have. In the Standard Model (Ch 3), we have seen that the top-quark is the most massive particle (Table 3.2), and we may now say that this maximum limit is dictated by the innate plasticity of spacetime. However, we may find more massive quarks when we build up bigger accelerators in the future, but this is the general rule!

Matter particle's maximum mass is dictated by
the innate plasticity of spacetime.

But what determines the *minimum* mass a particle may have in nature? Consider this: we have seen in Ch 9-D that spacetime curvature can only allow a certain magnitude of force to generate a ray of light. When the magnitude of force exceeds a critical value, then the negentropic–entropic equivalence is lost, and this causes the spacetime curvature to bend exponentially, causing it to curl upon itself tightly, thus creating a dense region of curled-up spacetime (which is the negentropic knot, Fig 9.5b and c). This means to say that there is a minimum magnitude of force below which it forms a photon, and only when it exceeds this limit would nature create a matter particle. But then the reader may realize that the minimum size of a negentropic knot is also determined by the spacetime's peculiar quality of plasticity (but not by the primordial force itself).

Matter particle's minimum mass is also
dictated by the spacetime plasticity.

In summary, it may be said that both the maximum allowable mass of a particle and its minimum mass are fixed by primordial spacetime. The reader may see that this understanding of minimum and maximum range of a particle in nature explains why we have unassailable constants in nature and why we have particles with least masses and particles with most masses, and by extrapolation, this also explains why we find atoms and molecules only in the ranges as we find them in nature and why we find only a limited number of elements in nature! And these limitations of atoms and molecules also set the maximum and minimum sizes of our living cells and, perhaps, the length of their lifespan!

This analysis of primordial entities concludes this chapter. Equipped with all this information about matter and energy, we will now go into the realms of the cosmos and see how the negentropic model works out to explain its intricacies.

☼ Carry-Over Snippets—Chapter 11 ☼

☼ **Spacetime** and **force** are two **primordial entities** of nature. Spacetime is all-pervading, infinite, and expanding. Force is localizing, finite, and contracting.

☼ The **most preferred state** in nature is for spacetime and force to exist independently, but this exclusive state is only mythical and is non-intuitive to the human mind. Failing this exclusive status, the **next-preferred state** is for the spacetime and force to exist in a state of negentropic—entropic equivalence (i.e. a ray of light).

☼ Negentropic knots don't have perfect synchrony, but nature prefers perfect synchrony. Hence, nature abhors matter and prefers free energy.

☼ **Wave—particle duality** of light manifests because its entropic phase behaves like a wave; negentropic phase behaves like a particle.

☼ A **matter particle** is a conglomeration of both spacetime and force. **Matter waves** happen because spacetime imparts the inevitable property of motion to any particle with mass in nature, and this folding and unfolding are exhibited as wavy motion.

☼ **Observer's effect** happens only when there is negentropic—entropic dissociation as in the case of a matter particle—but not with light.

☼ **Heisenberg's uncertainty** disappears in a ray of light. It appears once a matter particle forms because the quantum state of any particle can be defined only by a combination of states of both spacetime and force.

☼ All exotic quantum phenomena may be explained by the duality of matter particles.

☼ The innate nature of spacetime sets in the **speed of light**; there is no faster-than-light velocity in nature.

☼ **Zero** and **infinity** are two sides of the same coin.

☼ **Energy is defined** as the amount of spacetime curl, and **work** is a measure of uncurling spacetime.

☼ **Entropic** response initiates **motion**—the more the entropic response is, the more is the inertia, and the more is the mass of that object.

☼ **Force** can neither be created nor destroyed.

☼ **Spacetime plasticity** is spacetime's innate property. It sets in the maximum and minimum mass of a matter particle.

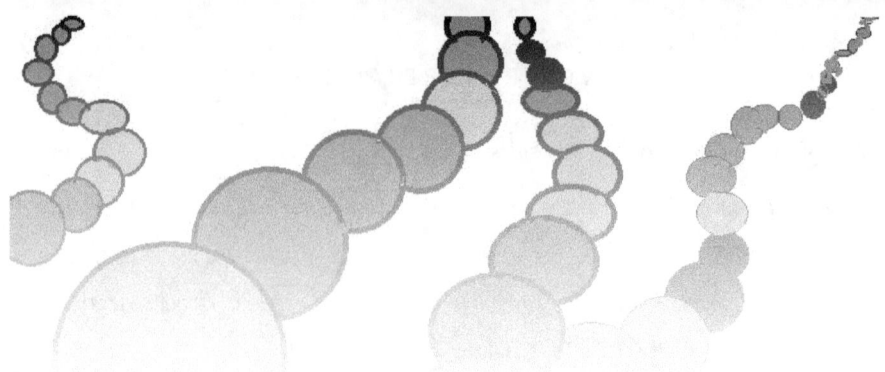

CHAPTER 12

The Differential Big Bang
And the Secret of Dark Matter and Dark Energy

Overview

S ome questions better remain unasked! As an apocryphal story goes, Saint Augustine, when asked 'What was God doing before creating the universe?' said to have replied tartly that 'he was preparing hell for people asking such questions'. Perhaps this might be true of the cosmologists who spend sleepless nights working upon to fathom the mystery of such riddles as 'What existed before the big bang?' But then it seems we may now be able to answer an assortment of such questions in cosmology with the help of our negentropic model!

Consequently, here in this chapter, we will apply the rules of the negentropic model hitherto discussed to answer the large questions that have bothered us in Ch 4 and Ch 5. First we will look at the infinite vastness of the universe and try to realize that it is actually a finitely infinite omniverse. Then we will start examining our own local universe and try to appreciate its vagaries and irregularities. In the subsequent sections, we will see the relationship between normal baryonic matter and dark matter, and this discussion soon plunges us into the darkest corners of black holes and big crunches. In this discussion, we will see how dark matter influences normal matter, and we will also see the origins of dark energy.

In the succeeding sections, we will discuss the enigma of singularity and how this leads to the birth of another universe. This discussion leads us to critically evaluate the events of the big bang and come out with a new concept of differential big bang, which explains almost all the vagaries of our universe. At the end, we will see a surprising similarity between the structure of the atom and the general structure of the universe!

Now enter the maze work of the omniverse to locate our own universe!

Section A
The Omniverse

All along the preceding chapters, we have seen that the negentropic model has efficiently answered many of the intricate problems related to light, matter, and gravity. In this chapter, we will see how this new model addresses the problems at the cosmic dimensions and answers many of the age-old questions that were hitherto left unanswered.

What we know about our universe pales into insignificance when we look at the puzzling voids of knowledge we have about the existence of the universe. We know neither of its real extent (the universe may be restrictively finite, or it may be expansively infinite) nor of its exact composition (the universe is utterly dominated by some unknown dark matter and dark energy), and at the same time, we are neither certain of the exact events at the time of its beginning nor do we have any indication of how it may end. We have seen in Ch 4-E and Ch 5-H and I that all these uncertainties have led the scientists to come out with many speculations. While some of them are products of learned guesswork based on known science, some run wild without any theoretical basis. In the following discussion, we will come up with a design of the universe with a new understanding of its origin, progression, and fate based strictly on the negentropic model hitherto studied.

In order to assist our understanding of the universe, we will make one simple assumption which is purely based on the characteristic features of spacetime and force we have discussed in Ch 11—the universe is *finitely infinite*. We will first explore the meaning of this phrase *finitely infinite*. We know that our universe is made up of essentially light, matter, and gravity (we will tackle the problem of dark matter and dark energy later in the chapter), and in the preceding chapters, we have seen that the primordial entities of spacetime and force interact to give rise to light, matter, and gravity. This means that every grain of matter in this universe, every unit of energy in this universe, and every event that is orchestrated in this universe are nothing but a manifestation of spacetime and force interacting in various ways. In short, we may say that spacetime and force are the precursors of the universe.

Spacetime and force are the precursors of our universe.

In Ch 11, we have discussed about the nature of spacetime and force and ascertained that spacetime is everlasting with an infinite spread, whereas force is localized and diminutive. We will now extrapolate the characteristics of these two primordial entities at the cosmic level and build a simple cosmic model of the universe. Basing on the above properties, we may hypothetically state that our universe, *and* the endless space beyond our universe, is composed of an infinite spread of spacetime. This indescribably expansive and infinitely boundless universe may be called the *omniverse* (called so for the lack of a better term!). In other words, the omniverse is an unlimited and unbounded sea of spacetime which spreads infinitely beyond our universe. In this context, we may define omniverse as an exquisite state of non-turbulence (or monotony) with no events happening in its serene ambience—that is, as long as it is left undisturbed!

Now imagine this situation: into this serene and changeless expanse of omniverse enters an agent called force. We have seen that this primordial force interacts with spacetime to trigger a phenomenon of change in the changelessness of spacetime by creating light and matter (we will worry about the details of cosmic creation later). Thus, the agent force may be said to initiate the formation of a finite and localized universe in the background of an infinite omniverse—just like the formation of a chunk of ice by the way of condensation of a small volume of water in a vast tub of water. This localized chunk of universe under the influence of force is called a *bubble universe*. It must be noted that each tiny bit of omniverse has a potential to generate a bubble universe under the influence of force, and one such bubble universe is our Mother Universe, the guardian of our Milky Way and the abode of our Mother Earth!

Each bubble universe is a finite entity because of the very property of primordial force being localized and diminutive. This means that each cosmic bubble that is generated in the omniverse tends to collapse at the end of its cosmic journey. Consequently, it may be presumed that each bubble universe has its own finite history with a discrete timeline in the omniverse—it has a finite beginning, a finite progression, and a finite ending (we will look into each of these phases in the later sections).

However, consider this argument: because force is a strictly conserved entity (as we have seen in Ch 11-E, force in nature can neither be created nor destroyed!), it may be envisaged that once the primordial force creates a bubble universe in the omniverse, it does not stop with the formation of a single bubble. Rather, it has a tendency to form a continuous chain of bubble universes for an indefinite period, one created after another! This means that as one bubble ends its life, the force continues on to give birth to another bubble in the omniverse, and this chain continues on and on! The reader may now see the meaning of the phrase *finitely infinite*—a finite bubble universe in the background of an infinite omniverse, and each bubble forming a chain consisting of infinite number of bubbles! Thus, our universe we live in, may be designated as a finitely infinite entity.

Our universe is a finitely infinite entity.

It may also be suggested that an infinite number of such chains of bubbles may exist in the omniverse, and the whole bunch of this multitude of bubble universes would make up a *multiverse*. All these chains represent bubbles which begin as points of space and time (under the influence of force) in the omniverse and grow up in size for some finite time, only to end up their lives ultimately—and to reappear in the form of a next bubble—and this sequence goes on and on forever (**Fig 12.1**)! In Sec-B, we will examine a typical bubble universe (i.e. our own universe) and see how it actually takes its birth in the omniverse, how it leads its cosmic journey, and how it may die out—but only to begin a next bubble!

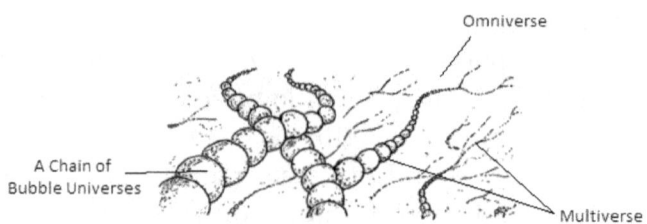

Fig 12.1: The omniverse and the multiverse.

To sum up, we may say that the omniverse is a nondescript infinite entity with no character and no events except that it has a potential to create an infinite number of bubble universes by splitting up spacetime into space and time in the event of its interaction with the agent force. In other words, the omniverse remains as a featureless entity in the background of all the known and unknown events that may happen in each of the bubble universes.

Now consider this intriguing question: do all the bubble universes that take origin in the omniverse have similar physical properties, such as constant speed of light, Planck's constant, second law of thermodynamics, gravitational laws, and sizes of atoms and molecules? Though we have asserted in Ch 11-C and E that these properties are dependent on the innate plasticity of spacetime, it is entirely possible (at least on hypothetical grounds) that different physical constants exist in other bubbles. We will not elaborate upon this argument any further here in this chapter.

With these ideas and presumptions in the background, we will now proceed with the description of the life cycle of a typical bubble universe such as our own!

Section B
The Present-Day Universe

From here on, we will exclusively confine our discussion to the study of our bubble universe (i.e. our own Mother Universe) and see how it might have taken its origin in the omniverse, how it could have expanded and arrived at this present state, and how it might end its life. And finally, we will also see how it may start the next cycle by forming another bubble!

Ours is a finite universe. The very fact that our universe has started off at the time of the big bang about 13.77 billion years ago bespeaks of its finite start, and since the universe has started at a finite point, it may be logically presumed, in all likelihood, to end at a finite point, as we will see below. But consider this: if there is a finite end to the universe (perhaps in the form of a big crunch), then why is the universe in a state of accelerated expansion at present, which actually diminishes the possibility of a recollapse into a crunch? In this chapter, we will look into a unique model which will explain these features of the universe along with a proper explanation of several of the intriguing features such as the nature of black holes, the composition of dark matter, the cause of accelerated expansion, and the existence of dark energy. The reader may note that all these issues are discussed within the ambit of the negentropic model without invoking any new hypotheses—Occam's razor in full application indeed!

The first and the foremost problem we have been tackling for a long time in fundamental science is the problem of the beginning of universe with questions like how the universe started and what the initial conditions of the universe were which led to the formation of stars, galaxies, black holes, and whatnot, all arranged in the manner as they exist today! And of course, the other problem that has been pestering us for a long time is the question of the ultimate fate of universe (Ch 5-I)! Now consider this: in order to answer the questions related to the beginning and ending of the universe, it would be profitable for us to first learn about the general cosmological features of our universe as it exists *today* and logically extrapolate these features backwards and forwards to understand its beginning and its ending (from here on, this universe as it exists today is called the present-day universe). Consequently, we will first make efforts to correctly understand the fundamental principles operating the present-day universe in order to unravel the mystery of the 'big beginning' and the 'big ending'.

Vagaries of the Present-Day Universe: We will first systematically analyze some of the general features of the present-day universe that could have a bearing on the beginning and the ending of the universe. Before proceeding further, the reader may be advised to revise the ideas presented in Ch 4-C and D and Ch 5-E-to-I because they come in handy in our present discussion. The general features of the universe that we are taking up for our study hereunder may be classified as 'cosmological', 'atomic', and 'operational' features.

First we will consider the *cosmological features* of our present-day universe. The present-day universe is irregular, with the vast expanse of space peppered with randomly distributed stars clumped together into galaxies under the influence of gravitational force. In fact, the very presence of matter in the universe makes it lumpy and bumpy (because if this were not the case, our universe would just be a uniform sheet of light with all photons monotonously spread over with unbroken blandness!). Interestingly, most of these galaxies contain all stages of stars in their life cycle ranging from nebular gas clouds (which are the precursors of stars Ch 4-B) to newly formed stars to middle-aged stars (including planets) to senescent and burnt-out stars (Fig 4.2). Apart from these, several galaxies contain exploding supernovae unevenly distributed in the galactic space along with a few black holes which are also randomly situated. If we come to think of it, all these features indicate that some sort of weird happenings are happening in the present-day universe, with some regions forming solid planets, some forming nuclear furnaces in the shape of stars, and some forming abysmal black holes. But then why doesn't the universe follow a pattern in these happenings? What could be the reason for this variegated presentation of our cosmos? Even when we think of the question of distribution of energy in the universe, it may be said that it is somewhat irregularly distributed when viewed at the smaller scale of galaxies. Though the baseline temperature of the general universe is 2.7 K, the temperature of the universe is much variegated with superhot regions, such as supernovae and supercool regions, such as the cores of black holes!

But then consider this feature of our universe: though the matter and energy distribution in the universe is irregular in the short range, it is usually assumed that the overall distribution of matter and energy in the universe is remarkably uniform, as evinced by the cosmological principle (homogeneity and isotropy, Ch 4-A) and by the extreme uniformity of CMB radiation on a larger scale (Ch 5-F). Thus, we may say that the universe has an irregularly regular distribution! But then how do we account for all these variegated events of nature which appear to have some short-range disorder in the background of some magnificent long-range order? How can these vagaries be explained by our negentropic model?

Our universe has a short-scale disorder with a large-scale order.

We will now consider the *atomic features* of our universe. All the known matter in our universe appears in the form of well-defined atoms (we will discuss dark matter later), all of which have a similar overall internal structure and constitution (for example, all atoms are composed of a diminutive nucleus consisting of least-massive quarks with least-massive leptons spinning around it). And also, all matter in nature exchange energy in a stereotyped sort of way, and thus, it may be said that all matter operates with grossly similar physical and thermodynamic properties. Even at the quantum level—though matter particles here are operated by the uncertainty principle—the events have some sort of predictable behaviour such as the behaviour of electrons in an atom and the phenomenon of stability of the nucleus. It is also interesting to note that our universe consists of all the possible natural elements in range—starting from the lightest hydrogen to the heavier and heaviest elements, such as uranium, neptunium, and plutonium, without any 'missing elements' in the periodic table in spite of the very odd and unstable atomic arrangement of certain elements (though it may be noted consistently that the odder elements have a rarer occurrence in nature). But then intriguingly, as has been adequately documented in the previous chapters, hydrogen and helium dominate the universe, and this preponderance has been explained, rather ambiguously, by the current big bang model (Ch 5-F). Now consider these questions: What could be the real cosmological significance of these atomic features of the present-day universe? Can the negentropic model do better in explaining these atomic vagaries?

Finally, we will consider the *operational features* of our present-day universe. By this phrase, *operational features,* we mean the details of interactions between matter and energy which are currently running the universe. A general survey of interactions of matter and energy in the universe gives us an interesting observation: matter in different parts of the universe behaves differently. In the intragalactic spaces (= empty spaces inside each galaxy, Ch 4-A), the gravitational force generally dominates, which helps in binding the stars together; however, in the intergalactic spaces (= large voids in between the galaxies), the gravitational force is far too weaker, allowing the galaxies to drift apart from each other, causing a generalized expansion of the universe. Putting it in a nutshell, we may say that the galaxies are collapsing, causing a localized contraction, whereas the general universe is expanding, causing clusters of galaxies to fly away from each other. These are weird happenings, and things get weirder still as we explore even further!

And yet at some other places within the galaxy, such as in the solar system, there is again a general slackening of gravity, causing the space within it to expand as evidenced by the observation that the earth is drifting away from the sun at a rate of about 15 cm per year, and the moon is drifting away from the earth at a rate of about 3.8 cm per year! Various reasons are propounded for expansion in the solar system, such as the play of tidal forces because of rotational effects, loss of mass of the sun because of nuclear reactions, change in the gravitational constant (G), effects of expansion itself, and the influence of dark matter and dark energy, but none of these explanations are entirely satisfactory. However, the fact remains that the universe behaves differently in different parts of the universe.

The universe expands at some places and contracts at some places.

It is also worth noting here that the behaviour of a galaxy, in general, cannot be accounted for by the calculated amount of mass constituting that galaxy (the reader may go to Ch 5-H for details). Our calculations clearly indicate that a significant amount of matter is found missing in a galaxy. In other words, the gravitational force exerted by the amount of mass in a galaxy is not adequate for the galaxy to hold down its constituent stars. Or we may say that with the magnitude of gravitational pull available in a galaxy, it must soon end up being ripped apart into the surrounding voids! But that's not happening! And this was explained by our scientists by invoking the presence of some mysterious matter called the dark matter, but then we have no inkling whatsoever of what this dark matter could be!

In continuation of the above discussion, we may state that the universe is accelerating at the periphery at a rate which may far exceed the speed of light (and this contradiction of special relativity has been vicariously attributed to the 'expansion of space itself'—without anybody really telling us the real meaning of this expression!). And the cause for this expansion is claimed to be due to the effects of some even more abstruse dark energy!

Most of these cosmological, atomic, and operational vagaries of the universe appear inconsistent to our present-day understanding, and they cannot be explained fully by using the current physical principles. We will now try to explain most of these oddities by employing the principles laid out in the negentropic model hitherto studied and try to piece together the great jigsaw puzzle of the universe in a new model called the *differential big bang model*. This model not only explains the current state of the universe but also unravels the concatenation of events that would lead to the mystery of its beginning and its ending. Before going into the new model, we will first systematically examine matter and energy in a different perspective, which allows us to unravel many secrets of nature, and these affairs would automatically lead us to the new concept of differential big bang.

SECTION C
Dark Matter – Heavy and Lean

We will begin our quest at the atomic level and step up to the cosmic level and see how the negentropic model explains the atomic connections to the cosmic behaviour of matter in the universe. In Ch 10-E, we have seen that the 'empty' spacetime arena around an atom or molecule in a substance is a complex conglomeration of various spacetime aberrations travelling in both directions (Fig 10.4). One set of spacetime aberrations mediated by the secondary matter waves, which tend to travel towards the nuclei of matter particles, thus bringing atoms/molecules together; and another set of spacetime aberrations mediated by the entropic blowout waves which tend to travel away from the matter particles, thus drifting them apart. Based on this, we may state that it is the difference in the magnitude of these two sets of spacetime aberrations which dictates the state of a particular region in the universe either to contract or to expand (the region of universe may be a bar of metal, a cloud of gas, the core of a star, or a chunk of a large galaxy in the cosmos). If the secondary matter waves (i.e. negentropic waves) dominate the intervening space, it would cause contraction of that particular region of the universe, and if the blowout waves (i.e. entropic waves) dominate, it would cause expansion of that region.

This simply means that the rate of expansion or contraction of a particular region depends on the overall magnitude of the total entropic forces or negentropic forces in that region. If the sum of entropic forces is in excess of negentropic forces, then that particular region tends to expand, and *vice versa*. But then we have already deduced from Ch 10 and 11 that the tendency to cause entropic excess is a function of primordial spacetime, whereas the tendency to cause negentropic excess is a function of primordial force. And thus, we may conclude that expansion of a region is basically a function of spacetime, and contraction is basically a function of force.

Expansion is spacetime dependent; contraction is force dependent.

Quite obviously, the above statement may be understood in a different perspective. In Ch 11-B, we have seen that force represents mass, and spacetime represents energy. And thus, the above statement may be translated into a simple statement in the following manner: excess of mass in a region of the universe causes it to contract, and excess of energy in a region of the universe causes it to expand. However, as already noted, scientists have identified that in most of the galaxies, the mass represented by their constituent stars could not explain the contraction of the galaxies, and this has been the very reason for the proposal of extra matter in the form of dark matter. And at the same time, scientists have failed to account for the extra amount of energy that is needed to cause an accelerated expansion of the periphery of the universe, and this has been the very reason for the proposal of dark energy.

With these ideas in the background, we will now proceed to explain the missing matter and energy of the galaxies basing on the fermionic negentropic model, which gives us a correct explanation of dark matter and dark energy, and this explanation would leave us at the doorsteps of the differential big bang model.

Dark Protons and Dark Matter: We have already seen the true nature of light/energy (as equivalent negentropic–entropic photons), matter (as dissociated negentropic knots), and gravitational force (as secondary spacetime aberrations), and now we will utilize these basic concepts to explain the dark matter and dark energy in the universe.

In Ch 9-E, we have seen that a negentropic knot, which is an embodiment of primordial force, is attractive in nature, and consequently, it may have an inexorable tendency to keep contracting so that, finally, it may collapse on to itself to end up into nothing! Nevertheless, this is not happening in nature. The negentropic knots (= quarks) combine to give rise to solid and stable baryons (protons/neutrons), which, in turn, form atoms of normal matter, which makes us and the things all around us! We have also learnt that this stability comes from the fact that different types of negentropic knots (with different masses) assemble in asymmetric configuration

(Ch 9-E; Fig 9.7b and c) to form baryons, which will not allow them to collapse on to themselves to disappear into nothing.

In Ch 9-E, it is also shown that when similar types of negentropic knots (with similar masses) combine to form baryons, it results in the formation of symmetric baryons, and they tend to spin in a negative direction incessantly until the whole assemblage of negentropic knots may ultimately disappear into a state of singularity (i.e. a state of nothing) (Ch 9-E; Fig 9.7a)! In fact, it may be supposed that in nature, the primordial force always tends to collapse into this state of nothingness, as we will discuss in Sec-F. At present, we will just state that formation of such perfectly symmetric baryons takes place in nature. These symmetric protons (= baryons) that are formed out of symmetrically spinning quarks, which tend to collapse on to themselves to a point of singularity, are called the *dark protons,* and the substance that is made up of these dark protons is called the *dark matter.*

Dark protons/dark matter arise out of baryons
of symmetric negentropic knots.

We will now study the characteristic properties of dark protons and see how these properties fit into the concept of dark matter currently understood! The dark protons differ from normal protons in several ways. First of all, the dark protons tend to have very tight adhesions among their constituent quarks because of their perfectly symmetrical geometric structure. In consequence to these tight adhesions, another unique feature arises: normal protons have relatively 'loose' quarks in them so that, as they spin, they 'trap' some amount of surrounding spacetime, which is released into the surroundings as an entropic blowout. However, in the case of dark protons, as the participating quarks' size become smaller and smaller, the adhesions become tighter and tighter, so much so that ultimately the dark protons do not allow any spacetime component to trap inside them as they spin in the surrounding spacetime continuum. The reader may draw this analogy: imagine a rotating fan with perfectly symmetrical blades making virtually no flutter of vibrations and sound and a fan with loose blades making a hell of a lot of sound and vibration (Ch 10-C; Fig 10.1)!

This means to say that, as we have seen in Ch 9-G, a slow-moving normal proton can generate entropic response in the form of an exuberant entropic blowout which goes on to form an electron cloud around it, thus attaining the status of a hydrogen atom (or any complex atom for that matter). But dark protons, on the other hand, by virtue of their tight adhesion of quarks and their perfect spin, will not generate much of any such entropic blowout, so they remain as dark protons without forming regular atoms.

> *Dark matter does not form regular atoms; they*
> *do not generate entropic blowouts.*

However, it is very important to note here that though dark protons do not generate entropic blowouts, they do generate primary matter waves (quark waves), and these primary matter waves surround the dark protons to make them attractive. In the sections below, we will study more of this essential property of dark protons.

Heavy and Lean Dark Protons: The other most important feature of dark protons is that they may exist in different mass ranges. This means that the symmetric negentropic knots that participate in the formation of dark protons may range from the least massive up-quarks (or perhaps even more lighter forms of quarks) to the most massive top-quarks (or perhaps even more massive forms of quarks) (the reader understands the mechanism of their generation as he/she reads along!). The dark protons formed out of lighter quarks (such as up-quarks, down-quarks, and strange-quarks) are called the *lean dark protons*, and the dark protons that are composed of heavy quarks (such as bottom-quark and top-quark) are called *heavy dark protons*. And of course, lean dark protons make up *lean dark matter*; heavy dark protons make up *heavy dark matter*.

Heavy dark matter differs very much in its properties from the lean dark matter. Heavy dark matter has a far too weaker attractive nature owing to the fact that the excessive mass of quarks in the heavy dark protons makes them redundant so that their quark–gluon cycles spin only sluggishly. Consequently, heavy dark matter has a gravitational pull which is far too weaker.

Another feature of heavy dark matter is its inert nature; it does not interact with either normal matter or light. Here the reader may note that though heavy dark protons are made up of massive quarks, their mass could not be measured because, as discussed in Ch 11-D, we measure mass of a unit of matter only when the constituent negentropic knots show up entropic response in the form of uncurling of spacetime. Since dark protons do not generate much entropic response, they remain massless and pass unnoticed to any of our gadgets.

Heavy dark matter is totally inert and is massless.

This perfect symmetry of heavy dark protons, their attractive nature, and their inability to generate adequate entropic response make the dark matter a totally inert substance in the sense that it will not react with normal matter or a passing photon in any way. Thus, heavy dark matter, though ubiquitously distributed in the universe (in Sec-G, we will see why dark matter exists amidst normal matter and learn about their common origin), has no effect on normal matter in any way, and light can pass through it as if it doesn't exist! It may be imagined as a dense transparent sago existing amidst normal matter without interacting in any way.* However, as already noted, heavy dark protons do generate primary matter waves (quark waves), and these negative waves generate weak gravitational waves, which make them somewhat attractive. Anyway, this weak gravity may be considered to cause certain mysterious effects on the normal matter, as we will see in Sec-D.

Lean dark matter, on the other hand, is a different stuff with different characteristics. The quark–gluon cycles of lean dark protons would be able to spin fast, which makes the lean dark matter exquisitely attractive. Nevertheless, this lean dark matter also is not detectable to us because the lean dark protons are also tightly wound and will not generate any blowouts for our detection. We will study lean dark matter further in Sec-E and F, where it becomes relevant.

---

* The reader may see that these properties of dark protons meet the requirements of WIMPs (weakly interacting massive particles as discussed in Ch 5-H).

We will now look into the two important implications of the above discussion. Because of the presence of huge amounts of dark matter amidst normal matter in a galaxy (this feature is fully explained in subsequent discussion), it exerts its gravitational effects on the galaxy, which actually contributes enormously to the total gravity of that galaxy. This explains the missing matter in a galaxy! In Sec-D, we will see how exactly this heavy dark matter is arranged in a galaxy, and in Sec-G, we will see how both normal matter and heavy dark matter are generated simultaneously at the time of the differential big bang.

Heavy dark matter contributes to the total gravity of a galaxy.

Lean dark matter has a different implication—it has all the characteristics to form a black hole. And thus, we may postulate that lean dark matter takes part in the formation of black holes in the universe. We will look into these affairs in Sec-D and E.

Lean dark matter forms black holes.

So finally, to sum up, we have seen two types of dark matter— heavy and lean—each with its own characteristic features and each with its own implications. We will now put forth the following relevant questions: Where exactly are these two types of dark matter located in the universe? How are they generated? What is the cosmological significance of these dark protons? Each of these questions is categorically answered in the following sections. First we will see how heavy dark matter assembles in nature to form the precursors of black holes called the primary black holes.

Section D
The Primary Black Holes

So now we will see how heavy dark matter, though inert in some ways, behaves and influences normal matter in a mysterious way. Because of the common origin of normal matter and heavy dark matter at the time of the origin of the universe (Sec-G), it may be envisaged that heavy dark matter spreads amidst normal matter of a galaxy nearly uniformly and homogeneously. But then in what form does this heavy dark matter exist in a typical galaxy?

We have seen that this heavy dark matter does not interact with normal baryonic matter in the usual way so that it allows light and normal matter to pass through it unaffected. But then more importantly, the heavy dark protons throw up primary, secondary, and gravitational waves into its surroundings, and it may be envisaged that this property of heavy dark protons makes dark matter 'sticky'. Thus, we may say that this attractive property of heavy dark matter leads to the formation of tiny aggregates, and these lumps may actually be considered to represent miniature black holes which are, in fact, the forerunners of the 'regular' black holes we observe in the universe. These miniature black holes are called the *primary black holes*. However, primary black holes may behave somewhat like black holes because of their attractive property, but they do not gobble up the surrounding matter owing to their weak gravity. Stephen Hawking had actually predicted the rampant occurrence of such 'primordial' black holes in the universe and had speculated some of their characteristics (Ch 4-C)!

Heavy dark protons transform into primary black holes.

However, it must be realized that these primary black holes will not stay put at one place amidst normal matter, but the heavy dark protons tend to travel in space along the arrow of time (as we will see in Sec-G). This relentless travel of primary black holes in space makes them present themselves as elongated tiny 'streaks'. With the passage of time, more heavy dark matter falls into these primary black holes (Sec-G) so that these tiny streaks of dark matter spread and ramify *amidst normal matter* all through the galaxy nearly uniformly and homogenously. In Sec-E, we will see how these primary black holes proceed to form streaks of 'capillary black holes', but for now, we will see how primary black holes affect the normal matter in the galaxy.

Thus, in summary, the primary black holes are thought to be made up of heavy dark matter, and these are supposed to be ubiquitous in nature, and they occur in great numbers amidst normal matter in a typical galaxy. The cumulative gravitational force of these innumerable primary black holes in a galaxy accounts for the total gravitational pull of the galaxy which holds its constituent stars. However, no gadget made up of baryonic matter can pick up the mass of this heavy dark matter for our study (because of the lack of entropic phase), and so it remains obscure to our observation!

Primary black holes contribute to gravity within a galaxy.

Now consider this: because heavy dark matter and normal matter have a common origin, it may be considered that primary black holes occur in a far greater proportion wherever there is normal matter. And this obviously means that dark matter is almost exclusively situated inside the galaxies (where normal matter is copiously seen), and dark matter may be said to be largely absent in the intergalactic voids. In consequence, this shows that dark matter exerts its gravitational influence chiefly within the galaxies but not in the intergalactic space, and this explains why the galaxies tend to contract, whereas the intergalactic voids themselves tend to expand, thus expanding the universe in general (we will look into dark energy below!).

There is no dark matter in intergalactic space, so it shows expansion.

Of course, it is the net amount of normal matter and dark matter which would finally determine the state of a particular region of the universe to either contract or expand, and this explains the variegated contraction and expansion of different regions of the universe as we have seen in the case of the expansion of the solar system showcased, for example, as drifting of the earth from the sun or the moon from the earth (Sec-B).

One important characteristic feature of primary black holes (containing heavy dark protons) is that they are *not* abysmal pits in space (like our massive black holes out there in the cosmos), and they neither attract the surrounding normal matter inexorably into their cores nor attract any photons into it. But then more importantly, primary black holes seem to have some discernible effects on the normal matter in nature. It may actually be said that these effects of heavy dark matter on normal matter are common and ubiquitous.

However, we fail to appreciate their significance owing to their very subtle nature. Now we will study these subtle effects on normal matter to realize their importance in our day-to-day activities.

The Problem of Activation Energy: First, recapitulate this from Ch 6: we have seen a queer and unexplained phenomenon in Ch 6-C. Though the second law of thermodynamics irrevocably asserts that nature always tends to favour *disorder*, we find innumerable examples in nature wherein *order* tends to happen. In fact, this building up of order is the essence of all matter formation in nature. An atomic nucleus is an embodiment of order; a boulder of rock with piled-up molecules is an instance of order in nature; our solar system with revolving planets is an essential example of order; all life processes going on in a cell are vitally important examples of order (Ch 6-D). Likewise, the phenomenon of photosynthesis, the growth of a highly orderly foetus inside a uterus, baby stars taking birth from nebulae, and even the formation of negentropic knots to create subatomic matter particles in various nuclear reactions are all examples of building up of order in nature. In short, all our known events in nature showcase the importance of order and negentropy. Though this order is ultimately compensated for by generating more disorder in the general universe, we have no clear explanation of why this order in nature should happen in the first place! In other words, the question of why nature should allow negentropy to happen in the first place remains to be explained!

Why should nature allow order to pile up in nature?

There is another common phenomenon in nature which illustrates the intriguing nature of negentropy, and this is the problem of *activation energy*, and this needs a proper explanation. Take the example of burning of a piece of paper. We have seen that paper is an ordered form of cellulose, and it tends to disintegrate into CO_2 and H_2O spontaneously (Ch 6-B). However, this process of disintegration will not start readily and automatically—to start this process, certain amount of initial energy input is often needed. This is called the *activation energy*. Here's another example. In all our living cells, there are several large molecules which naturally tend to disintegrate into smaller molecules (in the metabolic cascades). But then almost no chemical reaction that happen in the metabolic cycles of a cell happens spontaneously and rapidly. They all need an activation energy provided by certain agents called *enzymes* (which are biological *catalysts*). In other words, we may say that enzymes reduce the activation energy needed to push the reaction. Even nuclear reactions need some amount of initiating energy to run them (e.g. fast-moving neurons in a nuclear reactor, Ch 3-E)! Now why does nature need this input of energy? Why doesn't paper burn on its own, and why don't our cells run the metabolic steps spontaneously? But then we must realize that when reactions happen in nature without activation energy, then our lives would become too horrible. A sheet of paper whooshes out into thin air as soon as it is made, our cells exhaust and die out immediately as they are born, all radioactive atoms turn into instantaneous atom bombs, and so on! In fact, nothing much of value ever exists in nature (including us) if there were no activation energy. All material things in the universe become unimaginably unstable; in fact, the formation of anything in nature becomes questionable without activation energy!

But where does this activation energy spring out in nature? Why is normal matter stable? By the way of the following discussion, the reader will see that this stability of normal matter is due to the negentropic effects of dark matter! We have already seen that normal matter is very intimately associated with dark matter present in the primary black holes (which, of course, is unseen by us and is not picked up by any of our experiments). The weak negentropic (attractive) action of this heavy dark matter builds up a certain degree of order in normal matter, which is reflected in all the negentropic phenomena described above! Putting it in the other way, we may say that this order-building tendency of normal matter is because of the presence of heavy dark protons in the densely ramifying primary black holes present amidst normal matter. In consequence, in any phenomena of nature, this negentropic force of dark matter has to be overcome first in order to initiate the entropic process, and this negentropic force represents the activation energy which we have seen above! In short, we may conclude that dark matter 'stabilizes' normal matter by its weak attractive nature. Thus, we may say that the resistance of a piece of paper to burn spontaneously is due to the stabilizing effects of heavy dark protons amidst cellulose molecules.

Heavy dark matter stabilizes normal matter.

However, it appears that this negentropic influence of dark matter on normal matter is not very active and 'intrusive' (owing to its very weak nature) – rather, it acts on a 'standby mode', meaning that it is inactive in an ordinary sense but is active in the background in a vague sense (appearing instantaneously when it is just 'called for'!). This phenomenon of self-limiting activity of dark matter may be descriptively called the *negentropic supervision*. Thus, we may say that all the known normal matter in a given galaxy is constantly under the gentle supervision of heavy dark matter present in the ubiquitous primary black holes. Though this negentropic supervision is weak so as to render the primary black holes undetectable to direct human experiments, it has a significant negentropic influence on the normal matter as shown in the above examples!

All normal matter is constantly under 'negentropic supervision' of dark matter.

In other words, we may say that heavy dark matter confers normal matter some of its negentropic properties. However, it must be noted here that normal matter itself shows up some negentropic features owing to the presence of negentropic knots in the protons and neutrons of their nuclei, but this negentropic tendency is augmented by the heavy dark matter!

Speaking the other way, we may say that heavy dark matter makes the normal matter 'lethargic' to undergo entropic changes too quickly. If heavy dark matter were to be absent in nature, the normal matter would have been quick to disintegrate into entropic blowouts, and normal matter would not have existed for too long as it exists today, and so our earth, the solar system, and this universe would never have existed at all (Sec-G)!

Heavy dark matter makes normal matter lethargic to show rapid entropy.

Thus, we may conclude that this masterly negentropic supervision is responsible for the present-day universe, which would otherwise have been a tabula rasa with matter disappearing as it was created so that it's just a universe that is filled with only the maximally entropic light spreading across everywhere!

SECTION E
The Progression of Black Holes

We will now study the progression of primary black holes in the universe and see the effects they show as they pass along with the progression of time. In other words, we will study how heavy dark matter behaves in the gigantic stride of the arrow of time in our bubble universe. We will begin our discussion with two questions: How do the primary black holes navigate amidst normal matter in the present-day universe? And what is their ultimate fate? We will now categorically describe the events that take place in the universe in connection with the progression of primary black holes and see how they end up in the big ending of the universe called the *big crunch*.

Progression of Primary Black Holes: Thus, in continuation of the above discussion, we may say that the primary black holes (which contain heavy dark matter) are ubiquitous in nature, and they exist as infinitesimally microscopic structures permeating the very fabric of the cosmos spreading amidst normal matter in the form of a fine network (just as the microscopic capillary roots of a plant spread in between the particles of soil). In other words, the fine streaks of dark matter ramify into each and every region of the present-day universe amidst normal matter wherever it is present (i.e. spreading in the empty spaces in between the atoms of a galaxy). These streaks of primary black holes are called the *capillary black holes* (Fig 12.2).

Though the heavy dark matter in these capillary black holes is inert and unobtrusive to normal matter and light, it does generate primary and secondary matter waves. These negentropic waves naturally make the capillary black holes attractive, and so, as time proceeds, they tend to coalesce with each other to form bigger channels of dark matter (Fig 12.2). These capillary channels flow forward along with the passage of time in the direction of the arrow of time (Ch 6-E), and as they continue their journey up, they begin to join and rejoin to form bigger and bigger black holes (just as the capillary roots join to form bigger roots). These medium-sized black holes are called the *secondary black holes* (Fig 12.2). Here we may presume that the massive black holes that we observe in our galaxy which are situated amidst the stars are nothing but these secondary black holes.

Secondary black holes are the black holes we observe amidst stars.

And of course, numerous such secondary black holes may arise amidst stars of a galaxy to finally merge with each other, forming 'supermassive black holes' which are situated at the centres of many galaxies. These black holes may be named *galactic black holes* (Fig 12.2). These galactic black holes would proceed further and further along the arrow of time, and they coalesce with each other to form even bigger *celestial black holes* (which we haven't perhaps observed so far!) (Fig 12.2).

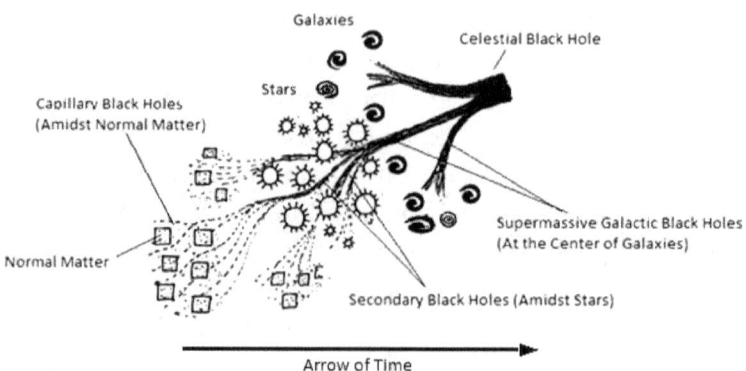

Fig 12.2: Progression of black holes in a bubble universe (a schematic representation).

Our scientists have already identified such merging black holes in the cosmos (see LIGO experiments, Ch 1-F), and it may be envisaged that the coalition of these secondary, galactic, and celestial black holes is a superlatively violent affair because of the huge gravitational attraction involved among them. And consequently, these gigantic cosmic collisions send powerful gravitational ripples into the surrounding space which were actually detected recently by the LIGO experiments (Ch 1-F). Now we will look into the fate of these gigantic celestial black holes and see what happens to the heavy dark protons inside the black holes as they transform into bigger and bigger black holes.

In the above discussion, we have seen how the microscopic primary black holes gradually evolve into gigantic celestial black holes. However, the reader may notice an astounding feature in this process of transition—the primary and capillary black holes situated amidst normal matter existing in between the atoms and molecules of the present-day universe exhibit only weak gravitational attraction, and they will not gobble up the surrounding normal matter and light. On the other hand, the secondary and galactic black holes present amidst stars and galaxies are abysmal pits into which normal matter and light get inexorably dragged into and destroyed. Now here's the question: what makes this phenomenal difference between the primary and secondary black holes?

The reader may realize that heavy dark protons are relatively unstable owing to the massive nature of their quarks, and this instability reflects on their tendency to transform into lean dark protons as time proceeds. The mechanism of reduction of mass of quarks in heavy dark matter is by the phenomenon of cleavage wherein the massive quarks of heavy dark protons, as they spin, are split up into less massive quarks. This cleavage of heavy quarks is an unrelenting process which continues until, perhaps, the quarks reach the lowest possible limit—that is, the dimensions of the Planck scale. In short, heavy dark protons continuously transform into lean dark protons.

Heavy dark matter cleaves itself into lean dark matter.

With this feature in mind, we may propose that as the capillary black holes bundle up to form bigger black holes, the heavy dark protons disintegrate into leaner dark protons so that the secondary, galactic, and celestial black holes are composed of progressively leaner and leaner dark matter. It must also be realized that as the mass of the quarks decreases, the spin of the quark–gluon cycles in the dark proton increases, which brings about some important changes in the behaviour of black holes, as we will see below. With this basic premise, we will work our way up to the ultimate fate of dark matter. The reader may see that the following discussion not only paves way to the concept of singularity but also gives us a better understanding of the phenomenon of event horizon, black hole radiation, dark energy, and other features!

Lean Dark Matter: We have already remarked in Sec-C that lean dark matter is a different stuff. This is because of the exclusive fact that the quark–gluon cycles of the lean dark protons would be able to spin faster and faster (as they become leaner and leaner), which makes the lean dark matter exquisitely attractive. However, it must be noted that even this lean dark matter is not detectable to us because the lean dark protons also will not generate any blowouts for our detection. Here, the reader may recapitulate that normal protons (i.e. baryonic protons) contain exuberant amounts of primordial spacetime in them so that, as they spin, they churn out spacetime lavishly out of them in the form of entropic blowouts, which actually makes their detection possible. In this context, it may be noted that heavy dark protons also may have some little remnants of primordial spacetime entrapped amidst their quarks, but as the dark protons become leaner and leaner, they tend to expel spacetime entirely from them so that by the time they reach Planck scales, it attains the status of utterly 'pure' dark matter (Sec-F).

Lean dark matter at Planck scales is entirely devoid of spacetime entity.

We have already seen that heavy dark protons spontaneously disintegrate into lighter ones because, quite obviously, the lean dark protons are more stable in nature than heavier ones. But then how much time does it take for the heavy dark matter to cleave into lean dark matter? This cleavage may be thought of as an exceedingly sluggish process so that it may take *billions* of years for it to happen. This is simply because of the fact that the heavy dark protons are also relatively very stable because of their symmetric composition.

We have some evidence in support of this sluggish disintegration. We know that it has taken about 13.7 billion years for the events at the big bang to reach the present state, and this indicates that the primary black holes (which were presumably formed at the beginning of the universe, as we will see in Sec-G) have taken as much time to form secondary and galactic black holes!

Cleavage of heavy dark protons is a sluggish process.

Thus, we may envisage that secondary black holes contain lean dark matter, and as they coalesce into galactic black holes, the dark protons get leaner and leaner, and by the time they form celestial black holes, the lean dark protons may attain nearly infinitesimally small quark–gluon cycles.

Now we will look into the exquisite properties of lean dark matter. As noted above, the smaller dark protons of lean dark matter have tighter and tighter adhesions between their quarks (negentropic knots), and thus, they trap lesser and lesser amounts of spacetime entity in them. In other words, we may say that as the lean dark matter gets smaller and smaller, it becomes darker and darker (with lesser and lesser amount of spacetime trapped in them). And as time proceeds further, lean dark protons lose sufficient mass to attain infinitesimally small quark–gluon cycles nearing the size of Planck length. Any further reduction of mass is theoretically impossible, and this state goes with the state of singularity. We will deal with the events of singularity later in Sec-F, but for now, we will concentrate on the events that take place in the secondary and galactic black holes wherein such an absolute singularity is not *yet* attained.

The significance of the above discussion is that we may now state that the term *dark matter* itself is *not* synonymous with singularity! In fact, we would have never observed any black holes in the universe if all dark matter were to achieve singularity instantaneously. This is because all heavy dark protons would have disappeared into singularity as quickly as they are born! But in reality, we have many black holes in our universe which appear to dutifully move along the universal arrow of time to end up slowly and eventually into supermassive black holes, which would, in turn, end up into galactic and celestial black holes at some point in the future. All these features mean that the black holes have a timeline of their own. In short, all black holes may be said to behave as though they have their own natural life cycle.

Each black hole has a timeline of its own!

We will now study secondary and galactic black holes and see how they behave like contracting cones with defined 'edges' called the event horizons.

Events at the Event Horizon—Fireworks at Black Holes: We know that all our known black holes are interminable pits in space which devour the normal matter and light in their vicinity. We have also seen that this area where normal matter and light disappear into a black hole is called the event horizon, and this edge of a black hole is one of the most violent regions of the universe. The events that occur at the brim of a black hole are so highly turbulent that they result in the creation of a bright halo in its vicinity (which actually makes a black hole's detection possible)! Why does a black hole behave this way, and why are the events turbulent? How could our negentropic model explain these features of a black hole?

We have already seen in the above discussion that even though absolute singularity is not attained in secondary and galactic black holes, they possess sufficiently lean dark protons to make them exquisitely attractive so that they attract normal matter and light unremittingly into their cores. Because secondary black holes are intensely attractive, they develop well-demarcated edges where normal matter begins to fall into them (whereas capillary black holes are not so attractive, so they are devoid of event horizons). Now we will study the weird happenings that occur at an event horizon where normal matter gets transformed into dark matter.

According to the negentropic model, event horizons are the places around black holes where the asymmetric fermionic negentropic knots constituting normal matter baryons (protons/neutrons) are converted into symmetric bosonic negentropic knots constituting dark protons. Hereunder, we will see a mechanism by which this process takes place at the event horizon.

Before proceeding further, we must note that there exists a superlatively powerful gravitational force in the form of spacetime aberrations surrounding the secondary black holes which are created by the continuous generation of primary matter waves from the lean dark protons in them. Now consider this: we have seen in Ch 11-B that negentropic knots of normal matter are composite units composed of both primordial spacetime and force. At the event horizon, the surrounding extremely powerful gravitational force of a black hole affects the negentropic knots in such a way that the negentropic force component is inexorably pulled towards the core of the black hole, leaving behind the primordial spacetime component at the event horizon. This means that the negentropic knots of normal matter are literally split into spacetime and force components. Or we may say that the normal matter is stripped off its primordial spacetime entity so that the primordial force entity exists in a pure state. The primordial force entities are dragged into the core wherein they are reassembled into lean dark protons (which are entirely devoid of spacetime entity).

At the event horizon, spacetime and force are split—
force enters in; spacetime is ejected out.

The primordial force that falls back into the black hole eventually leads to singularity; we will study these events in Sec-F. And the spacetime entity which remains outside the black hole takes part in the events at the event horizon, as we will see now. The reader may see that these events of spacetime that happen at the event horizon also explain the enigmatic dark energy and the exact cause of accelerated expansion we have seen in Ch 5-H.

The event horizon of a black hole is a complex region of space. It has two sets of spacetime aberrations in its ambience: one set are the powerful gravitational spacetime ripples which drag the normal matter into the black hole (as discussed above), and the other are the entropic blowout spacetime waves originating from the splitting of normal matter, which tend to spread out of the black hole into the surrounding vicinity. The reader may notice the analogy between these spacetime ripples around a black hole and the intermolecular spacetime aberrations that surround molecules of normal matter

(Ch 10-C; Fig 10.2), but then the only difference are their magnitudes (whereas the magnitude of intermolecular aberrations is minuscule, that of the black hole is expectantly superlative!).

The reader may recollect that these blowout spacetime aberrations are actually empty spacetime aberrations without any negentropic phase incorporated in them (Ch 10-C). In reality, they are virtual waves! However, at the event horizon, these empty blowout waves would acquire the status of matter particles owing to the influence of tremendous gravitational force which acts upon spacetime (Fig 12.3). This intense gravity existing outside the black hole warps the spacetime and incorporates (or 'drives in') the leftover negentropic force into the spacetime so that negentropic knots form many sorts of exotic particles (much like the formation of free electrons, as discussed in Ch 9-G). In simple words, the fanning blowouts at the event horizon acquire negentropic knots and become matter particles, such as leptons, quarks, and their antiparticles, and these charged particles, in turn, annihilate themselves to liberate huge flashes of electromagnetic radiation (Fig 12.3). Of course, in addition to this electromagnetic output from annihilation, a massive amount of gamma radiation is also liberated directly from the event horizon as a result of the incorporation of primordial force into entropic blowouts to form ultrashort waves of negentropic–entropic phases. In short, we may say that there is a fierce activity of light and matter generation at the event horizon, contributing to the formation of a magnificent bright halo around the black hole!

Massive amounts of matter particles and EM
radiation are generated at the event horizon.

There is another important consequence of this phenomenon of entropic aberrations around a black hole, and that is the phenomenon of dark energy.

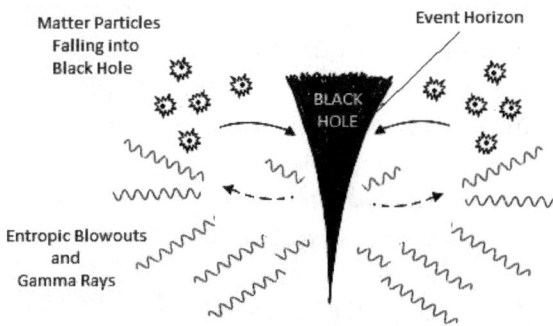

Fig 12.3: The event horizon.

Dark Energy and Accelerating Universe: Consider this: the magnitude of the above-mentioned events are such that though some of the entropic spacetime aberrations at the event horizon are converted into light and matter, a significant portion of these empty aberrations are left out as they are, and these empty aberrations spread out into the surrounding omniverse to finally merge with it. This phenomenon of spreading spacetime aberrations around a black hole results in a tremendous spacetime warpage, and when light rays pass through these aberrations, they create effects of refraction akin to gravitational lensing we have seen in Ch 5-H. However, we may presume that this sort of 'entropic lensing' diverges light (rather than converge it, as in the case of gravitational lensing), and so when an observer from within a galaxy makes an observation (such as us making an observation), the periphery of the surrounding universe appears to undergo exponential expansion.

Entropic aberrations create entopic lensing, which is more prominent in bigger black holes.

Moreover, it may be noted that the phenomenon of entopic lensing becomes more intense as the black holes become bigger so that at the event horizons of galactic and celestial black holes, they diverge light more significantly. This explains why we observe the phenomenon of accelerated expansion of the universe. As we make an observation of more distant objects, they appear to travel at increasing speeds, ultimately crossing the universal speed limit of the speed of light! Thus, finally, we may say that the phenomenon of accelerated expansion of the periphery of the universe is an illusion created by the entropic spacetime aberrations emerging from the bigger black holes. In short, we may say that the phenomenon of dark energy is only an aberration in our observation.

Accelerated expansion is an illusion created
by entropic spacetime aberrations.

The reader gets an overall picture of the universe in Sec-G, which will highlight these issues, but before going into these issues, we will peep inside these galactic and celestial black holes themselves and see the ultimate fate of the lean dark protons that are contained in them.

Section F
The Big Crunch Singularity

It may be said that the final event in a bubble universe is the big crunch singularity! But then what is this singularity, and how is it achieved in a black hole? How is this huge amount of dark matter in the black holes decimated to a single point? What are the terminal events in a big crunch? In this section, we will categorically examine the peculiar situations of the inside of a black hole, and this discussion would really pave way for us to understand the conditions which would create the next bubble universe.

In Sec-E, we have seen that the capillary black holes coalesce to form bigger and bigger black holes to attain the massive sizes of secondary and galactic black holes. These galactic black holes, in turn, coalesce to form celestial black holes. These supermassive celestial black holes may now be envisaged to join with each other to ultimately end up in a unified black hole which is of infinitesimal size with progressively leaner and leaner dark protons. As this progression happens, the event horizon of the bigger black holes becomes more intense so that, finally, the entropic spacetime aberrations released out of this unified black hole reach immense proportions, which ultimately loin the surrounding omniverse (Fig 12.3). In contrast, the infinitesimally large unified black hole keeps contracting as it progresses across the space, and finally, it collapses into a singular state. This contracting unified black hole is called the *big crunch*. In effect, we may say that the universe at this stage is split up into two divisions—one division consisting of spacetime ripples which ultimately merge with the surrounding general omniverse and the other division consisting of primordial force that would create a singular, ever-contracting terminal event culminating into the big crunch. The happenings inside this big crunch are described hereunder.

First consider this: as the dark protons travel from the periphery to the core, they keep losing their residual spacetime (i.e. whatever little spacetime entity that is left entrapped in the dark protons) and become leaner and leaner. And finally, when they reach the core, the lean protons attain Planck scales, and in this state, there is absolutely no spacetime so that the dark protons exist as the purest form of primordial force. Because of this total absence of spacetime inside the core of the big crunch, the passage of time stops altogether, and so time may be said to have come to an absolute standstill. And because of the same reason, inside the core of the big crunch, the concept of spatial orientation of this naked primordial force also becomes meaningless. Thus, a state of timelessness and spacelessness prevails in the core of the big crunch.

Timelessness and spacelessness prevail in the big crunch.

But then as we have observed in Ch 11-A, we, the humans, are not equipped with the perceptive abilities to understand the existence of the pure primordial force, and thus, we may not be able to completely comprehend the real happenings inside a black hole because therein exists the purest form of primordial force. Nevertheless, we will now proceed with the ultimate fate of this big crunch.

The Bosonic Catastrophe: In the above discussion, we have seen that the infinitesimally large unified black hole keeps contracting as it progresses across the omniverse to collapse finally into a singular state of big crunch. This situation of oneness is actually a *state of nothing* wherein matter may be thought to exist in a state of one-dimensional point of infinite mass and density with no meaning for time and space (Ch 4-C)! In other words, it may be said that the term *singularity* really signifies a state of nothingness.

Now consider this paradox: the fermionic negentropic knots of normal matter have an 'antisocial behaviour' in the sense that all fermions exist in their independent states with their own spatial orientation (Ch 3-D). As the heavy dark protons cleave repeatedly to become leaner and leaner dark protons, they lose their fermionic property, and they attain bosonic behaviour to fall at one place because of their newly acquired 'social behaviour'. However, we have seen in Ch 11-E that primordial force is a strictly conserved entity. Force can be neither created nor destroyed, so when all the lean dark protons fall at one place, they tend to disappear into nothing. This state of nothing is not allowed in nature because the primordial force has to keep account of its existence owing to the law of conservation of force. But even so, the big crunch must ultimately attain singularity because, at all costs, all the infinite number of bosons in the big crunch must be accommodated at a single point! Therefore, this is a 'paradoxical crisis' that exists inside a big crunch which bespeaks of conservation of primordial force on one hand and the necessity of nothingness on the other! This ambiguous condition is called the *bosonic catastrophe*—primordial force tends to disappear into nothing without ever disappearing!

Bosonic catastrophe is a paradoxical crisis!

How does nature orchestrate events inside a black hole in this paradoxical situation? Nature appears to follow a simple mechanism to circumvent this problem. The Planck-sized protons present inside the big crunch may be envisaged to undergo a process called the *bosonic conation* (Fig 12.4), which is nothing but a gradual collapse of all dark matter at a single point in the general space of the omniverse! And from this bosonic cone arises another interesting event which saves the day for the primordial force to maintain its conserved nature, and this is the phenomenon of the birth of another universe with yet another set of events which go on to build up a new world! In other words, we have arrived at the point of the creation of another bubble universe by the way of another set of events which may be described in a concept called the differential big bang, and this is our topic in Sec-G.

SECTION G
The Differential Big Bang

Finally, we have arrived at the doorsteps of a new universe. We have seen in the above discussion that the phenomenon of conservation of primordial force has led to the development of bosonic conation, which, in turn, would result in the creation of another bubble universe. We will now study the events that take place in the conversion of this singular bosonic status of big crunch back into a variegated fermionic world of a new cosmic bubble. The bosonic cone may be thought of as a cosmic pen with which the design of a new bubble universe is written on the clean monotonous sheet of the omniverse. Here the reader may notice that we have broken down one age-old concept. According to the new negentropic model, it is shown that there was singularity *before* the big bang, and that singularity is *broken* at the beginning of the universe to give rise to a pluralistic universe.

Singularity existed before the big bang; it
was broken down at the big bang.

The mechanism involved in the process of this grand transformation may be conveniently called the 'differential big bang', and in the forthcoming discussion, we will get a clarification as to why we should prefix big bang with the term *differential*!

The Point of Bosonic Cone: Presumably, the beginning of a new universe starts off at a point in space, and this point, obviously, is exactly where the bosonic cone impinges upon the omniverse to start writing down its story! This reference point at which the singularity of the big crunch ends and the plurality of a new universe begins is called the *point of bosonic cone* (PBC). And in the following discussion, it is shown that this plurality is the precursor of all the irregularities we observe in the present-day universe. Now we will look into the cascade of events that take place at the point of the bosonic cone which leads to the formation of our universe.

Before proceeding further, we may understand two important issues. First, the primordial force that exists in the bosonic cone is completely devoid of spacetime in the sense that it is in a state of pure negentropic force. At the point of the bosonic cone, this pure state of negentropy impinges upon the monotonous state of the omniverse with its infinite spread of spacetime. Second, at the PBC, the primordial force (or pure negentropic force) that is accumulated in the bosonic cone starts amalgamating (or mixing) with the primordial spacetime of the omniverse, resulting in the creation of smaller units of matter in the form of discrete negentropic knots which are nothing but mixtures of force and spacetime (Ch 11-B). This may raise questions in the mind of the reader: why does the negentropic force repeatedly break away from the PBC to form a multitude of pieces of discrete matter particles, and why does not the negentropic force form a continuous stretch of primordial force in a smooth manner until the big crunch is completely drained of its negentropic force, much like a naughty boy who squeezes off all toothpaste out of his toothpaste tube? To answer these questions *is* the start of our discourse on differential big bang!

Consider the very nature of the primordial spacetime. We have seen in Ch 11-E that the primordial spacetime can accommodate only a certain magnitude of primordial force in its warpage, which means that there is a maximum limit of primordial force that can be wrapped up by the spacetime beyond which its innate nature will not allow. It is, in fact, shown that the innate plasticity of spacetime sets in a maximum limit for the magnitude of primordial force that can be cloaked up into a negentropic knot (or quark)!

Spacetime sets in a maximum limit for force to form a negentropic knot.

But then by the above statement, we may conclude that all the matter particles that are broken off at the bosonic cone must be of the maximum possible units of mass as dictated by the above maximum limit, and in consequence, we may say that the entire universe must have a monotony of the same matter particles in it without diversity.

Instead, the universe is much more variegated with different sorts of matter particles with different masses. So how could this variegated nature of the universe become possible? Now enter the differential big bang!

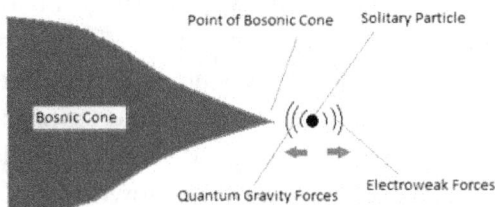

Fig 12.4: The bosonic cone and the differential big bang.

The Variegated Universe: To begin with, we will imagine that at the PBC, an exquisitely large magnitude of primordial force has fallen over the ambient primordial spacetime continuum of the omniverse, which has exerted a finite action on the infinite, all-pervading spacetime, resulting in the creation of a negentropic knot (or quark). This negentropic knot obviously takes the shape of an exquisitely large quark (perhaps a top-quark or an even heavier one). We may imagine that during the briefest initial moments of the universe (perhaps in the first few 'Planck seconds', Ch 2-F), this massive quark exists as a *solitary particle* in the omniverse. This solitary matter particle, by itself, has no characteristics of its own. It has no mass, no motion, or no other attributes of a matter particle simply because to have any physical dimension, a particle must be compared in relation to another companion particle (after all, we cannot compare a particle with itself—dimension is a relative concept, Ch 1-A!). However, it may be assumed that this solitary particle soon develops certain attributes *in relation* to the point of the bosonic cone itself (from which it originates) so that we may say that this solitary particle can now move *away* from or *towards* the PBC. The importance of this movement is discussed below. From here on, we will study the changes that this solitary particle brings about in the omniverse.

The first change the solitary particle brings about in the omniverse is a change in its disposition itself. It must be noted that

RIDING ON A RAY OF LIGHT

this solitary massive quark is the result of an amalgamation of both spacetime and force, and the natural tendency of this quark is to disintegrate so that spacetime and force are once again separated. This instability of this massive negentropic knot is because of the extreme negentropic burden it carries (Ch 9-H), and this instability manifests itself in the form of an entropic 'reaction' (the reader may note that the birth of a negentropic knot itself denotes 'action'). Putting it the other way, we may say that the folded primordial spacetime of the omniverse (under the influence of force) makes an attempt to restore its unruffled original state by unfolding, and this unfolding takes the form of an entropic blowout!

This process of unfolding of spacetime brings about a change in the position of the solitary matter particle (in relation to PBC), and this change in its position is nothing but its motion in space (Ch 11-D). But then this sort of motion may be designated as a manifestation of 'fundamental force' because in Ch 8-B, we have defined force as 'an agent which causes change in nature'. But more specifically, we may dub this force as electromagnetic force because it causes entropy (i.e. increased rarefaction) by the way of moving the particle *away* from the PBC. However, if we recollect our discussion in Ch 9-G, we may note that this massive quark with its fanning entropic response (or blowout) qualifies itself to be called a weak force particle. Putting these two together, we may dub this complex field force around the solitary particle as the *electroweak force*!

However, there is another type of change that is *simultaneously* taking place in the disposition of the solitary matter particle. Consider this: we have already seen in Ch 9-E that a quark disintegrates instantaneously (as it is born in nature) into a quark wave (or primary matter wave), and this quark wave, being attractive, tends to take the particle *towards* the PBC. In other words, apart from throwing a blowout wave, the solitary particle also demonstrates a tendency to give rise to primary matter waves! This change in the disposition of the solitary particle in the opposite direction may be considered to represent another fundamental force, and that is the strong force (because it is a quark wave which actually represents a gluon, Ch 9-E)! But we have also seen in Ch 10-E that these quark waves are the precursors of secondary matter waves, and these secondary matter waves, as they spread out, are represented in the long range as proto-gravitational waves. Putting these two together, we may dub this complex field force around the solitary particle as the *quantum gravity*!

> *Electroweak force and quantum gravity are created*
> *instantaneously and simultaneously.*

Therefore, at the beginning of the universe, the solitary particle is subjected to a complex field force created by these two opposing forces—the quantum gravity pulling the particle towards PBC and the electroweak force pushing the particle away from it. Now consider this: we know that the creation of matter particles at the PBC is a continuous process—i.e. an array of matter particles are created at the PBC and released into the omniverse on a continuous basis. As these new particles are released into the ambience of our solitary particle, they become influenced by the complex field of opposing forces created in its vicinity. This executes a *differential action* of forces upon the forthcoming matter particles which dictates a complicated course for them to take, and this, in turn, results in the generation of a variegated universe! This is the essence of the differential big bang. We will now briefly study the various events that may take place in the early universe because of this differential action.

At the outset, it may be envisaged that the continuous array of massive matter particles that are generated at the PBC take two chief courses *simultaneously*: *one*, the *negentropic course* by which the order of the universe is increased, and *two*, the *entropic course* by which the order of the universe is decreased. As we may expect, the quantum gravity plays a role in the former events of the developing universe, and electroweak forces plays a role in the latter events. However, it must be reiterated that even though we study these two courses separately, both of these courses happen *at the same time* in the early universe! Moreover, it must also be noted that these two courses interact freely with each other, which is the actual reason why a variegated universe is created! We will now look into these two courses which would explain many happenings of our present-day universe in a logical manner.

The Negentropic Course: We will first look into the negentropic course. At first, we may presume that a majority of the matter particles released from the PBC are massive quarks of similar type. It is a natural consequence that because of the extreme negentropic field force created by these massive quarks, they generate strong quark waves which tend to pull them together. These assembled quarks soon form quark–gluon complexes which take the shape of heavy dark protons because of the symmetrical nature of quarks. These heavy dark protons, in turn, coalesce to form primary black holes.

Negentropic course chiefly ends up in primary black holes.

The reader may see that this negentropic course is the most preferred course of primordial force. It will not allow primordial force to mix up with the spacetime so as to create normal matter; rather, it takes a course which leads to attain singularity at the earliest possible time! This propensity is the reason why heavy dark protons and primary black holes populate the early universe freely and rampantly.

Negentropic course is the most preferred action at the big bang.

We have already looked into these miniature primary black holes in Sec-D and learnt that they progress along the arrow of time to become capillary black holes which, in turn, form bigger and bigger black holes which ultimately terminate in the big crunch singularity! We will now go into the discussion on the entropic course of events, and as we move along in the discussion, we will observe how this negentropic course interferes with the other events of the universe and participate in shaping the present-day universe.

The Entropic Course: It may be noted at the outset that only a minority of matter particles released from PBC take this entropic course. We have already seen that nature prefers primordial force to take a negentropic course so that majority of negentropic knots are consumed in the process of becoming heavy dark protons and black holes, and this leaves only a few massive negentropic knots that are created at the PBC to take part in the entropic course. But then it may be said that the entropic course of events leads the early universe to take a much more variegated path so that it makes our universe a much more interesting place!

The chief difference between negentropic course and entropic course is that in negentropic course, the matter particles does not mix up with the spacetime but assemble together and transform into heavy dark protons and black holes, whereas in entropic course, the primordial force mixes up freely with the ambient spacetime so as to develop a strong and turbulent entropic reaction. This mixture of spacetime and force results in the creation of an infinitesimally massive number of various forms of smaller matter particles, force particles, and electromagnetic radiation as discussed below.

Entropic course mixes spacetime and force
entities to develop turbulent events.

It may be presumed that, in the entropic course of events, some of the massive negentropic knots that are liberated at the PBC may disintegrate directly into equivalent negentropic–entropic waves— or in short, they disintegrate to form photons. But alongside this event, some of the large quarks also disintegrate into smaller negentropic knots (matter particles), and this phenomenon of creation of matter particles is presumably as follows: the large negentropic knots released at the PBC acquire huge entropic blowouts in their surroundings as a part of exuberant entropic response, and these complex structures with fanning entropic blowouts and large negentropic knots are nothing but weak force particles in the form of W particles (Ch 9-G). And we know that these complex W particles are quite unstable, and they soon disintegrate, giving rise to numerous smaller particles, such as quarks, antiquarks, electrons, positrons, neutrinos, and antineutrinos, apart from a huge amount of electromagnetic radiation. However, many of these smaller particles eventually annihilate each other, generating more electromagnetic radiation. As may be guessed, all this radiation are in the range of powerful gamma rays which travel relentlessly along the arrow of time and eventually lose their energy as they travel in the cosmic timescale to transform into microwaves, and this radiation ultimately presents itself as the cosmic microwave background radiation (CMB) in the present-day universe (Ch 5-F).

Entropic course ends up chiefly in CMB radiation.

In summary, we may say that the majority of large negentropic knots released at the PBC transform into black holes, which continue to spread across the early universe, forming an all-pervading network of capillary black holes which join and rejoin to form bigger and bigger black holes as time progresses; and a minority of the large quarks disintegrate into smaller particles which, in turn, annihilate each other, thus generating massive amounts of radiation which ultimately give rise to CMB radiation. All this is well, but how is normal baryonic matter formed in the universe?

Formation of Normal Matter, Stars, and Galaxies: Consider this: in the turbulence of the entropic events described above, it may be presumed that some of the smaller quarks that are released are left out in the ambient universe. These quarks exist in different mass ranges because during their formation, they are exposed unequally to numerous intervening forces composed of different ranges of electromagnetic radiation created by weak forces and different ranges of matter waves created by quantum gravity. All these assorted forces muddle up the ambient universe so that, as the arrow of time proceeds forward, many smaller particles in the range of up-quarks and down-quarks are created. Hitherto, the universe is in a great entropic state with high ambient temperatures, and as it cools down, the smallest possible negentropic knots in nature (i.e. up-quarks) *chance* to assemble with down-quarks (next smallest particles) and form stable baryons (protons and neutrons) (Ch 9-E) which further take part in the formation of atoms. But here is an enigma: it is justifiable that the reader may get a legitimate doubt here as to why the up-quarks do not assemble themselves (as nature prefers!) so that they can form the leanest dark protons and fall into black holes to disappear into abysmal black holes. Or why do they escape their fate of annihilation with their antiparticles to become EM radiation?

The answer to this intriguing question is the peculiar phenomenon of 'negentropic supervision' discussed in Sec-D. We will now see how this phenomenon assists nature to build up our bountiful universe which we observe today!

We have already seen in Sec-D that negentropic supervision is a phenomenon generated by the presence of heavy dark matter, which exerts some sort of weak background influence on the matter particles in its vicinity. This background influence is shown to slow down the entropic response of normal matter so that it becomes stabilized without rapidly disintegrating into a flash of light. In the same way, it may also be envisaged that negentropic supervision plays a role in building up normal matter in the early universe. In this connection, it may be suggested that the negentropic supervision, led by the heavy dark protons in the capillary black holes, may *actively* mediate the asymmetric binding of up-quarks with down-quarks, resulting in the formation of stable baryons. Or in the least, this negentropic supervision may *passively* prevent the randomly assembled asymmetric baryons from breaking down. In short, we may say that the heavy dark protons influence the ambience of the early universe to allow normal matter to exist!

Negentropic supervision allows the creation of baryons.

And furthermore, this negentropic supervision allows normal matter to assemble into nebulae, stars, and galaxies in the present-day universe, and this negentropic supervision also bespeaks of the underlying mechanism of all the vagaries of our universe, such as red giants, white dwarfs, supernovae, planets, satellites, and all such stuff of the universe! This phenomenon may presumably explain many vagaries of nature, such as the existence of very heavy atoms in nature in the form of radioactive elements or the unequal distribution of some elements in the universe. However, it must be reiterated that once normal matter forms, the struggle of nature starts once again so that this baryonic matter tends to disintegrate and to finally achieve the maximum entropy!

It may also be stated here that these turbulent events of the entropic course has set in a state of generalized expansion in the universe, thus driving all the galaxies out into the peripheries of the bubble universe.

'Universe in an Atom': The reader my find a striking similarity between the gross structure of the universe and the gross structure of an atom. If we take a cross section of our bubble universe at the level of the bosonic cone, it would have a dense core of contracting dark matter which is surrounded by an exuberant area of spacetime aberrations. We know that an atom has just exactly the same arrangement—an ever-contracting nucleus representing negentropy and an expansive electron cloud representing entropy!

Both universe and atom have contracting
cores with exuberant peripheries!

Our Irregularly Regular Universe: Having studied differential big bang, we may also account for the irregularly regular distribution of the universe (as discussed in Sec-B)—i.e. a short-range disorder (of our planets, stars, and galaxies) in the background of a long-range order (homogeneity of CMB radiation). Simply put, the entropic course of differential big bang gives rise to irregularities in the short range, and the negentropic supervision of the negentropic phase gives rise to long-range order.

Explanation for Olbers' Paradox: It is interesting to note that the differential big bang model of the universe offers an explanation for the Olbers' phenomenon (Ch 5-F) wherein the enigma of why the night sky should be darker is discussed. The big bang theory has stated that in an expanding universe, the light from distant stars becomes progressively red-shifted, which would render light dimmer and dimmer, and may eventually disappear. However, the differential big bang gives us a more valid explanation. We have seen in Ch 11-D that all events in nature are directed by the cosmic principle of complete segregation of spacetime and force. A light ray passing in the cosmos for indeterminate periods of time is also subjected to this cosmic tendency of total spacetime–force separation, and consequently, the negentropic phases of light get elongated and elongated (because of the inexorable tendency of spacetime to straighten), eventually getting separated from the spacetime altogether, which causes the disappearance of light! But then what becomes of the primordial force in the separated negentropic phases? It may be said that these negentropic phases disappear into the ubiquitous network of black holes pervading the cosmos!

A light ray undergoes negentropic separation
and disappears into black holes.

We will now go into Sec-H, which is merely a re-evaluation of what we have already discussed above in order to answer the big question of 'What is this universe?' and 'What is the ultimate purpose of life?'

SECTION H
The Cosmic Purpose

Now that we have seen the beginning and ending of our universe in a new light, we will engage ourselves in looking at the universe from a different vantage point. Differential big bang is a point in space wherein all these laws of nature come back to life sequentially so that it results in the birth of another universe with all the vagaries and irregularities which, after a period of time, once again undergo the culminating events described above to end up in another big crunch. However, the events that happen inside a particular cosmic bubble depend on the very initial conditions that have prevailed at the time of the differential big bang. Consequently, it may be envisaged that there are a few new twists and turns in each bubble so that we may have a variety of bubble universes in this omniverse. And presumably, the story of this birth and death of universes goes on and on forever because of the indestructible nature of force itself in the omniverse! We will now see how this model explains a variety of miscellaneous phenomena we have discussed in the preceding chapters.

Arriving at the Cosmic Purpose: We will now examine another important question and see how we may discuss it in accordance with our negentropic model and the differential big bang. Why should there be an arrow of time at all in the universe, and why doesn't this admixture of force and spacetime behave in an erratic way so that this cosmic entanglement of primordial entities travel in all directions to drive the events of the universe? Or in other words, why is time travel in one direction but not the other way? The explanation offered by the negentropic model would be simple: the purpose, or the drive with which the universe moves forward, is single—the primordial force, having got entangled with the primordial spacetime, will continually always strive for a total disentanglement. This is the cosmic principle we have seen all along in the above discussion! All the events of the universe are driven by this single *cosmic purpose*. The only way to reach this state of total disentanglement is to achieve the big crunch singularity as we have discussed in Sec-F, and the only way by which it can be achieved is by the way of forming a bosonic cone! And the simplest way for nature to achieve a bosonic cone is by the way of differential big bang!

Differential big bang is the simplest way to achieve singularity.

Thus, the universe as we see with normal matter is caused by the differential big bang. And this is the purpose of all movement, all matter, and all inertia/entropy and other events of nature from big bang to big crunch, and this is the purpose of all life and strife in the universe, and this is the purpose of existence of the universe.

Thus, spacetime constantly attempts to disembark the force impinged upon it, and force constantly attempts to liberate itself from the spacetime entanglement. This struggle not only represents the life cycle of our universe but also constitutes the essence of our life on the earth too! We have already seen in thermodynamics of life (see Ch 6-D for discussion on life) that the birth and growth of an organism are an negentropic process which is followed by the decay and death of that organism in a timescale, and life represents this period wherein the struggle continues for the mixture of spacetime and force to separate themselves!

And finally, it may be said that the birth of a bubble universe is a *cosmic accident,* and its end is a *cosmic necessity.* In the same way, birth of a life in nature is a cosmic accident with a lot of negentropy incorporated into it, and death is a cosmic necessity with the inevitable entropy taking over life's negentropy!

> *Birth of a bubble universe is a cosmic accident;*
> *its end is a cosmic necessity.*
> *Likewise, life is a cosmic accident; death is a cosmic necessity.*

This is the 'big cycle' of nature at the heart of all the activities of the universe. And this is the 'big clock' which has set its universal time, and our universe has spent about 13.77 billion years of it already, and this cosmic time is ticking away constantly to reach its end by the way of a big crunch. We, the humans, are the transient custodians of this cosmic purpose to serve the same purpose as that of the universe. This universal time was considered a gigantic clock, ticking away eternally in a cosmic heartbeat with which all other clocks may be set, as Newton had envisaged centuries ago!

Perhaps the span of life of each bubble universe depends not only on the magnitude of the primordial force injected into it at the time of the big bang but also on certain mathematical factors which operate between the negentropic and entropic forces at the initial moments of the differential big bang, as we have seen in Sec-G. And this time period is further influenced by the existence of dark matter in the form of miniature black holes which meddles with the affairs of normal matter.

This concludes our great saga of the universe from the point of the big bang to the big crunch with a logical explanation of all its matter and energy!

☼ *Carry-Over Snippets—Chapter 12* ☼

☼ *Omniverse is an infinitely boundless spread of spacetime. A* **bubble universe** *forms when primordial force acts on spacetime. An infinite number of bubble universes make a* **multiverse.**

☼ *Our Mother Universe (i.e. **present-day universe**)* has a variegated distribution of matter which expands at some regions and contracts at some regions.

☼ **Dark matter** *is composed of* **dark protons** *which are symmetrical baryons that end up in black hole formation.* **Heavy dark protons** *have massive quarks, and* **lean dark protons** *have lighter quarks.*

☼ **Heavy dark matter** *forms* **primary black holes** *and* **capillary black holes**, *which are inert and massless but contribute to the total gravity of a galaxy. However, heavy dark matter influences normal matter in a subtle way of* **negentropic supervision**, *which* **stabilizes normal matter**.

☼ *Capillary black holes join to form* **secondary black holes, galactic black holes**, *and* **celestial black holes**, *which contain lean dark matter. These black holes have event horizons where the normal matter is broken down. Negentropic phases form bosonic dark protons which fall in to black holes; entropic phases cause emission of several matter particles along with massive amounts of radiation.*

☼ *Massive entropic aberrations at the event horizons give us, as observers inside a galaxy, the illusion of* **accelerated expansion**, *which is interpreted by us as* **dark energy**.

☼ **Bosonic catastrophe** *of the* **big crunch** *gives rise to a* **bosonic cone** *wherein singularity exists.*

☼ *Bosonic cone impinges on the spacetime of the omniverse to create another bubble universe. The concatenation of events lead to a* **negentropic course** *giving rise to dark matter and black holes and an* **entopic course** *giving rise to dark energy and CMB radiation.*

☼ *Interaction between negentropic and entropic courses gives rise to* **normal matter**.

☼ *All events of the universe are driven by the* **cosmic principle** *of achieving total segregation of primordial spacetime and force.*

EPILOGUE

Big Questions, Brief Answers!

H aving come to the end of our discussion on the negentropic model, I am at liberty to assume that several issues would have popped up in the minds of the readers during the course of their perusal of this work. I will now try to answer some of the anticipated questions in this question and answer session. At the outset, I will run a bird's-eye view over the newly described principles, and in passing, I will also deal with the hard question of what is 'ultimate reality' and see if our negentropic model qualifies itself to represent reality. However, I am sure that there would be many more questions in the minds of the readers than that are presented here, and expectantly, such questions are discussed on other platforms post-publication!

1. How is the universe born, and how does it end?

We will discuss this question by briefly outlining the events that take place in the lifetime of a typical bubble universe such as ours. The omniverse is made up of a monotonous sheet of infinite spacetime which is bespeckled with areas of irregularities in the form of bubble universes resulting from the action of force upon spacetime. At the beginning of each bubble, the primordial force acts on spacetime of the omniverse, resulting in the formation of matter in the form of massive quarks. However, the cosmic principle states that spacetime and force always tend to exist separately, and this tendency makes the generated matter unstable. Thus, the majority of matter takes a negentropic course attempting to allow matter to exist in heavy dark matter, which transforms itself into a pure state of primordial force in the form of lean dark matter as time proceeds. Nevertheless, a minority of matter would take an entropic course (because of the innate property of spacetime), which orchestrates the formation of normal matter and light. This differential activity is the essence of differential big bang.

The innate property of primordial spacetime and force tending to exist separately dictates the arrow of time. The formed matter at the big bang being an amalgamation of both spacetime and force, it always attempts to separate, and this sets in a time span for each bubble universe (our universe has passed about 13.77 billion years so far and, presumably, has about that much of time to pass on before ending in a big crunch!). This arrow of time orchestrates all the events in a bubble universe, and this explains the inevitability of entropy that we observe in nature.

The dark matter and normal matter mix freely as the time proceeds. The normal matter (which is unstable because of the huge negentropic burden it carries) has an inexorable tendency to disintegrate into rays of light because a ray of light with equivalent negentropic–entropic phases is the next-preferred state in nature. However, normal matter achieves some stability owing to the stabilizing effect of the intervening dark matter, and this negentropic supervision of dark matter over normal matter fine-tunes all the events of nature we see around us. This explains the stability of a piece of paper or a lump of coal without burning down immediately as they form, and this negentropic supervision also slows down our metabolic cycles, making life in nature possible!

But then ultimately, the primordial force manages to separate completely from primordial spacetime to form a big crunch to achieve singularity, which takes the shape of a bosonic cone. The separated spacetime creates the effects of dark energy before merging away with the surrounding omniverse. The bosonic cone finally impinges upon spacetime of the omniverse to start over another bubble universe once again!

2. How are light, energy, and matter related? And what is their relationship to work, entropy and inertia?

Energy, matter, and their interactions are merely the result of spacetime warpage caused by force. Light is composed of equivalent proportions of primordial spacetime and force travelling with a fixed and finite velocity and spreading infinitely. The phenomenon of light illustrates an absolute relationship between spacetime and force which together dictate the absolute speed of light, and this relationship between these two entities in a photon represents the Planck constant. Thus, all features of light stand explained using the negentropic model.

Matter, on the other hand, is the incorporation of excessive force in a unit of spacetime. Formation of matter (in the form of negentropic knots) brings about a change in the changeless spacetime by segregating space and time so that a unit of matter can now exist with a spatial orientation as well as with a temporal orientation (which represents local time). This segregation further brings about several changes in nature, which orchestrates all the events of the universe. These changes bespeak of the cardinal features of matter, such as mass and motion. But then here is a caveat—because matter is represented by both spacetime and force, a matter particle's quantum state regarding its spatial orientation and its time lapse in motion cannot be understood separately and simultaneously, and this allows the Heisenberg's uncertainty to come into play as soon as a matter particle is born.

Now how to understand the phenomena of energy, work, entropy, and inertia? Energy is simply the amount of spacetime warpage at a given locale in the universe—the more the curl, the more is the energy stored in there! Consequently, a radio wave has less energy, gamma ray has more energy, and a unit of matter (with its constituent complex curling of spacetime in the form of negentropic knots) has the most energy (which follows the equation $E = mc^2$). Because this curl in spacetime is due to the influence of primordial force upon it, the magnitude of force indicates the amount of energy. The unimpeachable property of spacetime to revert back to its original state is manifested as the work done in nature—the more the spacetime curl, the more is the energy and the more work done. A moving object tends to move at the speed of light, and this tendency is an attempt of the object to make its constituent spacetime attain its original status, and this is depicted as the object's inertia. And consequently, more the mass of the object, more is the inertia. Entropy of a thermodynamic system represents a measure of the same tendency of spacetime to revert back to its original status.

3. What became of the classical fundamental forces? How do we understand electromagnetic force and gravity now?

The notion of 'four fundamental forces' has changed completely. The four fundamental forces are merged into two workable forces— negentropic force and entropic force. Any change that brings about an increase in order in nature and cause attraction is called a negentropic force, and any change that brings about an increase in disorder and cause repulsion is called an entropic force.

Strong force and gravity were shown to be one and the same force, and they are merely empty spacetime aberrations propagating in the negative directions, causing an increased order in nature. Strong force acts in the immediate vicinity of quarks within the nucleus, and this represents the strongest force—though at larger distances it is represented by gravity, which becomes progressively diminutive and weaker at increasing distances outside the nuclei. Quantum gravity has turned out to be nothing but strong force, and of course, this miniature negentropic force had a pivotal role at setting up events of differential big bang at the beginning of the bubble universe and in shaping the initial events of the universe.

Light (i.e. electromagnetic force), with its equivalent negentropic– entropic phases, has attained a special status and is no longer considered a force! One surprising outcome of the negentropic model is the proper understanding of the medium of light propagation; spacetime is the medium, and the medium itself contributes to the structure of light!

But then what would be the precise definition of a *force field*? A force field is the warpage of spacetime around a point of force by which the point of force can exert its effect on the surrounding objects—just like the field force of gravity of the sun influencing the earth!

4. What is the exact meaning of arrow of time? Why is nature so fastidious about the second law?

4D light, which is nothing but the spacetime of the omniverse, has neither time nor space. However, 3D light, which is our ordinary daylight, has an absolute time dimension and an absolute space dimension, which together has fixed the speed of light at 299,792,458 metres per second! Therefore, it may be said that force has created space and time. But when matter formed in nature, the absolute relationship between space and time is lost, and nature always prefers to gain back its original status. This 'urge of nature', this natural tendency of nature, sets forth a direction and a purpose in all the events of universe, which becomes manifested as an inevitable tendency of nature to increase entropy. This overall tendency of matter sets in the gigantic arrow of time which drives the bubble universe from the point of the big bang to the big crunch and fixes a lifespan to it.

5. Is the universe finite or infinite? Is the universe regular or irregular? Is the universe flat or curved?

The universe is finitely infinite, and it is irregularly regular. This means that each bubble universe is a finite unit, but these finite units form infinite chains of multiverse in an infinite omniverse! The present-day universe is an irregular place with vagabond planets and stars waddling across the galaxy in different directions. However, the universe has an overall homogeneous distribution as indicated by the CMB radiation. Differential big bang has shown that during the interactions between entropic and negentropic courses, irregularities arise in the short range, leading to the formation of lumps of normal matter transforming into stars and galaxies, but at the same time, the background negentropic supervision of dark matter gives rise to a long-range order across the universe in general!

Flatness is the property of spacetime, whereas curvature is the property of force! Therefore, absolute linearity is only a rule of the omniverse, which is absolutely flat. All bubble universes have a curvature. After all, bubble universes are bound by gravitational force originating from its constituent stars, black holes, and planets.

6. Does this model, with its primordial entities of spacetime and force, represent the 'ultimate truth'?

The negentropic model does not qualify to represent the *ultimate truth* (or the *ultimate reality*). The problem of the ultimate truth is that it really is not a scientific concept (though, historically, science has taken its roots from the quest for absolute truth!) – rather, it falls in the realm of metaphysics, which is a quasi-scientific branch of philosophy. To examine the concept of ultimate truth in its correct perspective, the reader must first understand the meaning of 'truth' in a philosophical context. In order to do that, we have to first comprehend the philosophical concept of *noumenon*, as defined by Immanuel Kant in his work *Critique of Pure Reason*. Noumenon is a *thing-in-itself* which exists independently of human sensory perception or thought. This term is in contrast to *phenomenon* (= phenoumenon), a term used to describe an object as perceived by the human senses. Noumenon, or a thing-as-such, is the true nature of a thing (or an object), which is outside any of our senses or thinking. And thus, noumenon, by definition, should be an 'idea' beyond human comprehension (the reader may notice that an 'idea' itself is a human conception, and so this definition is merely a paradox).

Now consider this: does the concept of the primordial entities of spacetime and force represent noumenon? The answer would be a yes and no. Yes, because both of these entities, by themselves, are inconceivable to the human mind, and hence, they are unknowable! No, because a thing-in-itself, by its sheer definition, has no attributes and thus cannot exist in two entities. It should have an absolutely singular status! Therefore, to be called a noumenon, we must be able to say that both spacetime and force are one and the same—an impossible feat for us who, having learnt of their role in the creation of light, matter, and gravity, look at them as two utterly contrasting entities. Thus, the negentropic model, along with our two primordial entities, is strictly a scientific working model for our practical understanding of matter and energy, and it cannot be treated in terms of metaphysics and philosophy – hence, for us to engage in any further interpretation of the true nature of these entities is beyond the realm of physical science. And so we have stayed away from such a metaphysical interpretation all along the preceding chapters; rather, we have adhered to the hard facts of physical science and tried to understand nature as we probe with our senses!

Thus, the negentropic model is just one of the simplest empirical ways to understand our nature and its phenomena. Perhaps it is the best model so far proposed in the history of science! In other words, nature is best understood by the way of primordial entities of spacetime and force because, as we have seen in Part II, this model has efficiently explained almost all the phenomena of nature, such as light, matter, gravity, and our universe. This comprehensive model has the fewest number of assumptions to sew all the known laws into one grand fabric—a perfect example of Occam's razor indeed! Nevertheless, we may expect that, with further advances in science, this model may undergo some significant transformations, or we may hope that even better models may replace this model.

7. Can the negentropic model be experimentally tested? How can we rectify the theoretical inconsistencies that have crept in the negentropic model?

So far, we have discussed the negentropic model only as a concept; all concepts need experimental validation for them to be incorporated into the mainstream science! However, such experiments may be either prospective (when we conduct experiments to actively search for evidence) or retrospective (when experiments already conducted stand in their support). We have one such outstanding retrospective experiment which comes in support of the negentropic model. Alexander Hartung, a doctoral student at the Goethe University, Frankfurt, Germany, has studied the problem of distribution of conservation of momentum of photon in the case of photoelectric effect. It has already been shown that, in the case of photoelectric effect, some of the photon's energy is utilized to break up the atom, and the remaining energy is transferred to the released electron, which flies away with certain velocity. The question of which reaction partner (electron or atom nucleus) conserves the momentum of the photon has not been clear to the physicists. Hartung's experiments showed that 'as long as the electron is attached to the nucleus, the momentum is transferred to the heavier particle, i.e. the atomic nucleus, and as soon as it breaks free, the photon momentum is transferred to the electron'. This is an interesting finding which incidentally stands in support of the negentropic model! Below is the explanation.

If we recapitulate the mechanism of heat transfer (as presented in Ch 10-D), we have seen that the negentropic phase of a photon gets incorporated into the atomic nucleus in the form of non-valence quarks, which increases the mass of the nucleus (without disturbing its atomic number). Of course, a little bit of the negentropic force (in the negentropic phase of a photon) is simultaneously transferred to the surrounding electron cloud, and this negentropic force is then incorporated as a negentropic knot preparing it for take-off as a free electron flying away from the parent atom. This clearly explains the photoelectric effect: a photon preferentially transfers its energy to the nuclei in a substance first, thereby increasing its mass by heating – and only if the energy is in excess, then it is incorporated into the electron cloud, which results in the photoelectric effect! Thus, UV rays give only heat to a slab of metal, whereas X-rays induce photoelectric effect.

Several aspects of the negentropic model need to be tested by prospective studies, and it is my yearning hope that this model is tested for the benefit of posterity!

8. What are the pitfalls of the negentropic model?

It must be acknowledged that the negentropic model is only an approximate hypothesis in the sense that it simply indicates a general approach to the problems in theoretical physics. Neither is it a complete model because, though based firmly on empirical grounds, it is not treated with mathematical formulations as it has become necessary in the present times! For this reason, the author acknowledges any of the inadvertent inconsistencies and incongruities that might have crept in the discussion. Such defects must have resulted from my own faults and shortcomings, and I humbly plead the learned academics for their scholarly excuse – this model is still in its infancy, and it has to face the test of time for its validation or invalidation!

9. What are the biological implications of the negentropic model?

Life, as noted in Ch 6-C, is a thermodynamic cycle composed of birth and growth of an organism (which represent increase in order and negentropy) and its decay and death (which represent increase in disorder and entropy). Clearly, our negentropic model has shown, in Ch 12-H, that birth of life is a cosmic accident and death is a cosmic necessity. At the same time, it was noted that, on a larger scale, the birth of a bubble universe also is a cosmic accident and the end of universe is a cosmic necessity because once the primordial force starts its play by building up a bubble universe, the primordial spacetime always attempts to restore its original status, thus finally causing separation of spacetime and force (which leads to the end of the bubble universe).

This shows that life is just a mere local sport played by the big universe, and Mother Nature always laughs at each instance of birth and death of an organism unconcerned and moves on ahead in the giant orchestra of the cosmic bubble guided by the ruthlessly unstoppable arrow of time! It must be realized that it is only our human interest that we attempt to preserve and sustain life on the earth—our universe has no cosmic concern to sustain life! Nature will not care what you do with the ecology on the earth. It will not bother, for example, whether you wastefully drive your gas-guzzling car to the grocery to buy your daily ration or if you walk down to save on the fuel – to get to the grocery is merely doing some work in the universe, and to do that nature has to spend energy, and the universe doesn't care if you spend it on your fuel or your calories – it is rather interested in only managing to get the arrow of time going by upholding the inevitable entropy!

It may interest the reader to consider that the effect of dark matter (in the form of negentropic supervision) may perhaps be responsible for the various intriguing properties of life, such as awareness, memory, consciousness and mind, intellect, and even free will.

10. Of what use is the study of hypothetical fundamental science and cosmology? Are there any practical uses of the negentropic model?

A bit of wisdom is a joy forever. The correct understanding of governing principles of the universe at the cosmic and atomic level gives us an immense scholastic pleasure and satisfaction. And the eternal struggle of mankind to find out the ultimate truth, however impossible that may seem at present, gives us a sense of intellectual triumph. Apart from these sublime reasons, we may realize that a scientific concept is always the backbone of all technology – today's science is tomorrow's technology! It may be easily conceived that a concrete knowledge of the structure of photon and matter, or a correct understanding of the nature of our governing principles of the universe, certainly paves way for great technological advances to serve humanity better.

Also, it may be presumed that, equipped with a better understanding of the governing principles of the cosmos, we may go ahead with our space exploration more efficiently, which may turn out to be to the best advantage of humanity in the near future!

POTPOURRI OF
NEWER CONCEPTS

☼ All *matter* and *energy*, and all events in the universe, are composed of two primordial entities—*spacetime* and *force*.

☼ The impingement of primordial force over primordial spacetime gives rise to *negentropic event*, which curves the spacetime, and this is accompanied by an *entropic event* which attempts to straighten the spacetime.

☼ *Equivalence* of negentropic and entropic phases gives rise to *light*.

☼ *Dissociation* of negentropic and entropic phases gives rise to *matter*.

☼ All the phenomena that occur in nature, such as *motion*, *mass*, *entropy*, *inertia*, *charge*, the *arrow of time*, and all such others, are driven forward by the *cosmic principle*, which states that the primordial spacetime and force always tend to exist in an *independent state*. The *next-preferred state* in nature is the state of negentropic–entropic equivalence (which is a ray of light).

☼ Once matter is created by the way of negentropic–entropic dissociation, nature always tends to bring it back to a state of equivalence (by the way of generating photons) and thence into a state of independent existence.

☼ *Energy* is defined as the magnitude of spacetime curvature, and *work* is a measure of the effort of spacetime to uncurl itself.

☼ *Negentropic knots* form *quarks*, which are the *basic units of matter*. Motion of quarks happens in *quark–gluon cycles*, which represent strong force and are manifested as *primary matter waves*.

☼ The *anatomy of an atom* is defined, and the characteristic properties of electrons, neutrinos, and other particles are explained. *Negentropic burden* explains the stability of the nucleus.

☼ *Dark protons* form when symmetric quarks take part in the formation of baryons.

☼ *Dark matter* constitutes *black holes*.

☼ Matter generates primary matter waves from their negentropic knots, which travel out in a negative direction manifested as *secondary matter waves* and *proto-waves* at short distances and as *gravitational waves* at longer distances. All these negative waves are empty spacetime aberrations.

☼ All the weird *quantum phenomena* may be explained by spacetime–force interactions.

☼ *Omniverse* is a monotonous sheet of primordial spacetime. Primordial force starts a bubble universe at a given locale by the way of *differential big bang*.

☼ Negentropic course of the differential big bang forms *dark matter* and *black holes*; entropic course ushers in *dark energy* and *cosmic radiation*. Interaction between negentropic and entropic courses gives rise to *normal matter*.

☼ *Negentropic supervision* stabilizes normal matter and fine-tunes all the events of nature.

☼ Black holes merge to form *big crunch singularity*, which terminates in a *bosonic cone*, which once again initiates another bubble universe.

ACKNOWLEDGEMENTS

have presented a lot of scientific stuff throughout the book, and this academic exercise was possible only because we live in the Age of Information – and my first acknowledgements must go to all these dedicated science writers who have made their invaluable contributions to the encyclopaedias in the net such as Wikipedia and Britannica with a single aim of promoting the cause of science. Whereas the computers and the net have allowed me to make a comprehensive study of all the mysterious mechanisms operative in the universe, it was really the vast number of books written by specialists in the field which have helped me comprehend the complex theories in a better way, and this has enabled me to compile all the information into an organized form of a book. Here I must acknowledge that I have gained some crucial information provided by the following books in particular: *Relativity Simply Explained* by Martin Gardner is one of the best books available on relativity, and *Big Bang* by Simon Singh is a great repository of information which is expertly written – these books are very convenient sources of information for any beginner. *The Elegant Universe* (Brian Greene), *The Grand Design* (Stephen Hawking and Leonard Mlodinow), *A Universe from Nothing* (Lawrence Krauss), and *Hyperspace* (Michio Kaku) are excellent sources of recent advances which make you ponder over the hard problems in theoretical physics!

Writing a book is something, and publishing a book is a quite different thing – and I am fortunate that Partridge Publishers have very gracefully accepted to publish this huge academic work – I am grateful to Antoniet Saints for her sincere initial efforts in my enrolment! I profusely thank Kathy Lorenzo for her efficient editorial skills in guiding me through the multitasking enterprise of publishing – she has, in fact, shaped the book to the present form in the shortest time possible! My sincere thanks must go to Emman Villaran and Clyde Pontillas for their meticulous copy editing which has really fine-tuned my presentation and brought in a great glint and elegance to the book. I am thankful to Julius Artwell for his ebullient marketing schemes which would surely make this otherwise drab scientific output into a bubbling enterprise. I also profusely thank the Partridge book-designing team for their artistic efforts in bringing out this elegant volume.

I must admit that this work would not have been possible but for the efforts of my wife Sathya, who was my first filter through whom the final sketch of chapters took their shape! I extend my special thanks to Ramvara Prasad, our ENT surgeon and computer giant, for his assistance in formatting the images in the book and the front cover illustration. I am grateful to Sitaram Kotike, advocate, High Court of Andhra Pradesh, my friend and mentor, for his constant advice and encouragement throughout this period. I am deeply indebted to Prof. Sudhakar Penagamuri, retired principal of the Adoni Arts and Science College, my guru and guide, for his help and guidance in compiling the text. I must thank Manasa, my daughter, a techie and a busy buzz, for bringing me out of a few conceptual stalemates and for her revision of crucial chapters.

And lastly and more importantly, I must thank all the readers for their patience and perseverance in pursuing the complex topics of this book.

REFERENCES

Abdul Kalam, A. P. J., and Srijan Pal Singh (2015). 'Going All Out for Neutrino Research'. *The Hindu* (June 17).

Barnett, Michael, and Andria Erzberger (2013). 'The Particle Adventure' at Lawrence Berkeley National Laboratory. www.particleadventure.org.

Becker, Adam (2018). *What is Real?: The Unfinished Quest for the Meaning of Quantum Physics*. United Kingdom: John Murray Publishers.

Carroll, Sean (2010). *From Eternity to Here*. London: Oneworld Publications.

Choi, Charles Q. (2013). 'Something from Nothing? A Vacuum Can Yield Flashes of Light'. *Scientific American* (February 12).

Cox, Brian, and Jeff Forshaw (2011). *The Quantum Universe: Everything That Can Happen Does Happen*. United Kingdom: Penguin Books.

Creighton, Jolene (2015) 'Science Explained: Do Atoms Last Forever?' (August 18). futurism.com.

Desikan, Shubashree (2015). 'Hubble: Towards Resolving the Age of the Universe'. *The Hindu* (April 23).

Durant, Will (2006). *The Story of Philosophy*. New York: Pocket Books.

Feynman, Richard P. (1988). *What Do You Care What Other People Think?*. United Kingdom: Penguin Books.

—— (2011). *Six Easy Pieces*. New York: Basic Books.

—— (2011). *Six Not-So Easy Pieces*. New York: Basic Books.

Gardner, Martin (1997). *Relativity Simply Explained*. New York: Dover Publications.

Greene, Brian (2000). *The Elegant Universe*. London: Vintage Books.

Hardin, Jeff, Gregory P. Bertoni, and Lewis J. Kleinsmith (2012). *Becker's World of the Cell*, 8th edition (International Edition). San Francisco: Pearson Education.

Hawking, Stephen (1988). *A Brief History of Time: From the Big Bang to Black Holes*. London: Bantam Books.

——. (2007). *The Theory of Everything*. Mumbai: Jaico Publishing House.

—— and Leonard Mlodinow (2010). *The Grand Design*. London: Bantam Press.

Hossenfelder, Sabine (2018). *Lost in Math*. New York: Basic Books.

Javadi, Hossain (2017). *Beyond the Standard Model*. USA: Supreme Century.

Kajita, Takaaki, and Arthur McDonald (2015). 'For the discovery of neutrino oscillations which shows that neutrinos have mass', (October). www.Nobelprize.org.

Kaku, Michio (1995). *Hyperspace*. New York: Anchor Books.

Kane, Gordon (2006). 'Are Virtual Particles Really Constantly Popping in and out of Existence? Or Are They Merely a Mathematical

Bookkeeping Device for Quantum Mechanics?' *Scientific American* (October 9).

Kanjilal, Pratik (2016). 'In Fact, Understanding Micius, Beijing's Big Push for Quantum Security'. *The Indian Express* (August 22).

Krauss, Lawrence M. (2007). *Fear of Physics*. New York: Basic Books.

——. (2012). *A Universe from Nothing*. New York: Atria Paperback.

Nalawade, Sharad B. (2012). *The Speed of Time*. Madison: Wordizen Books.

——. (2014). *Physics for Backbenchers*, Backbenchers series of books. Bangalore.

——. (2019). *Journey to the End of the Universe*. ISBN 978-93-5382-864-6, Printed at Bengaluru.

Natarajan, Priyamvada (2016). *Mapping the Heavens*. Noida: HarperCollins Publishers.

Penrose, Roger (2011). *Cycles of Time*. London: Vintage Books.

Piccioni, Robert L. (2013). *Einstein for Everyone*. Mumbai: Jaico Publishing House.

Pimbblet, Kevin (2015). 'The Fate of the Universe—Heat Death, Big Rip, or Cosmic Consciousness?' Phys.Org (September 4).

Prasad, R (2016). 'Listening to the Chirps of Gravitational Waves'. *The Hindu* (February 15).

Prasannan, R. (2012). 'Fast Train to the Past? Holy Ghost!' *The Week* (January 22), pp. 39–43. Kottayam: Malayala Manorama Press.

Reucroft, Stephen, and John Swain (1998). 'What Is Casimir Effect?' *Scientific American*, (June 22).

Rovelli, Carlo (2014). *Seven Brief Lessons on Physics*. United Kingdom: Penguin Books.

Russell, Bertrand (1925). *The ABC of Relativity*. New York: Routledge Classics.

Schrödinger, Erwin (1958). *What Is Life*. Cambridge University Press.

Singh, Simon (2005). *Big Bang*. London: Harper Perennial.

Strassler, Matt. 'Of Particular Significance: Conversations on Science with Theoretical Physicist Prof Matt Strassler'. https://profmattstrassler.com.

Thomas, Andrew D. H. (2013). *Hidden in Plain Sight 2: The Equation of the Universe* (e-book).

—— (2014). *Hidden in Plain Sight 3: The Secret of Time* (e-book). Aggrieved Chipmunk Publications.

Thomas, Andrew D. H. (2016). *Hidden in Plain Sight 5: Atom* (e-book).

—— (2016). *Hidden in Plain Sight 6: Why Three Dimensions?* (e-book).

University of Nottingham (2016) Press release: 'There are at least two trillion galaxies in the universe, ten times more than previously thought' (October).

Zhang, Lei (2013). 'The van der Waals Forces and Gravitational Force in Matter'. *General Physics* (March 14), arXiv:1303.3579.

Zorn, Carl. 'How Long Is the Lifespan of an Atom?' Jefferson Lab education at jlab.org/qa/radelement.

Zyga, Lyka (2015). 'Why Do Measurements of Gravitational Constant Vary So Much?' Phys.Org (April 21).

INDEX

light:
    behaviour of 8-9, 11, 463
    bending of 38, 53, 156, 221, 417,
        419-20, 448, 451-2, 454
    cosmic 316, 328
    features of 52, 60, 340, 543
    finite 331-7, 339-40, 342-3, 346-7
    generation of xx, 269, 316-19,
        334-5, 340, 345, 347, 349-
        50, 355, 368-9, 381, 394, 455
    infinite 330-5, 337, 339-40, 342-3
    infrared 486
    propagation of 327, 335, 341, 544
    properties of 9, 13, 50, 53-4, 316,
        318, 330, 356, 436, 465
    refraction of 8, 417, 449
    wave nature of 54
    wave-particle duality of 45,
        463-8, 470
'light clock' experiment 16
light years 184, 186
Linde, Andrei 169
locomotion 250, 254, 292-3
Lorentz (physicist) 11, 15-16

M

McDonald, Arthur 101
mass:
    gravitational 19
    inertial 19, 385, 390
    proper see mass, rest
    relativistic 20, 22, 42, 122, 291
    rest 19-22, 97, 231, 385
matter, normal 127, 415, 494, 505,
    508-15, 517, 520-1, 526, 532, 535,
    538-40, 542
matter particles:
    elementary 94
    fundamental 134-5
    solitary 529-30
maximum entropy, state of 266, 476

Maxwell, James Clerk 54, 57, 90, 116
mesons 41, 98, 105, 108, 126, 377,
    392-3, 398
Messier, Charles 183, 185
Michelson, Albert 11, 13
Michelson-Morley experiment 9,
    11, 330
microwaves 82, 200, 205, 349, 534
Milky Way 5, 139, 151, 159, 173-4,
    183, 185-6, 192, 201, 210, 216-
    17, 497
momentum, conservation of 101,
    113, 286-8, 290, 395, 548
Morley, Edward 11, 13
Mother Universe 497, 499, 540
motion:
    absolute 5-6, 9-12, 25, 28-9, 31,
        39, 278, 280, 359, 483
    accelerated 26, 29, 271, 283
    inertia of 272-3, 282, 300,
        303, 314
    Newton's laws of xv, 26, 257, 475
    first law 270-1, 314
    second law 46, 226-7, 230-3,
        236-7, 239, 243-4, 256, 271,
        283, 545
    non-uniform 6, 25, 31, 39, 270-4,
        281, 283-9, 291, 298, 300-1,
        303-4, 314, 333, 335, 338,
        343-4, 347, 353, 386, 390,
        396-7, 400, 446, 459, 466,
        473, 476, 488
motion, relativistic 280, 282
motion: simulated uniform 273-4,
    282, 288, 314, 333, 335, 356, 400
multiverse 169, 497, 545
multiverse theory 168-9
muons 22, 101, 106, 117, 126, 361,
    376-7, 391, 406

N

nebula 145, 419, 440

negentropic burden 412-15, 437-8, 483-6, 488

negentropic course 532-3, 540

negentropic force 322-3, 325-7, 342-4, 346-7, 349, 351-5, 379-80, 386, 389-90, 402, 404, 407-8, 425-6, 431, 436, 438, 440, 486, 513, 521, 528, 544, 548

negentropic knot formation 380, 393

negentropic knots 379-85, 387, 389-90, 398-9, 402-11, 415, 417, 433, 437-41, 446, 460, 468-70, 472, 474, 476, 480, 482-7, 490-2, 505-6, 512, 514, 519, 521-2, 528-30, 532, 543, 548

negentropic model xviii, xx-xxi, 267, 316, 410, 413, 425, 428, 440, 455, 463-5, 468, 474, 483, 494-5, 499, 501-4, 520, 538, 541, 543-4, 546-50

negentropic phase 343-7, 351-2, 354-6, 367-9, 379-83, 385, 399-401, 403, 415, 433, 438, 447, 450, 453, 459-60, 465-7, 470, 473, 492, 522, 536-7, 548

negentropic supervision 514-15, 535-6, 540, 542, 550

negentropic waves 443-4, 504, 515

negentropy 233, 239-40, 242-4, 255-6, 265-6, 322, 379, 381, 398, 400-2, 413-15, 433, 436, 460, 512, 528, 539, 549

neutrinos 23, 101-2, 104, 106, 113, 115, 126, 150, 159, 166, 211-12, 313, 359-60, 371-2, 377, 406-7, 409, 415, 533

neutron stars 150-1, 154

neutrons, free 372, 411-12

Newton, Isaac 2, 8, 16, 18, 26-8, 38, 46, 48, 53, 187, 271-3, 282-3, 286, 294, 298, 303, 314, 321, 416, 539

nothing, state of 223, 506, 525-6

nuclear fission 313-14, 348, 371
  induced 114
  spontaneous 112, 114

nuclear forces 104, 154

nuclear reactions 93, 99, 102, 109, 115, 138, 145-7, 149, 153, 161-2, 164-6, 296, 313, 337, 348, 371-2, 375, 377, 390-1, 401, 404, 406, 409, 419, 436, 502

nuclear reactors 48, 114, 513

nucleons 98, 105, 108, 111-12, 122, 371, 385-6, 393, 410, 429-31

nucleosynthesis 149-50, 213
  primordial 149, 198-9, 208, 213, 369-70
  stellar 149, 199, 201, 369-70

O

observer's effect 64, 69, 71, 455, 465, 469-70, 492

Olbers, Wilhelm 204

Olbers' paradox 204

omniverse 495-9, 525-31, 537, 539-42, 545-6

orientation, spatial 34, 329, 360, 362, 383, 405, 484, 525-6, 543

Origin of Species (Darwin) 181

Ørsted, Hans 54

Ostriker, Jerry 217

outer space 142-4, 148, 205, 274, 321, 376

P

parallel worlds 135, 168

particle accelerators 103, 119, 125-6, 135, 160, 372

particle annihilation 371, 374

waves:

    blowout 431-4, 437-9, 441, 450-2, 454, 481, 485, 504, 530

    electromagnetic 57, 92, 342, 345, 367, 387, 453, 481, 485

    entropic blowout 432, 450, 454, 504

    gravitational xvi, 2, 43, 443-8, 452-4, 456, 510

    negative 443-4, 446, 508

    primary matter 387, 389, 392, 430-1, 444, 446, 448, 453-4, 507-8, 521, 530

    proto-gravitational 442-3, 445, 481, 531

    quark 387, 389-93, 411-12, 415, 430, 444-6, 453, 490, 507-8, 530

    secondary 431-4, 441, 443, 450-1, 454

    secondary matter 431-4, 441-2, 444, 446-50, 452-4, 481, 504, 515, 530

    virtual 433, 443, 447-8, 454, 481, 522

Weinberg, Steven 118

Wheeler, John 39, 132, 154, 171

Wieman, Carl 110

Wilczek, Frank 104

Wilson, Robert 205

Witten, Edward 135

work-energy theorem 298

wormholes 138, 168, 170-1, 175

Y

Young, Thomas 10, 53

Z

Zweig, George 94

AUTHOR'S BIOGRAPHY

And the Disclaimer of a Non-specialist!

What do you call a chap who writes about cells, cell membranes, and lipid rafts at one moment and then swaps over to touch upon the intricate details of atoms, quarks, and black holes at the next moment and then again switches over to pen upon the myriad philosophical aspects of the brain, mind, and consciousness in the next instant? Whatever he might be—mad as a crazy nut or odd as a loopy egghead—one thing he could *not* be is a *specialist* in any field! And by Jove, that's exactly what I am—a non-specialist!

But then as it turns out, I have started my career as a specialist in the highly specialized field of medical science, and that was my walk of life to earn my bread, so to speak, for my life's entire working tenure. Here I should mention that I graduated in medicine from Kurnool Medical College, Kurnool, and MS in general surgery from Kasturba Medical College, Manipal, and took my training in vascular surgery at Nizam's Institute of Medical Sciences, Hyderabad. And all along, I have been practicing surgery at Adoni in India for nearly three decades.

However, this specialty being a purely applied field of science, it could not quench my thirst to study a more subtle and sublime branch of science—the field of fundamental science! But then what do we mean by 'fundamental science'? What is its scope, and who would specialize in it? Simply put, basic science, as it is also called, is a search for the basic mechanism that is operative at the fundamental level in any branch of science. Thus, all branches of natural science, such as physics, chemistry, and biology, at their core deal with discovering the governing principles that are operative at the most fundamental level. And one should realize that at this fundamental level, the boundaries among all our branches of science tend to become smudgy with similar forces operating in the background. However, many academic specialists feel uncomfortable to go into these intricate depths in their respective fields (perhaps justifiably, for this being a practically impractical affair), and thus, these aspects are generally left untouched for a long time in the history of science! But nowadays, there are many specialists who would try and venture out into these unknown fields with an unbridled mind of a non-specialist who can come out of the lock-stepped approach of a specialist to reach for higher goals, and the reader may find many such specialized non-specialists in the galaxy of writers of science among us today!